Food Science and Technology

nova
science publishers

www.novapublishers.com

Food Science and Technology

Proceedings of BIOSPECTRUM: The International Conference on Biotechnology and Biological Sciences: Biotechnological Intervention Towards Enhancing Food Value
Sanket Joshi, PhD (Editor)
Susmita Mukherjee (Editor)
Moupriya Nag, PhD (Editor)
2022. ISBN: 978-1-68507-985-7 (Hardcover)
2022. ISBN: 979-8-88697-069-2 (eBook)

Nutraceuticals: Food Applications and Health Benefits
Anita Kumari, PhD (Editor)
Gulab Singh, PhD (Editor)
2022. ISBN: 978-1-68507-488-3 (Hardcover)
2022. ISBN: 978-1-68507-512-5 (eBook)

A Closer Look at Polyphenolics
Peter Bertollini (Editor)
2022. ISBN: 978-1-68507-480-7 (eBook)
2021. ISBN: 978-1-68507-434-0 (Hardcover)

Food Processing: Advances in Research and Applications
Myriam Huijs (Editor)
2022. ISBN: 978-1-68507-570-5 (Hardcover)
2022. ISBN: 978-1-68507-581-1 (eBook)

Cheeses around the World: Types, Production, Properties and Cultural and Nutritional Relevance
Ana Cristina Ferrão, Msc (Editor), Paula Maria dos Reis Correia, Raquel de Pinho Ferreira Guiné, PhD (Editor)
2019. ISBN: 978-1-53615-418-4 (Hardcover)
2019. ISBN: 978-1-53615-419-1 (eBook)

More information about this series can be found at https://novapublishers.com/product-category/series/food-science-and-technology-series/

Wing-Fu Lai
Editor

Food Packaging

Safety, Management and Quality

nova science publishers
www.novapublishers.com

https://doi.org/10.52305/SEPV4757

Library of Congress Cataloging-in-Publication Data

ISBN: 979-8-88697-249-8

Published by Nova Science Publishers, Inc. † New York

Contents

Preface

Packaging can help retard deterioration of food products and prolong their shelf life [1, 2]. Food products sold in the market are generally put inside packages, which play an important role in protecting the packaged food from environmental attacks (such as oxygen, light, and water vapor) and microbial contamination. Over the years, different types of materials (ranging from wood and glass to a wide variety of polymers) have been adopted to generate food packages. Various strategies such as supercritical fluid impregnation have also been incorporated into the process of film production to enhance the functionality and versatility of the generated films in food protection. Despite this, till now none of the materials can meet every single need in food packaging. Each type of material has its pros and cons. For example, while glass is transparent, thereby allowing the appearance of the food product inside the package to be seen and hence is appealing from the product design perspective [3, 4], glass-generated packages are comparatively heavy and easy to be broken. On the other hand, polymers are known to show high flexibility in structures and functions [5-7]. Polymers can also be easily generated as films [8, 9], which can then be used to generate thermosealable and microwavable packages in different dimensions and shapes to meet the needs of different food products [10-14]. Yet, many of the polymer-based food packaging films are highly permeable to gases and other low-molecular-weight compounds. This reduces the capacity of the packages in combating the adverse effects caused by the surroundings of the packaged food. Clearly developing an ideal package for food products is not a simple task.

The objective of this book is to present a snapshot of the latest advances in the development, characterization and quality management of films for food packaging purposes. This book consists of three sections. The first section aims at presenting an introduction to materials selection and design for food packaging. Different types of materials, ranging from polysaccharides to graphene oxide nanocomposites, for film development will be discussed. The second section reviews the current status of engineering and optimization of food packaging films. Applications of the films for packaging foods such as green vegetables and mushrooms will also be presented. The last section is devoted to delineating practices useful for quality control and assurance in food packaging. Concepts and implementation of HACCP and other quality control measures will be discussed. Strategies to reduce human error in quality control of fresh produce labels will also be covered for food production purposes. It is hoped that this book can serve as a reference guide for those working in the academic and industrial sectors to get an overview of the latest concepts and techniques in the development, application, and quality management of food packaging films.

Here I would like to express my thanks to the contributors of different chapters of this book. I have enjoyed excellent support from the staff in Nova Science Publishers. The quality of this book, and its value to its readers, depends in large measure on the care and promptness with which submitted manuscripts are reviewed. I would like to thank the reviewers for taking their time to assess and select the best manuscripts for inclusion in this book, and for providing critical comments to the contributors of the selected chapters. Thanks are extended to Obireddy Sreekanth Reddy from the Department of Chemistry at Sri Krishnadevaraya University for his editorial and administrative assistance in the development of this book. Finally, Yau-Foon Tsui and Noah Zhao are acknowledged for their helpful suggestions and support in bringing out this publication.

Wing-Fu Lai, PhD
Editor

References

[1] Parhi, A.; Tang, J. M.; Sablani, S. S., Functionality of ultra-high barrier metal oxide-coated polymer films for in-package, thermally sterilized food products. *Food Packag Shelf Life* 2020, *25*, 100514.

[2] de Carvalho, A. P. A.; Conte, C. A., Green strategies for active food packagings: A systematic review on active properties of graphene-based nanomaterials and biodegradable polymers. *Trends Food Sci Tech* 2020, *103*, 130-143.

[3] Lai, W. F.; Zhao, S.; Chiou, J., Antibacterial and clusteroluminogenic hypromellose-graft-chitosan-based polyelectrolyte complex films with high functional flexibility for food packaging. *Carbohydr Polym* 2021, *271*, 118447.

[4] Lai, W. F.; Wong, W. T., Edible clusteroluminogenic films obtained from starch of different botanical origins for food packaging and quality management of frozen foods. *Membranes-Basel* 2022, *12* (4), 437.

[5] Hu, K. Q.; Huyan, Z. Y.; Ding, S. X.; Dong, Y. Y.; Yu, X. Z., Investigation on food packaging polymers: Effects on vegetable oil oxidation. *Food Chem* 2020, *315*, 126299.

[6] Fernandes, S. S.; Romani, V. P.; Filipini, G. D.; Martins, V. G., Chia seeds to develop new biodegradable polymers for food packaging: Properties and biodegradability. *Polym Eng Sci* 2020, *60* (9), 2214-2223.

[7] Diez-Pascual, A. M., Antimicrobial polymer-based materials for food packaging applications. *Polymers-Basel* 2020, *12* (4), 731.

[8] Lai, W. F.; Wong, W. T., Edible clusteroluminogenic films obtained from starch of different botanical origins for food packaging and quality management of frozen foods. *Membranes-Basel* 2022, *12* (4), 437.

[9] Lai, W. F., Design of polymeric films for antioxidant active food packaging. *Int J Mol Sci* 2021, *23* (1), 12.

[10] Topuz, F.; Uyar, T., Antioxidant, antibacterial and antifungal electrospun nanofibers for food packaging applications. *Food Res Int* 2020, *130*, 108927.

[11] Nesic, A.; Cabrera-Barjas, G.; Dimitrijevic-Brankovic, S.; Davidovic, S.; Radovanovic, N.; Delattre, C., Prospect of polysaccharide-based materials as advanced food packaging. *Molecules* 2019, *25* (1), 135.

[12] Muller, P.; Schmid, M., Intelligent packaging in the food sector: a brief overview. *Foods* 2019, *8* (1), 16.

[13] Huang, T.; Qian, Y.; Wei, J.; Zhou, C., Polymeric antimicrobial food packaging and its applications. *Polymers-Basel* 2019, *11* (3), 560.

[14] Chen, X.; Chen, M.; Xu, C.; Yam, K. L., Critical review of controlled release packaging to improve food safety and quality. *Crit Rev Food Sci Nutr* 2019, *59* (15), 2386-2399.

Section I: Materials Selection and Design for Food Packaging

Chapter 1

Polysaccharides-Based Food Packaging Films: An Overview

Dinusha L. Peramune, Sandali Ranaweera, Wasana N. Marasinghe, Danushika C. Manatunga, K. G. L. R. Jayathunge and Rohan S. Dassanayake*

Department of Biosystems Technology, Faculty of Technology,
University of Sri Jayewardenepura, Homagama, Sri Lanka

Abstract

Over the past few years, there has been increasing attention and research on biopolymers, including polysaccharides, among the scientific community as an alternative to petroleum-based packaging materials. Polysaccharides have recently transpired as edible food packaging materials due to their inherent properties, including renewability, relative abundance, nontoxicity, edibility, high mechanical strength, excellent barrier, organoleptic, antimicrobial and antioxidant activity. As the first chapter of Section I, we will cover the structure, properties, and applications of the most used polysaccharides in edible food packaging. This chapter also discusses the advantages, challenges, current trends and the role of active packaging, intelligent packaging, and nanotechnology in polysaccharide-based edible food packaging.

Keywords: edible food packaging, active packaging, intelligent packaging, nanotechnology, polysaccharides

---

* Corresponding Author's Email: rdassanayake@sjp.ac.lk.

In: Food Packaging
Editor: Wing-Fu Lai
ISBN: 979-8-88697-249-8

1. Introduction

According to the Food and Agriculture Organization (FAO) of the United Nations 2021 report, approximately 11 billion tonnes of food and many non-food products, including 32 million tonnes of natural fibres and 4 billion m^3 of wood, are produced by the world's agri-food systems annually (FAO 2021b). FAOSTAT 2021 reports that the estimated gross value of agricultural output in 2018 was USD 3.5 trillion (FAO 2021a). In 2020, approximately 10% of the global population suffered from hunger, which is an increase of around 120 million compared to 2019 (FAO 2020). Even though there is an escalating demand for food, adverse weather conditions, including droughts, floods, climate variability, anthropogenic activities such as deforestation and construction, pandemic situations, economic inequalities, price instability, human conflicts, and post-harvest losses have immensely impacted the food production and agri-food systems globally. Among those many issues, post-harvest losses due to handling, storage and distribution have significantly affected the agri-food systems because of improper packaging. Moreover, foodborne illnesses that result in improper packaging and food processing have also become a severe public health threat, causing over 400,000 annual deaths worldwide (Halonen et al. 2020). Therefore, food packaging is considered one of the vital aspects of addressing post-harvest losses and has been gaining interest among the scientific community in recent years.

Food packaging is the proper packaging of the food product that plays an essential role in the modern Food Industry. Food packaging is imperative to limit the addition of food preservatives, commercialize the food product, and reduce food waste. Most importantly, food packaging protects food from chemical (oxygen, carbon dioxide, moisture, temperature, radiation such as ultraviolet, infrared, visible) and physical (shock and vibration during distribution and handling) and biological (pathogens, rodents, insects, and other animals) damages and preserves the quality, while enhancing the shelf-life of the food product (Amaraweera, Gunathilake, Gunawardene, Fernando, Wanninayaka, Manamperi, et al. 2021, Geueke, Groh, and Muncke 2018, Marsh and Bugusu 2007). Food labelling used in packaging helps food manufacturers inform customers about food ingredients and nutritional values and promote the product. Therefore, many food industries outsource packaging operations to contract packagers to reduce costs, improve quality and innovation, and speed up the packaging and distribution process.

Most widely used food packing materials include plastics, glass, metals, and paper and paperboard. The packaging material and the container type are dependent on the type of food. Various container types, including aseptic processing, cans, bags, corrugated boxes, paper (cartons, coated), flexible packaging, pallets, trays, and wrappings, are commonly used for food packaging (Alamri et al. 2021). However, Petroleum-based plastics are the dominant materials in food packaging industries because they are versatile, cheap, and easy processability. A wide variety of these plastics are used in both flexible and rigid forms. Moreover, petrochemical plastics have found the most usage in food packaging industries due to their lightweight and excellent tensile properties. They also act as an effective barrier against oxygen, carbon dioxide and moisture. During the last few decades, the utilization of petrochemical plastics has increased with the rapid popularization of single-used plastic containers (Geyer, Jambeck, and Law 2017). Low-density Polyethylene (LDPE), linear low-density polyethylene (LLDPE), polypropylene (PP), polystyrene (PS), polyethylene terephthalate (PET), polyvinyl chloride (PVC) are major types of petroleum plastics employed as food packaging materials (Pongrácz 2007, Geyer, Jambeck, and Law 2017). However, many plastics, including single-use plastics used in food packaging, are non-biodegradable and not environmentally friendly, causing enormous environmental and health burdens. Moreover, microplastics which are fragments of any plastic type in micro-scale, notoriously known as silent killers, can bioaccumulate and adversely affect the environment and human health and is still a relatively unexplored area. Therefore, increasing environmental and health concerns and the generation of large amounts of waste have scrutinized plastic packaging in recent years.

Besides plastics, paper and paperboard are the second most popular primary food packaging materials, representing 31% of the global packaging market (Deshwal, Panjagari, and Alam 2019). There are different types of paper packages: kraft paper, parchment paper, greaseproof paper, and paperboards (Goswami 2019). Surprisingly, several environmental and health issues are also associated with paper and paper boards as food packages. Numerous additives are added to the papers and paperboards during the manufacturing process to improve the quality of the product. These chemical additives, including organic and inorganic dyes, mineral oils, phthalates, polyfluorinated substances (PFAS), can interact with food products, causing health effects, thus limiting the usage of paper and paperboards as primary food packaging materials (Trier, Granby, and Christensen 2011, Fierens et al. 2012).

To address many issues related to petroleum plastics and paper food packaging materials, significant attention has been paid to biopolymer-based raw materials as renewable alternatives. Biopolymers are long-chain macromolecules produced by living organisms or derived from biomass. Three main types of biopolymers include polysaccharides, polypeptides and polynucleotides. Biopolymers can be extracted directly from biomass such as cellulose, starch, chitin; produced by chemical synthesis using monomers such as polylactic acid (PLA); produced by microorganisms or genetically modified microorganisms such as polyhydroxyalkanoates (PHAs); and by chemical synthesis using monomer units of biopolymers and petroleum-based monomers such as poly(butylene succinate) (Ncube et al. 2020). Biopolymers and their derivatives exhibit intriguing properties, including relative abundance, renewability, biodegradability, nontoxicity, environmental friendliness, and tunable surface properties. Like conventional plastics, biopolymers can also be processed using casting, electrospinning, melt-mixing, hot pressing and extrusion to produce a diverse spectrum of food packing products (Mangaraj et al. 2018). Therefore, biopolymers have widely been considered for food packaging, coating, and edible food packaging (Fabra, López-Rubio, and Lagaron 2014). Interestingly, biopolymer-based edible food packaging has recently emerged as an innovative solution in the food packaging industry, giving value-addition to packed food. Edible food packaging is a type of food packaging developed to be consumed and can effectively biodegrade similar to the food it contains. Edible food packing currently comes in many forms and is being widely improved from polysaccharides and their derivatives.

This chapter provides comprehensive coverage of the structure, properties, and applications of polysaccharides and their derivatives in edible food packaging. Moreover, the advantages, challenges, current prospects, and the involvement of nanotechnology in polysaccharide-based edible food packaging have also been exclusively discussed.

2. Edible Food Packaging

The key to sustainable food packaging is to develop food packages using innovative materials that can protect the food and environment, substitute fossil-based plastics, reduce waste accumulation, and limit the negative impact on human health. Notably, considerable research has recently been employed to investigate renewable alternatives to conventional food packaging materials, mainly due to increased consumer awareness and demand for health, food quality and safety, and environmental sustainability (Boz, Korhonen, and Koelsch Sand 2020). Interestingly, edible food packaging materials have recently transpired as advanced packaging materials, eliminating many concerns about conventional food packing materials. Edible biopolymers, including polysaccharides, polypeptides, lipids, waxes, and other edible compounds derived from renewable sources, have been considered in edible packaging. These edible packaging materials are intended to be integral parts of food products and be eaten with the products or disposed to the environment, resulting in efficient biodegradation (Janjarasskul and Krochta 2010a). Commonly used edible food packing materials include films, coatings, sheets, gels and pouches (Janjarasskul and Krochta 2010a). Edible films and sheets are stand-alone structures separately formed as a thin layer and then wrapped on or placed between food products (Jeevahan and Chandrasekaran 2019). Several polysaccharides, including cellulose, chitosan, starch, pectin, alginate, and carrageenan, have been explored for their potential to form edible films (Cazón et al. 2017). The edible coating is performed as a thin layer directly on the surface of the food product (Jeevahan and Chandrasekaran 2019). Edible packaging provides excellent protection and barrier properties, structural integrity, enhancement of the quality and the safety of the food. Besides barrier properties, edible packaging has many added advantages such as edibility, biodegradability, nontoxicity, low cost, higher chemical stability, processing adaptability, antimicrobial and antifungal properties with extended shelf-life (Dhumal et al. 2019, Benbettaïeb et al. 2016, Umaraw et al. 2020). Polysaccharides and their derivatives have been employed in edible food packaging due to their characteristic features. Also, different edible composite materials can be developed by combining polysaccharides with other biopolymers to improve the properties of the packaging material (Adzaly et al. 2018).

2.1. Structure and Properties of Polysaccharides in Edible Food Packaging

Polysaccharides are ubiquitous and structurally the simplest biopolymers in nature. Polysaccharides consist of monosaccharide units linked by glycosidic bonds. Polysaccharides are the most abundant carbohydrate found in food. Almost all the natural polysaccharides display qualities, including relative abundance, non-toxicity, renewability, biodegradability and eco-benignity, which lead them to be applied in many applications sustainably. Due to their unique properties, polysaccharides and their derivatives have been widely considered in food packaging and edible food packaging applications. The following section discusses the chemical and physical properties of the most used polysaccharides in edible food packaging.

Starch is one of the extensively used plant-based polysaccharides in developing food packaging. Many studies have shown the use of starch from various sources such as corn flour, cassava, potato, red rice, and wheat in producing edible coatings with different properties (Pelissari et al. 2019). It consists of two types of molecules: linear amylose and branched amylopectin. The amylose to amylopectin ratio of starch varies depending on the source. Generally, starch comprises about 20-25% amylose and 75-80% amylopectin by weight (Amaraweera, Gunathilake, Gunawardene, Fernando, Wanninayaka, Dassanayake, et al. 2021, Amaraweera, Gunathilake, Gunawardene, Fernando, Wanninayaka, Manamperi, et al. 2021). These molecules are formed by α-condensation of D-glucose monomer units. Figure 1 shows the chemical structures of linear amylose and branched amylopectin molecules of starch. In the purest form, starch is a white, odourless polysaccharide. Starch can be categorized into A-and B-Type. A-Type starches contain an orthorhombic crystal structure, while B-type contain a hexagonal structure (Nilsen-Nygaard et al. 2021). In plants, starch is usually found as semi-crystalline granules, and their sizes and other properties depend on the plant type. Typically, the diameter of starch granules of different plant types ranges from 1-150 μm. Starch granules undergo swelling in water reversibly or irreversibly depending on the temperature. Starch possesses excellent properties, including oil resistivity, oxygen impermeability, nontoxicity and biodegradability, leading to suitable material for edible packaging (Menzel 2020, García et al. 2020). One of the setbacks of starch in food packing is the amorphous structure generated by amylose, causing poor mechanical properties, particularly brittleness, leading to moisture migration through these regions (Sadeghizadeh-Yazdi et al. 2019).

Cellulose is the most abundant biopolymer on earth and is widespread in plants, algae, tunicates, and bacteria. It is generated from D-glucopyranose monomer units linked by β-1,4-glycosidic bonds (Dassanayake et al. 2021). The abundance of hydroxyl groups on the monomer units builds up strong intra- and inter-molecular hydrogen bonding between cellulose chains, constructing microfibrils with randomly distributed crystalline and amorphous regions. Cellulose exhibits low density, high stiffness, thermal and mechanical stability, qualifying it as edible food coatings and food packaging (Atta et al.

2021, Jancy, Shruthy, and Preetha 2020, Souza, Gottschalk, and Freitas-Silva 2020). Cellulose is insoluble in water and most other organic solvents. Nevertheless, methylcellulose (MC), carboxymethylcellulose (CMC), hydroxypropyl cellulose (HPC), and hydroxypropyl methylcellulose (HPMC) are water-soluble derivatives of cellulose produced through etherification reactions. They are also widely considered in edible food packaging materials owing to their characteristic physical and chemical properties such as oxygen transfer, fat resistance, nontoxicity, odorless, tasteless and flexibility.(Li et al. 2020) Wettability is another function that plays an important role, especially in edible food coatings. Out of these derivatives, the wettability of MC is amphiphilic, whereas CMC, HPC and HPMC are hydrophilic (He et al. 2021). Figures 2A and B show the structure of cellulose chains showing inter-and intra-molecular hydrogen bonding networks and (B) the structure of cellulose derivatives (MC, CMC, HPMC), respectively.

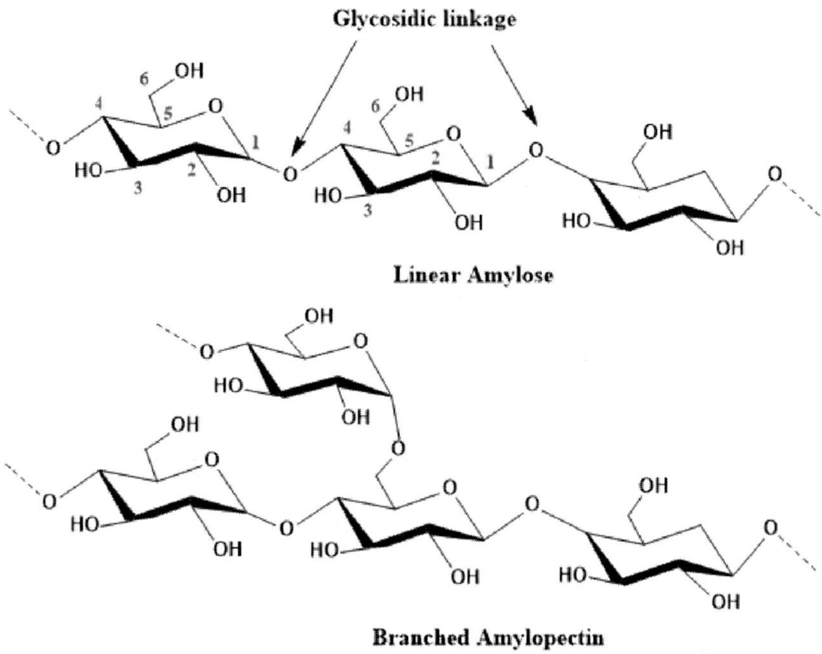

Figure 1. Chemical structures of amylose and amylopectin molecules of starch.

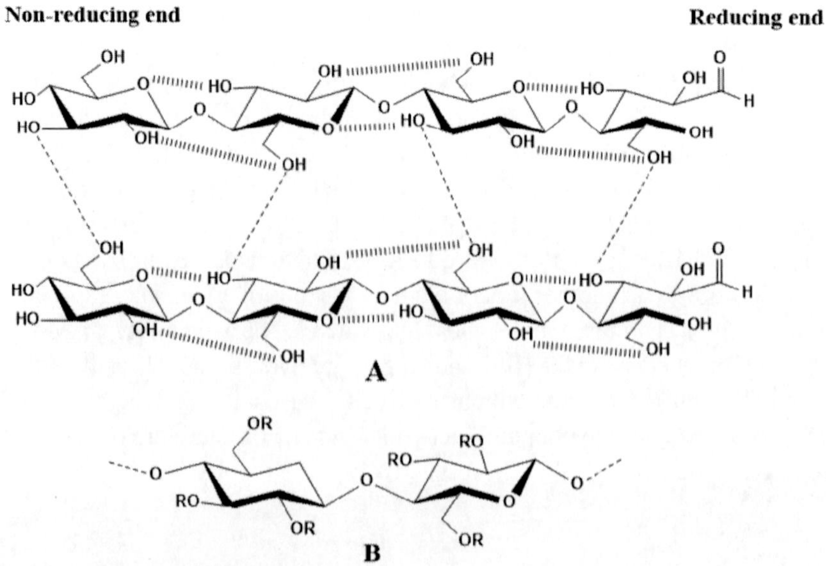

MC: R = H or CH_3
CMC: R = H or CH_3COOH
HPC: R = H or $CH_2CHOHCH_3$
HPMC: R = H or CH_3 or $CH_2CHOHCH_3$

Figure 2. (A) Structure of cellulose chains showing inter- and intra-molecular hydrogen bonding network and (B) structure of the cellulose derivatives (MC, CMC, HPMC).

Hemicellulose is the second most abundant polysaccharide component and is often associated with cellulose in plant cell walls. Due to their excellent properties, including biocompatibility, biodegradability and nontoxicity, hemicelluloses have gained attention in numerous applications, including food packaging. Hemicelluloses are made of various monosaccharide units called heteropolymers, whereas cellulose is exclusively comprised of glucose monomers. Hemicellulose chains contain fewer monomer units (300-5000) than cellulose (7000-15000) and are branched. Xylan and glucomannan are examples of hemicellulose types widely employed in edible food packaging (Nechita, Mirela, and Ciolacu 2021). Both xylan and glucomannan consist of pentose, hexose and arabinose sugar units as monomers. Xylan is a branched polymer of β-1,4-linked xylose residues branched with α-arabinofuranose and α-glucuronic acid residues (Hsieh and Harris 2019). In contrast, glucomannan is a more linear polymer with less branching, containing D-mannose and D-glucose as the monomer units (Hashemi and Jafarpour 2021). Figure 3 shows the chemical structures of xylan and galactomannan. Hemicelluloses are distinguished from cellulose by their crystallinity, where they are mostly less strong amorphous polymers. Most hemicelluloses are water-soluble and easily hydrolyzed by dilute acids or bases (Hsieh and Harris 2019).

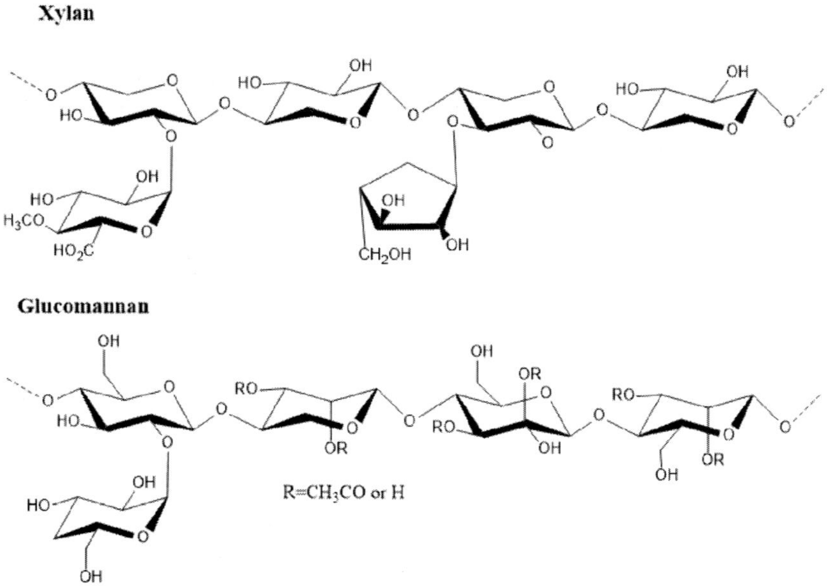

Figure 3. Chemical structures of xylan and glucomannan.

Chitin, the second most abundant biopolymer on earth after cellulose, is a polysaccharide found in the exoskeletons of insects, including arthropods and crustaceans (Mohan et al. 2020, Borić et al. 2020). While structurally analogous to cellulose and polymerized with (1,4)-β-N-acetyl glycosaminoglycan monomer units, chitin is highly crystalline due to its linearly arranged hydroxyl and acetamide groups. Chitosan, the N-deacetylated derivative of chitin, is a linear polysaccharide with randomly distributed deacetylated units of D-glucosamine and N-acetyl-D-glucosamine linked with β-1,4 linkages. Figure 4 depicts the chemical structures of chitin and chitosan. Both chitin and chitosan express similar properties such as nontoxicity, biodegradability, selective gas permeability and good tensile strength, making them good candidates for edible food packaging materials. As an added advantage, chitosan shows excellent antimicrobial properties due to its polycationic nature arising from the protonated amino groups (Rodrigues et al. 2020). However, the main drawback of chitosan-based packaging is its high permeability to moisture.

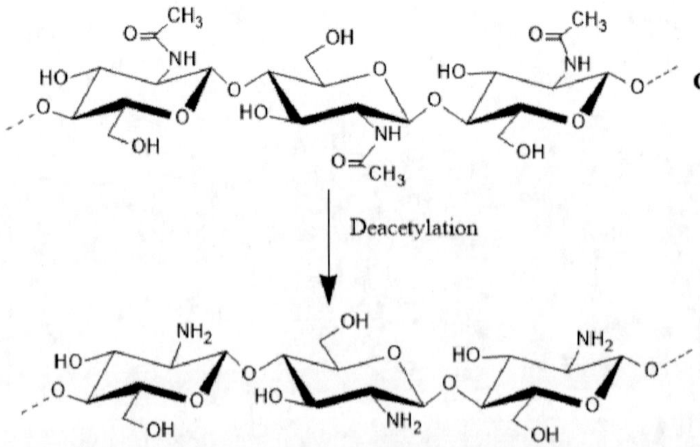

Figure 4. Chemical structures of chitin and chitosan.

Polysaccharide gums, also known as hydrocolloids, are a group of polysaccharides soluble in water and show a gelation ability. Additionally, they possess excellent properties, including thickening, moisture retention, stabilization and emulsification (Diego Aires da Silva 2020). According to their origin, gums can be classified into several categories, including algae extracts (alginate, carrageenan, agar), plant seed endosperm (guar gum), and plant cell wall extracts (pectin). Algae extracts are the most popular type of polysaccharide gums that are used for edible food packaging. Among these polysaccharides, alginate is the metal salt of alginic acid, see Figure 5, a naturally occurring polysaccharide that can be extracted from seaweeds. It is also abundantly used in edible food packaging due to its biocompatibility, film-forming ability, nontoxicity and relatively low cost. Alginate is a linear copolymer containing α-L-guluronate (1-4) linked with β-D-mannuronate arranged in an irregular, block-wise pattern. The block-wise pattern includes three types of residues of α-L-guluronate residues, β-D-mannuronate residues and a residue with a combination of the two monomers (Rahmani et al. 2017). Carrageenan, another polysaccharide type extracted from marine algae, is also recently used in developing edible food packaging. It is a group of linear polysaccharides made of partially sulfated galactans. The water-solubility of carrageenans usually depends on their ester sulfate content and the associated cations, including Na^+, K^+, Ca^{2+} and Mg^{2+}. There are three types of carrageenans, kappa-carrageenan (κ-carrageenan), iota-carrageenan (ι-carrageenan) and lambda-carrageenan (λ-carrageenan), see Figure 5. The emulsion stabilizing and low oxygen permeability of carrageenans lead it to have good mechanical properties and to be able to protect food surfaces from dehydrating (Dwivany et al. 2020).

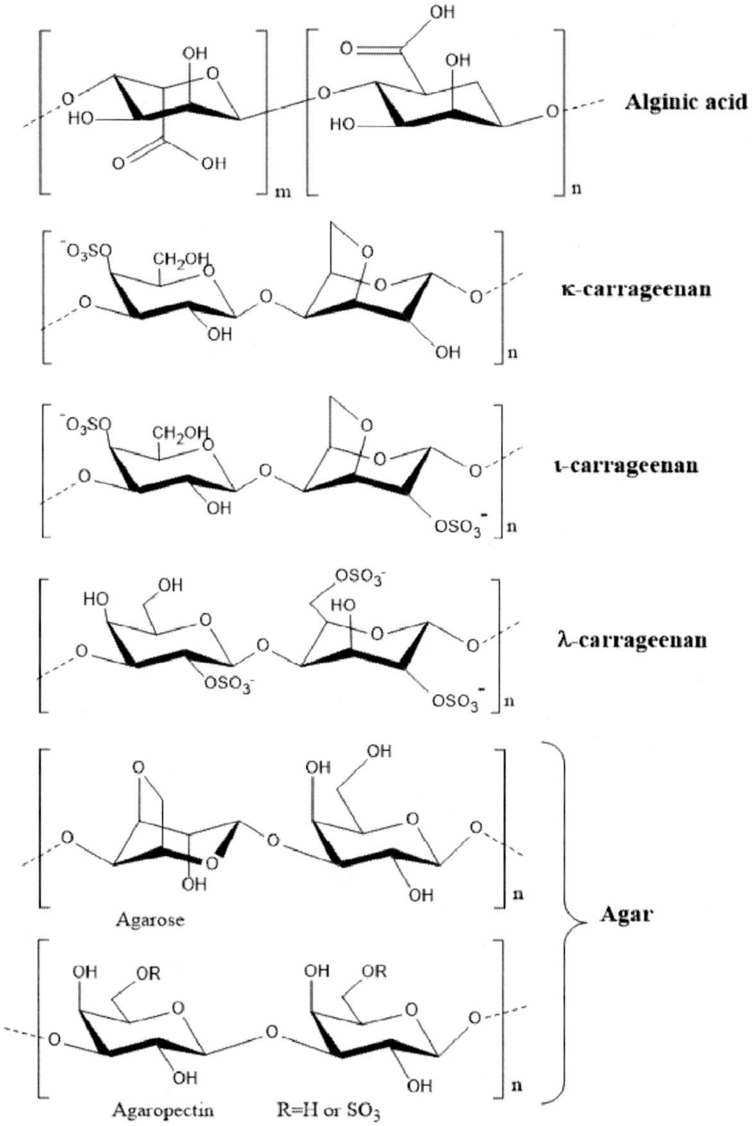

Figure 5. Structure of alginic acid, carrageenans (κ, ι, and λ), and agar.

Agar is composed of two polysaccharides, agarose and agaropectin-like starch. Agarose is a linear biopolymer made of several agarobiose residues, whereas agaropectin is a complex polysaccharide with uronic and sulfuric acid residues, see Figure 5. Agarose is mainly responsible for gelation properties. Gelation of agar occurs due to the aggregation of the molecular chains by hydrogen bonding between agarobiose units forming gels, making them good candidates for edible food coating and packaging (Park and Khan 2009). For guar gum, it is a polysaccharide extracted from cluster beans, is chemically categorized as a galactomannan polysaccharide polymerized with (1-4) β-D mannopyranosyl linear units bound with (1-6) α-D galactopyranose (Liyanage et al. 2015). Figure 6 shows the chemical structure of guar gum. Like some other polysaccharides, hydroxyl groups of guar gum develop hydrogen bonds with the water molecules leading to increased viscosity and complete water solubility. Gelation, emulsifying, binding properties along with nontoxicity, biodegradability and overall pH stability of guar gum make it a favorable candidate for edible food packaging (Yavari Maroufi, Ghorbani, and Tabibiazar 2020, Aydogdu et al. 2020). Finally, pectin is an anionic polysaccharide that is present in plant cell walls. It is built with β-1,4-linked α-D-galacturonic acid residues containing completely or partially methyl esterified uronic acid carboxyl groups, making the two main types of pectin. Those are high methoxy pectin and low methoxy pectin. Pectin can be found in fruits and vegetables such as citrus peel and apple (Janjarasskul and Krochta 2010b). Figure 7 depicts the chemical structure of pectin. Among the other edible food packaging materials, pectin has received much attention due to its plasticizer ability. Besides that, pectin possesses nontoxicity, biodegradability, and gas barrier properties (Jahromi et al. 2020). However, poor moisture barrier properties limit the usage of pectin-based edible films (Lazaridou A. 2020).

Microbial exopolysaccharides are a class of polysaccharides secreted by microorganisms and have recently gained attention food packaging industry. The most common exopolysaccharides include pullulan and kefiran. Pullulan is the most often studied microbial polysaccharide in edible food coatings. It is composed of α-1,6-linked-maltotriose units to form a linear polysaccharide. Pullulan is extracted from the yeast-like fungus *Aureobasidium pullulans*. Pullulan exhibits properties such as oil resistance, heat-insulating, and low oxygen permeability (Kumar et al. 2021, Barcelos et al. 2020). Kefiran is another exopolysaccharide extracted from the kefir grains, which contains a complex combination of lactic acid bacteria, acetic bacteria, yeasts, and fungi used in edible coatings (Prado et al. 2015). This polysaccharide is made of D-glucose and D-galactose in a linear manner with roughly equal participation. It comprises a backbone with 1,6-linked glucose, 1,3-linked galactose, 1,4-linked galactose, 1,4-linked glucose and 1,2,6-linked galactose residues and branches attached to O-2 atom of galactose and glucose residues. Hydrolysis of kefiran is difficult due to its complex linkage system (Moradi and Kalanpour 2019). Kefiran is a water-soluble polysaccharide with many exciting properties, including biodegradability, biocompatibility, non-toxicity, emulsifying and stabilizing effect, and good mechanical strength (Moradi and Kalanpour 2019, Motedayen, Khodaiyan, and Salehi 2013). However, the water vapor permeability and many other mechanical and physical properties of kefiran depend on the structure of the molecule and the extraction conditions, including the temperature (Pop et al. 2021). The antimicrobial property of kefiran is an added advantage in edible food packaging applications (Hasheminya and Dehghannya 2020). Figure 8 shows the structures of pullulan and kefiran.

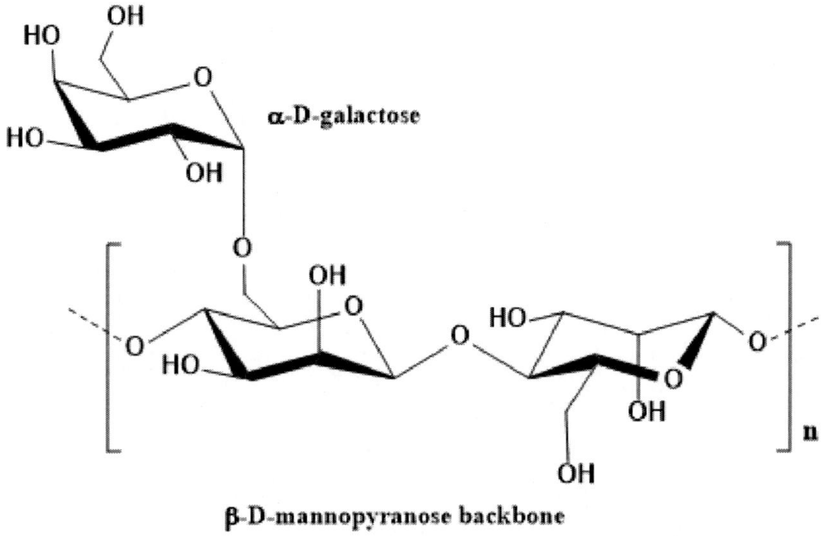

Figure 6. Structure of guar gum.

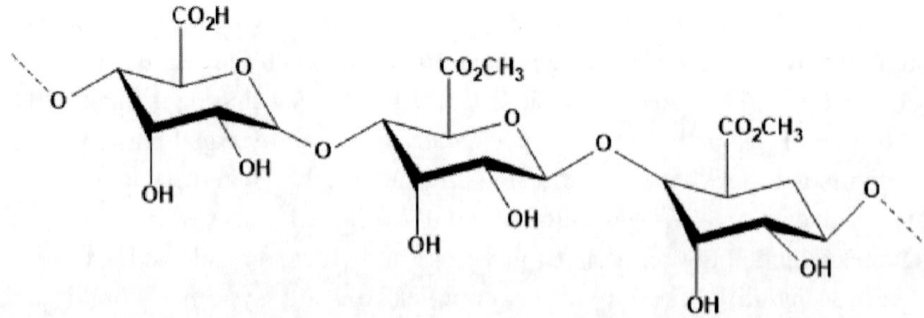

Figure 7. Structure of pectin.

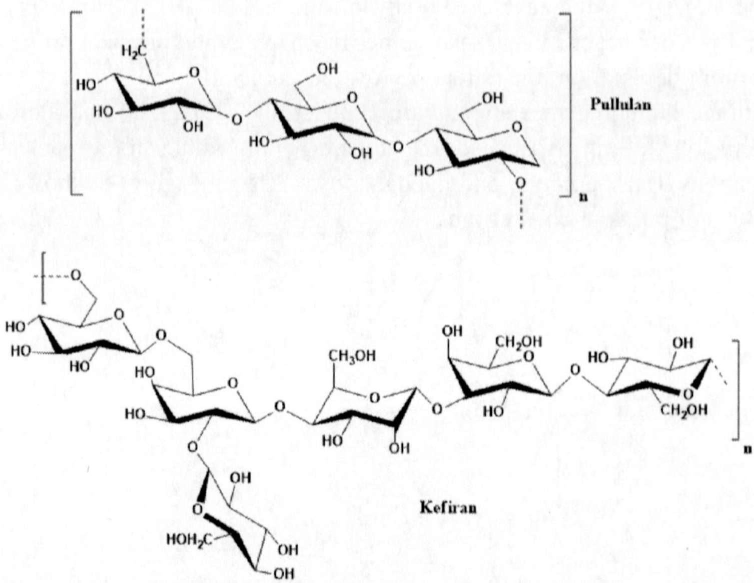

Figure 8. Structures of pullulan and kefiran.

2.2. Polysaccharides-Based Edible Food Packaging

Polysaccharides are excellent raw materials for edible packaging due to their characteristic properties (Hassan et al. 2018). The reliance on petroleum resources can be successfully substituted using polysaccharide-based materials while reducing the carbon footprint of the food industry. Hence, they have been swiftly industrialized and ingrained into edible food packaging. The following section summarizes several important polysaccharides, advantages and limitations and their applications in edible packaging reported in the literature over the last few years. Tables 1 and 2 show the most prevalent polysaccharides used in the production of edible packaging and their common sources, and their benefits and drawbacks.

Table 1. Categorization and sources of commonly used polysaccharides in edible packaging

Polysaccharide	Examples	Sources	Refs.
Cellulose and its derivatives (Plant-based/ microbial-based structural compound)	Cellulose, Methylcellulose (MC), Carboxymethylcellulose (CMC), Hydroxypropyl cellulose (HPC), Hydroxypropyl-methylcellulose (HPMC)	Wood, cotton, husks, bagasse, vegetables, some algae and fungi, some Gram-negative bacteria.	(Nešić et al. 2020, Tayeb et al. 2018)
Hemicellulose (Plant-based structural compound)	Xylan	Hardwoods, gramineous plants, sugar cane, cereal straw, corn stalks and cobs.	(Mikkonen and Tenkanen 2012, Ebringerová 2005)
	Glucomannan	Tubers, Softwoods, tubers and seeds of *Amorphophallus konjac* plants, bacterial and yeast cell wall.	(Chen 2021, Nešić et al. 2020)
Starch (Plant-based energy storage compound)	Amylose	Rice, corn, cassava, wheat, potatoes, tapioca, breadfruit, jackfruit, lotus seeds, pineapple stems and other starchy fruits and vegetables.	(Zhu 2020, Wang et al. 2018, Nakthong, Wongsagonsup, and Amornsakchai 2017)
	Amylopectin	Potato, corn, wheat, barley, tapioca and other starchy fruits and vegetables etc.	(Ai and Jane 2016)
Chitin (Animal-based structural compound)	Chitin, Chitosan, N-(2-azidoethyl) chitosan, and N-(3- azido-2-hydroxypropyl) chitosan	Shells of crustaceans (shrimps, crabs, insects)	(Elieh-Ali-Komi and Hamblin 2016, Wang et al. 2020)
Polysaccharide gums (Plant- and animal-based binding compounds)	Pectin	Fruit and vegetable (apple peel, citrus peel, sweet potato residue, and beet residue)	(Pérez-Jiménez and Saura-Calixto 2018)

Table 1. (Continued)

Polysaccharide	Examples	Sources	Refs.
Polysaccharide gums (Plant- and animal-based binding compounds) (Cont'd)	Agar	Marine red algae (asparagus, ferns, laver, *Gelidium, Gracilaria)*	(Armisén and Gaiatas 2009)
	Alginate	Cell wall and intercellular mucilage of brown algae (*Laminaria, Sargassum, Kelp, Durvillaea potatorum*)	(Mohamed, El-Sakhawy, and El-Sakhawy 2020, Senturk Parreidt, Müller, and Schmid 2018a, Abraham et al. 2019)
	Carrageenan	Cell walls of marine red algae (*Eucheuma, Chondrus, Gigartina,Gelidium, Hypnea*)	(Delattre, Andrea, and Michaud 2011, Mohamed, El-Sakhawy, and El-Sakhawy 2020)
Microbial exopolysaccharides (EPSs)	Pullulan	Fungus (*Aureobasidium pullulans)*	(Farris et al. 2014)
	Kefiran	Lactic acid bacteria	(Moradi and Kalanpour 2019, Piermaria, Canal, and Abraham 2008)

Table 2. Benefits and limitations of polysaccharide-based packaging materials

Advantages	Disadvantages
Relative abundance	Moisture sensitivity
Renewability	Microbial attack
Nontoxicity	Complex Processing steps
Cost-effective	Long processing time
Biodegradable	High operational and chemical cost
Excellent immobilizers	Energy extensive
Excellent chelating agents	Poor thermal stability
Good chemical, physical and mechanical properties	Properties depend on source
Light weight	Limited shelf-life
Low abrasive native	Limited reusability
Less emission of greenhouse gases	Some polysaccharides may contain toxins
Minimal waste production	
Easy functionalization	

An earlier study prepared bio-based polymer composite films using cellulose and spent tea leaf powder (STLP) and displayed their potential for packaging and dye adsorption applications (Duan et al. 2016). The authors also reported high tensile and thermal properties of their biocomposite films (Duan et al. 2016). Pirozzi and coworkers summarized the usage of nanocellulose (NC) in edible coating (EC) for extending the shelf-life of highly perishable fresh and fresh-cut fruits and vegetables reported in the literature (Pirozzi, Ferrari, and Donsì 2021). Tumbarski and et al. investigated the shelf-life and the preservation quality of fresh strawberries using an edible coating prepared using carboxymethyl cellulose (CMC) and bacteriocin (B) produced by *Bacillus methylotrophicus* BM47 (Tumbarski et al. 2019). The authors revealed a considerable decrease in decay incidence and the inhibition of the fungal growth on the coated fruit (Tumbarski et al. 2019). De Dicastillo and coworkers studied the antioxidant activity of glutaraldehyde (GA) cross-linked biocomposite films prepared from methylcellulose (MC) and native Chilean murta fruit extract (MU) and exhibited the potential of their biocomposite film in edible food packaging applications (López de Dicastillo, Bustos, et al. 2016). They also developed antioxidant and antimicrobial active food packaging materials using glutaraldehyde (GA) cross-linked biocomposite film prepared from methylcellulose (MC) and maqui (Aristotelia chilensis) berry fruit extract (López de Dicastillo, Rodríguez, et al. 2016). Barty-King et al. synthesized the edible and cost-effective hydroxypropyl cellulose (HPC)-gels and described their ability to process industrially using extrusion, casting, and high-speed rotary-spinning techniques (Barty-King et al. 2021). They reported edible HPC-hydrogels in mechanochromic materials for colorant-free food decoration and sensors in biodegradable "smart labels" for food packaging applications (Barty-King et al. 2021).

Recently, the xylan-coated paper and film samples were prepared and tested for barrier properties, water-solubility, mechanical strength, and antimicrobial activity against pathogenic bacteria by Petronela et al. (Nechita, Mirela, and Ciolacu 2021). The authors demonstrated the application of xylan-coated paper and films for food packaging. Carboxymethyl xylan (CMX) films cross-linked with Agar (Ag), Ammonium zirconium carbonate (AZC) and linoleic acid (La) were prepared by Queirós and coworkers (Queirós et al. 2017). They also investigated the mechanical, barrier, thermal, optical and antimicrobial properties of CMX.Ag, CMX.AZC, and CMX.La films and evaluated their ability to use as edible films (Queirós et al. 2017). Liu et al. developed edible films based on konjac glucomannan (KGM) loaded with thyme essential oil (TEO) and investigated their mechanical, hydrophilic, antioxidant and antimicrobial properties (Liu et al. 2021). Due to the elongation and high tensile strength, starch-based edible coatings and films have been widely investigated for food packing, as well. For instance, Go et al. exhibited the wrapping of chicken breasts with rye starch films containing rosehip extract (RHE) (Go and Song 2019). The authors use rye starch films incorporated with RHE as active packaging materials to retard lipid oxidation in foods (Go and Song 2019). Yang and coworkers developed active packaging films using foxtail millet starch (FMS) that contains clove leaf oil as an antimicrobial and antioxidant agent (Yang et al. 2018). FMS films containing clove leaf oil also exhibited an improved shelf-life of queso blanco cheese (Yang et al. 2018). Baek et al. reported the prevention of lipid oxidation of salmon using edible films prepared from cowpea starch and maqui berry extract (Baek, Kim, and Song 2019).

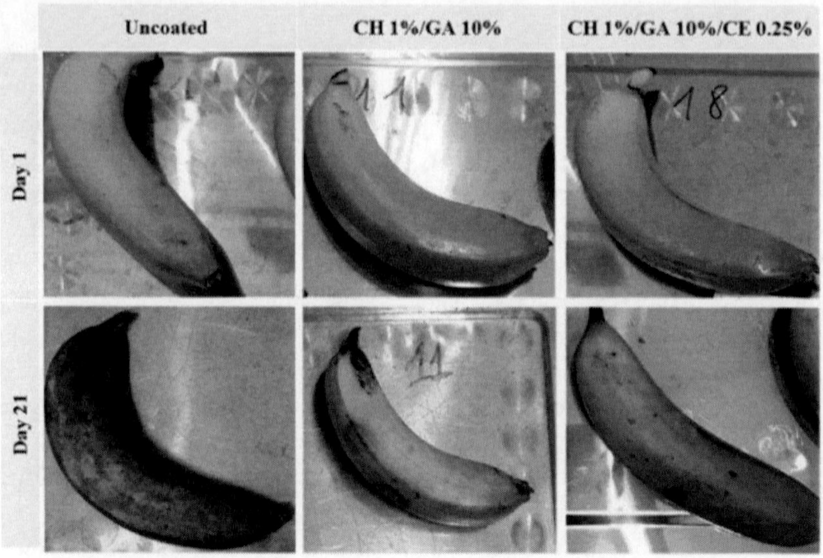

Figure 9. Physical appearance of bananas with and without coatings (CH-Chitosan, GA- gum Arabic CE- an antioxidant additive called Roxb.Extract) before and after 21 days of storage at 25°C and 70% relative humidity (Reprinted with permission from Le et al., 2021. Copyright 2021 Elsevier Ltd).

Chitosan is well-known for its antifungal and antibacterial characteristics and hence tested for edible food packaging. For instance, Le et al. prepared an edible coating using chitosan (CH), gum arabic (GA) and the antioxidant additive, Cleistocalyx operculatus (Roxb.) (CE). The CE/CH/GA coating was tested for the preservation of bananas under ambient storage conditions (Makvandi et al. 2021). The authors reported that the freshness of bananas treated with CH/GA/CE edible coating could last for 21 days of storage at room conditions, making them promising for preserving fruits. Figure 9 shows the physical appearance of bananas with and without CE/CH/GA coating before and after 21 days of storage at 25°C and 70% relative humidity. Malm et al. investigated the mechanical, barrier, water-resistance and light barrier characteristics of edible Cricket chitosan films and showed their potential as bio-based packaging material for food and pharmaceutical applications (Malm et al. 2021). Based on the previous studies reported, the potential use of chitin and chitosan in biodegradable functional films for food packaging applications has been summarized by Priyadarshi and coworkers (Priyadarshi and Rhim 2020).

Polysaccharide-based gum, pectin, is generally recognized as safe (GRAS) and commonly used in the food industry as gelling, stabilizing, and thickening agent. Pectin has widely been considered for edible food packaging. For example, Espitia and coworkers summarized the main methods for preparing edible pectin films, principal characterization techniques and applications of pectin edible films as antimicrobial food packaging based on past studies (Espitia et al. 2014). Valdés and coworkers reviewed the previously reported innovations in food packaging by using pectin-based edible films (Valdés et al. 2015). Agar and other gums have been effectively employed as edible coatings in food products. Mostafavi reviewed the applications of agar-based edible films in food packaging (Mostafavi and Zaeim 2020). Wardana and coworkers developed bioindicator edible films (BEF) from tapioca starch (TS), agar, and red cabbage juice (RC) and demonstrated its performance on sausage deterioration detection (Wardana and Widyaningsih 2017). Kumar et al. exhibited that the shelf-life of green grapes could extend with agar-ZnO nanocomposite coatings (Kumar et al. 2019). Pavli et al. displayed the antipathogenic activity of oregano essential oil (OEO) incorporated in Na-alginate edible films against *Listeria monocytogenes* strains with high-pressure processing (HPP) (Pavli et al. 2019). Parreidt and coworkers evaluated the previous literature on alginate-based edible films and coatings for enhancing the shelf-life of fresh-cut fruits and vegetables, meat, poultry, seafood, and cheese (Senturk Parreidt, Müller, and Schmid 2018b). The nanohybrid-based alginate-ascorbic acid-layered double hydroxide (alginate-AA-LDHs) coating was used to preserve the quality of the psychochemical parameters of strawberries by Madhusha et al. (Madhusha et al. 2020). As shown in Figure 10, the alginate-AA-LDHs dipped strawberry sample shows the highest preservation compared to other control experiments after 18 days. Semi-refined kappa-carrageenan (SRC)-based edible films plasticized with glycerol or sorbitol at 20, 25 and 30% (w/w) were prepared by Farhan et al. They demonstrated the applicability of SRC as an economical edible packaging material with improved mechanical and barrier properties (Farhan and Hani 2017).

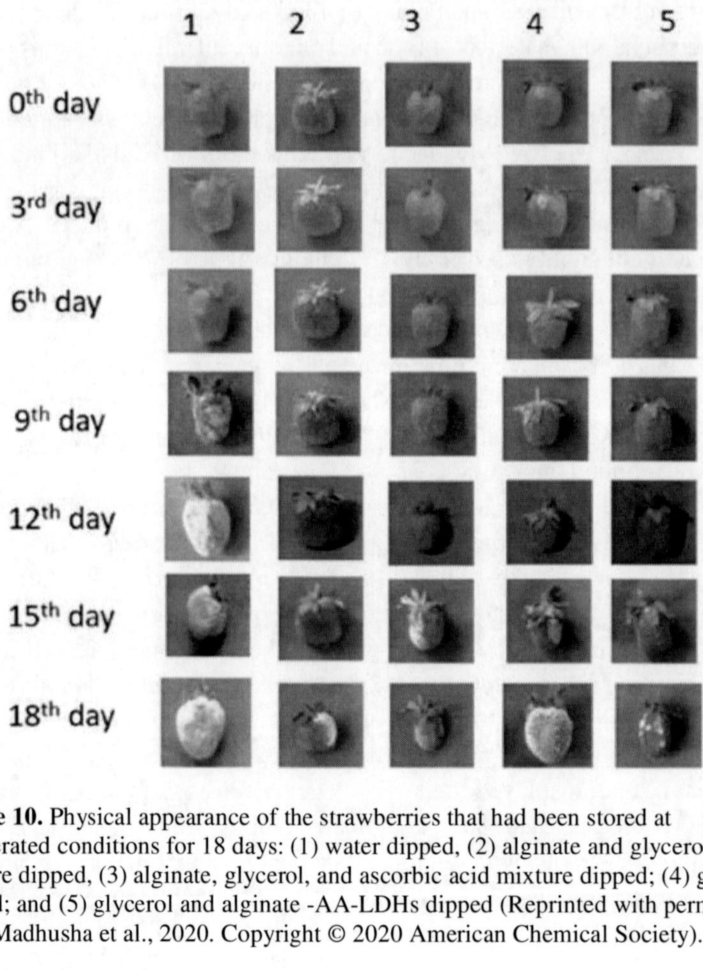

Figure 10. Physical appearance of the strawberries that had been stored at refrigerated conditions for 18 days: (1) water dipped, (2) alginate and glycerol mixture dipped, (3) alginate, glycerol, and ascorbic acid mixture dipped; (4) glycerol dipped; and (5) glycerol and alginate -AA-LDHs dipped (Reprinted with permission from Madhusha et al., 2020. Copyright © 2020 American Chemical Society).

Pullulan, a microbial exopolysaccharide, has been widely studied for edible coating due to its higher mechanical strength. For example, coatings made from pullulan/laminaria japonica oligosaccharides are known to extend the shelf-life of cherry tomatoes and pullulan coatings supplemented with *Bergenia crassifolia* extracts could produce a strong antibacterial activity against a variety of gram-positive and gram-negative foodborne pathogens and extend the shelf-life of peppers and apples during post-harvest storage (Wu, Lu, and Wang 2016, Kraśniewska et al. 2015). Kefiran, a food-derived microbial exopolysaccharide extracted from kefir grains, has also been extensively studied as edible food packaging material. For example, Tan and coworkers summarized the properties of kefiran as a food-derived biopolymer for agri-food and biomedical applications based on the studies reported in the literature (Tan, Chamundeswari, and Loo 2020).

2.3. Advantages and Challenges of Polysaccharide-Based Edible Food Packaging

The main advantage of edible food packaging is its edibility, and therefore no landfilling or recycling process is necessary (Cazón et al. 2017). Moreover, even if the edible film is not consumed, it does not contribute to environmental pollution because of efficient degradation. Many edible polysaccharides can be derived from renewable and abundant natural resources, and many of them are considered biomass waste or by-products (Kim et al. 2006). The edible polysaccharide packaging enhances organoleptic characteristics, including color, sweetness, and aroma. Due to their surface tunability, Polysaccharides have the ability to incorporate active compounds such as flavors, colorants, antioxidants, nutrients, and antimicrobial agents into the packaging material (Barbosa et al. 2021, Saklani et al. 2019). Active compounds can release or absorb substances into or from the food product, or the surrounding environment, expanding the shelf-life and improving the quality of packaged food (Barbosa et al. 2021). Polysaccharide-based edible packaging is also used to fortify natural coating at the exteriors layer of the product to prevent moisture losses, gas, aroma, and solute movement from the food (Kocira et al. 2021). Edible coating or film can also be used as a boundary among the layers of different foods to inhibit deterioration.

Despite the advantages mentioned above, one of the key issues associated with polysaccharide-based edible packaging materials is their poor thermal and chemical stability (Nešić et al. 2020). Weak mechanical properties, including brittleness, wrinkling, and shrinking are major drawbacks of polysaccharide-based edible packaging (Pal et al. 2021). Furthermore, the availability of hydroxyl, amino, or carboxyl functional groups in their structure, triggers moisture absorbance, diminishing the mechanical strength and barrier properties (Nešić et al. 2020). It has also been observed that some polysaccharides such as starch, agar, cellulose, and hemicellulose, have aided microbial growth by providing the required nutrients, thereby negatively impacting the shelf-life and the quality of the food product (Nešić et al. 2020). The life cycle assessment of the polysaccharide-based edible packaging materials has not been thoroughly evaluated to foresee the applicability, economic and social impacts. Safety risk assessment of many polysaccharides-based edible food packaging materials is yet to be fully examined (Zhao et al. 2021). Moreover, the enzyme activity in polysaccharide-coated fruits and vegetables on regulating genes associated with ripening and senescence is still under investigation (Li et al. 2017, Li et al. 2021).

3. Current Trends in Polysaccharide-Based Edible Food Packaging

The recent focus on using edible polysaccharide coatings as food packaging materials is to maintain the quality, safeguard the freshness, preserve the organoleptic properties, inhibit enzymatic browning, and postpone senescence of fresh-cut fruits and vegetables. Numerous studies are currently underway to produce food with high nutritional value and extended half-life using polysaccharide-based advanced materials.

3.1. Active Packaging

Current trends in polysaccharide-based edible food packaging include the preparation of micro and nanoencapsulation of active compounds, including anti-browning agents, antioxidants, enzymes, flavorings, vitamins, antimicrobials, essential oils and colorants, controlled release of active compounds, nutritional value enhancement, protection over extreme storage conditions and the development of bioactive delivery systems via the edible coating approach (Salgado et al. 2021). This process is called active packaging. The interaction between active material present in the package and food enhances the quality and safety of food. Currently employed active systems in food packing include oxygen scavengers, carbon dioxide emitters, moisture absorbers, anti-fogging and anti-sticking films, ethanol emitters, ethylene scavengers, time-temperature indicators, flavor releasing and absorbing films and antimicrobial films (Jafarizadeh-Malmiri et al. 2019, Kumar et al. 2020). Moreover, bioplastics developed from biomaterials, including proteins, polysaccharides, lipids, microorganisms, and bio-based monomers, have been widely considered for edible food packaging (Trajkovska Petkoska, Daniloski, Kumar, et al. 2021). In addition, nanotechnology and intelligent packaging have recently emerged as promising techniques to improve the functionality of edible food packaging (Trajkovska Petkoska, Daniloski, D'Cunha, et al. 2021).

3.2. Intelligent Packaging

Intelligent packaging materials are applied to monitor and provide the condition of packaged foods, including the quality and freshness during transport and storage. Generally, intelligent packaging contains external or internal sensors to detect freshness, gas, and temperature attached to the surface of headspace or packaging, providing specific qualitative information about the food product. Interestingly, polysaccharides have also been recently investigated for intelligent food packaging (Perera et al. 2021). For instance, Wu and coworkers demonstrated an electrochemical writing methodology on chitosan/agarose hydrogel film by embedding anthocyanin (ATH) (Wu et al. 2018). The authors observed a distinct color change upon the spoilage of crucian fish for their written chitosan/agarose/ATH packaging film. They further reported that written film could be used as an intelligent food packaging material to visualize and detect the spoilage of fish (Wu et al. 2018). Yao and coworkers developed different polysaccharides (chitosan, κ-carrageenan and locust bean gum)/polyvinyl alcohol (PVA) intelligent packaging films containing betacyanin-rich red pitaya flesh extract (RPFE) (Yao et al. 2022). They investigated light transmittance, moisture content, water vapor permeability, tensile strength, elongation at break, pH-sensitivity and ammonia-sensitivity for their polysaccharide/PVA-RPFE films (Yao et al. 2022). The locust bean gum/PVA/RPFE film effectively monitored shrimp freshness over other polysaccharide/PVA-RPFE films. Films and coatings prepared from polysaccharides and polyphenol compounds have widely been investigated for intelligent packaging of foods (Zhu 2021).

Recently, Qin et al. developed active and intelligent packaging films by loading different amounts of *Lycium ruthenicum* anthocyanins (LRA) into the starch matrix (Qin et al. 2019). The authors investigated the structural, physical, antioxidant and pH-sensitive properties of starch-LRA films (Qin et al. 2019). They also reported the application of starch-LRA films for monitoring the freshness of pork. Liu and coworkers prepared intelligent colorimetric films by incorporating anthocyanin-loaded ovalbumin-carboxymethyl cellulose nanocomplexes (OVA-CMC-ACNs) into corn starch/polyvinyl alcohol matrix (SP) (Liu et al. 2022). The authors demonstrated the utilization of SP/OVA-CMC-ACNs film for the freshness monitoring of mushrooms. Furthermore, electronic nose (EN) and electronic tongue (ET) devices are employed for intelligent packaging, too. ENs and ETs mimic human smell and taste sensors, detecting the smell and taste of food products (Baldwin et al. 2011). The artificial electronic eye (EE) is devised for visual inspection and acquiring color- and aspect-related information from food products (Calvini and Pigani 2022). Thermochromic inks are developed as freshness indicators in intelligent food packaging applications. These thermochromic inks change color based on temperature, indicating to the consumer whether a food is hot or cold (Salgado et al. 2021). Interestingly, intelligent packaging help also protects against tampering, counterfeiting and theft of the packing materials (Kumar et al. 2020).

3.4. Role of Nanotechnology in Polysaccharide-Based Edible Food Packaging

Nanotechnology involves the manipulation of matter at dimensions in the nanometer scale of 1 nm to 100 nm to produce new structures, devices and materials. Nanomaterials exhibit district characteristics that are significantly different from their bulk counterparts due to their high surface area to volume ratio and quantum effects at the nanoscale. Nanomaterials also possess unique features such as high surface area, high mechanical strength, solubility, diffusivity, nontoxicity, magnetic, optical, and thermodynamic properties. Therefore, nanotechnology promises scientific developments in various fields, including healthcare, medicine, agriculture, food, energy, materials and manufacturing. Nanotechnology is becoming increasingly important in the food packaging sector as a method for incorporating biological materials into nanoartifacts. Polysaccharides including cellulose, chitosan, starch, pectin, alginate, carrageenan, pullulan, and kefiran are commonly utilized in nanotechnology-integrated food packaging (Cazón et al. 2017). Polysaccharides are frequently combined with other materials to reduce the permeability and improve the mechanical properties of coating films for food packaging applications (Dogra, Verma, and Fortunati 2022). Nanomaterials in food packing enhance the mechanical strength, water and gas barrier qualities, and antibacterial and light-blocking properties. Moreover, the nanomaterials also improve the shelf-life and the food quality (Singh et al. 2017, Primožič, Knez, and Leitgeb 2021).

The application of nanotechnology in edible food packaging involves the development of packaging materials at the nanoscale or the addition of nanomaterials to improve the quality of the matrix (Chawla, Sivakumar, and Kaur 2021). Numerous studies have been reported on the use of nanomaterials in tandem with polysaccharides for preparing edible food packing materials. For instance, Trifol and coworkers developed a cellulose-based food packaging material using a polylactic acid (PLA)/cellulose nanofibers/nanoclay (C30B) composite. The authors reported reduced water vapor absorption and transmission over PLA material for their PLA/cellulose nanofibers/nanoclay (C30B) composite (Trifol et al. 2020). Electrospun fibrous composite synthesized from chitosan, nanolayers of Ag nanoparticles, chitosan, and polyvinyl alcohol (PVA) was successfully employed by Pandey and coworkers as meat packaging material with increased bioactivity and prolonged meat shelf-life by Pandey and coworkers (Pandey et al. 2020). Zhang et al. prepared a chitosan-gelatin bio-based edible coating infused with nano-encapsulated tarragon essential oils to preserve pork slices (Zhang et al. 2020). Their edible coating exhibited improved antioxidant, antibacterial, and sensory qualities (Zhang et al. 2020). Chitosan nanoparticles with diverse essential oils such as cinnamon, assai pulp, thyme, lemon, Zataria multiflora, and tarragon were used to develop active food packaging by Rehman et al. (Rehman et al. 2020). They showed reductions in oxidation and microbiological developments in a variety of food products (Rehman et al. 2020). Further, an edible film made with chitosan nanoparticles embedded in the starch film was successfully applied on cherry tomatoes to extend their self-life by hindering the development of E. coli and Staphylococcus aureus (Shapi'i et al. 2020). Moreover, the addition of inorganic nanofillers including TiO_2, ZnO, and ZnS, and carbon nanotubes, have improved the appearance, stability during shipment and storage, flavor, acid, sugar, color, and texture while reducing the spoilage (Bratovcic et al. 2015). It has also been observed that the thermal and mechanical stability of polysaccharide films and nanocomposites could be improved by adding reinforcing agents such as silver

(Ag) nanoparticles and cellulose nanostructures such as cellulose nanocrystals and nanocellulose (Pal et al. 2021).

The immobilization of nanomaterials helps improve the antibacterial properties of edible packaging materials. Metal-based nanoparticles have been widely exploited as antimicrobial agents in food packaging (Kraśniewska, Galus, and Gniewosz 2020, Primožič, Knez, and Leitgeb 2021). For example, Silva and coworkers reported effective antibacterial protection for their starch-based edible coatings containing silver nanoparticles against E. coli and Salmonella spp., especially storing food items such as fresh cheese (Silva et al. 2021). Similarly, grapes coated with a starch-based edible coating containing carrot cellulose nanofibers have provided excellent antibacterial protection against *E. coli*, *Penicillium spp*., and *L. monocytogenes* like bacterial spp (Yuan et al. 2021). Salarbashi et al. prepared a soybean polysaccharide (SSPS)/Cloisite 30B antimicrobial nanocomposite films and showed the inhibition the growth of *Salmonella typhi* PTCC 1609, *Staphylococcus epidermis* PTCC 1114 (ATCC 12228) and *Listeria monocytogenes* PTCC 1165 (Salarbashi et al. 2018).

Perishable foods including, vegetables, fruit, meat, and frozen foods, must be kept in optimal environmental conditions to avoid their deterioration. However, during food processing, packaging, transporting, and storing, the quality, freshness and shelf-life of food can be affected. Detection of food spoilage is often difficult to identify at the early stage. Interestingly, incorporating nanosensor technology in food packaging allows the detection, identification, and quantification of food freshness, toxins, adulterants and contaminants (Hernández-Muñoz et al. 2019, Kumar, Guleria, and Mehta 2017, Trajkovska Petkoska, Daniloski, Kumar, et al. 2021). Nanosensors ensure fast detection of food quality at any time, subsequently preventing the food-related diseases caused by the ingestion of spoiled food. Despite the fact that nanotechnology utilized in food packaging processing allows for significant improvements in packaging material and environmental qualities, more research and development are required for a better understanding of nanotechnology in food packaging materials (Primožič, Knez, and Leitgeb 2021). The risk of nanoparticle migration from packaging into food has been highlighted as a major concern. Since nanomaterials are very minute, they can enter human cells and cause damage to the life substances in cells. Therefore, it is always important to get approval from food regulatory bodies before using nanoparticles (Chawla, Sivakumar, and Kaur 2021). However, stating the benefits and risk assessment are ambiguous, they must be defined well, especially due to the shortage of additional data from the clinical trials (Chaudhary, Fatima, and Kumar 2020).

Conclusion

Edible polysaccharides have been widely explored in food and non-food packaging applications due to their renewability, degradability, non-toxicity, excellent mechanical and barrier properties, antioxidant and antimicrobial activity, and edibility. Polysaccharide-based edible packaging plays an imperative role in the safety, quality, transportation, and storage of perishable and processed food products. In this chapter, the structure, characteristic properties, and applications of polysaccharide-based materials in edible food packaging materials have been extensively discussed. Polysaccharides extracted from different origins, including plants, animals, algae, and microorganisms, have been described. Major advantages, challenges, current prospects and the role of active packaging, intelligent packaging and nanotechnology in polysaccharide-based edible food packaging have also been summarized. In this next chapter, our colleagues will discuss more deeply on the development of films by using cellulose for food packaging applications.

References

Abraham, Reinu E., Peng Su, Munish Puri, Colin L. Raston, and Wei Zhang. 2019. "Optimisation of biorefinery production of alginate, fucoidan and laminarin from brown seaweed Durvillaea potatorum." *Algal Research* 38:101389. doi: https://doi.org/10.1016/j.algal.2018.101389.

Adzaly, Noor Zainah, Norzaleha Kasim, Siah Moey, M. N. Anvarali, Mohd Romainor Manshor, K. H. Halim, A. Akbar, and H. Mohamad. 2018. "Shelf life study and sensory evaluation of culinary product ('Asam pedas' cube) using Chitosan film." *Malaysian Applied Biology* 47:79-87.

Ai, Y., and J. l Jane. 2016. "Starch: Structure, Property, and Determination." *Encyclopedia of Food and Health*. doi: 10.1016/B978-0-12-384947-2.00657-7.

Alamri, M. S., Akram A. A. Qasem, Abdellatif A. Mohamed, Shahzad Hussain, Mohamed A. Ibraheem, Ghalia Shamlan, Hesham A. Alqah, and Ali S. Qasha. 2021. "Food packaging's materials: A food safety perspective." *Saudi Journal of Biological Sciences* 28 (8):4490-4499. doi: https://doi.org/10.1016/j.sjbs.2021.04.047.

Amaraweera, Sumedha M., Chamila Gunathilake, Oneesha H. P. Gunawardene, Nimasha M. L. Fernando, Drashana B. Wanninayaka, Rohan S. Dassanayake, Suranga M.

Rajapaksha, Asanga Manamperi, Chakrawarthige A. N. Fernando, Asela K. Kulatunga, and Aruna Manipura. 2021. "Development of Starch-Based Materials Using Current Modification Techniques and Their Applications: A Review." *Molecules* 26 (22):6880.

Amaraweera, Sumedha M., Chamila Gunathilake, Oneesha H. P. Gunawardene, Nimasha M. L. Fernando, Drashana B. Wanninayaka, Asanga Manamperi, Rohan S. Dassanayake, Suranga M. Rajapaksha, Mahinda Gangoda, C. A. N. Fernando, Asela K. Kulatunga, and Aruna Manipura. 2021. "Preparation and characterization of biodegradable cassava starch thin films for potential food packaging applications." *Cellulose* 28 (16):10531-10548. doi: 10.1007/s10570-021-04199-6.

Armisén, R., and F. Gaiatas. 2009. "4 - Agar." In *Handbook of Hydrocolloids (Second Edition)*, edited by G. O. Phillips and P. A. Williams, 82-107. Woodhead Publishing.

Atta, Omar Mohammad, Sehrish Manan, Abeer Ahmed Qaed Ahmed, Mohamed F. Awad, Mazhar Ul-Islam, Fazli Subhan, Muhammad Wajid Ullah, and Guang Yang. 2021. "Development and Characterization of Yeast-Incorporated Antimicrobial Cellulose Biofilms for Edible Food Packaging Application." *Polymers* 13 (14):2310.

Aydogdu, Ayca, Clayton J. Radke, Semih Bezci, and Emrah Kirtil. 2020. "Characterization of curcumin incorporated guar gum/orange oil antimicrobial emulsion films." *International Journal of Biological Macromolecules* 148:110-120. doi: https://doi.org/10.1016/j.ijbiomac.2019.12.255.

Baek, Su-Kyoung, Sujin Kim, and Kyung Bin Song. 2019. "Cowpea starch films containing maqui berry extract and their application in salmon packaging." *Food Packaging and Shelf Life* 22:100394. doi: https://doi.org/10.1016/j.fpsl.2019.100394.

Baldwin, Elizabeth A., Jinhe Bai, Anne Plotto, and Sharon Dea. 2011. "Electronic noses and tongues: applications for the food and pharmaceutical industries." *Sensors (Basel, Switzerland)* 11 (5):4744-4766. doi: 10.3390/s110504744.

Barbosa, Cássia H., Mariana A. Andrade, Fernanda Vilarinho, Ana Luísa Fernando, and Ana Sanches Silva. 2021. "Active Edible Packaging." *Encyclopedia* 1 (2):360-370.

Barcelos, M. C. S., K. A. C. Vespermann, F. M. Pelissari, and G. Molina. 2020. "Current status of biotechnological production and applications of microbial exopolysaccharides." *Crit Rev Food Sci Nutr* 60 (9):1475-1495. doi: 10.1080/10408398.2019.1575791.

Barty-King, Charles H., Chun Lam Clement Chan, Richard M. Parker, Mélanie M. Bay, Roberto Vadrucci, Michael De Volder, and Silvia Vignolini. 2021. "Mechanochromic, Structurally Colored, and Edible Hydrogels Prepared from Hydroxypropyl Cellulose and Gelatin." *Advanced Materials* 33 (37):2102112. doi: https://doi.org/10.1002/adma.202102112.

Benbettaïeb, Nasreddine, Jean-Pierre Gay, Thomas Karbowiak, and Frédéric Debeaufort. 2016. "Tuning the Functional Properties of Polysaccharide–Protein Bio-Based Edible Films by Chemical, Enzymatic, and Physical Cross-Linking." *Comprehensive Reviews in Food Science and Food Safety* 15 (4):739-752. doi: https://doi.org/10.1111/1541-4337.12210.

Borić, Milka, Filipa A. Vicente, Damjan Lašič Jurković, Uroš Novak, and Blaž Likozar. 2020. "Chitin isolation from crustacean waste using a hybrid demineralization/DBD plasma process." *Carbohydrate Polymers* 246:116648. doi: https://doi.org/10.1016/j.carbpol.2020.116648.

Boz, Ziynet, Virpi Korhonen, and Claire Koelsch Sand. 2020. "Consumer Considerations for the Implementation of Sustainable Packaging: A Review." *Sustainability* 12 (6):2192.

Calvini, Rosalba, and Laura Pigani. 2022. "Toward the Development of Combined Artificial Sensing Systems for Food Quality Evaluation: A Review on the Application of Data Fusion of Electronic Noses, Electronic Tongues and Electronic Eyes." *Sensors* 22 (2):577.

Cazón, Patricia, Gonzalo Velazquez, José A. Ramírez, and Manuel Vázquez. 2017. "Polysaccharide-based films and coatings for food packaging: A review." *Food Hydrocolloids* 68:136-148. doi: https://doi.org/10.1016/j.foodhyd.2016.09.009.

Chaudhary, Pallavi, Faria Fatima, and Ankur Kumar. 2020. "Relevance of Nanomaterials in Food Packaging and its Advanced Future Prospects." *Journal of Inorganic and Organometallic Polymers and Materials* 30 (12):5180-5192. doi: 10.1007/s10904-020-01674-8.

Chawla, Rekha, S. Sivakumar, and Harsimran Kaur. 2021. "Antimicrobial edible films in food packaging: Current scenario and recent nanotechnological advancements- a review." *Carbohydrate Polymer Technologies and Applications* 2:100024. doi: https://doi.org/10.1016/j.carpta.2020.100024.

Chen, X.; Pang, J.; Wu, C. 2021. "Fabrication and Characterization of Antimicrobial Food Packaging Materials Composed of Konjac Glucomannan, Chitosan and Fulvic Acid." *Food Science* (7):232–239.

Dassanayake, Rohan S., Niwanthi Dissanayake, Juan S. Fierro, Noureddine Abidi, Edward L. Quitevis, Kiran Boggavarappu, and Vidura D. Thalangamaarachchige. 2021. "Characterization of cellulose nanocrystals by current spectroscopic techniques." *Applied Spectroscopy Reviews*. 1-26. doi: 10.1080/05704928.2021.1951283.

Delattre, Cédric, Fenoradosoa Andrea, and Philippe Michaud. 2011. "Galactans: An Overview of their Most Important Sourcing and Applications as Natural Polysaccharides." *Brazilian Archives of Biology and Technology* 54:1075-1092. doi: 10.1590/S1516-89132011000600002.

Deshwal, Gaurav Kr, Narender Raju Panjagari, and Tanweer Alam. 2019. "An overview of paper and paper based food packaging materials: health safety and environmental concerns." *Journal of Food Science and Technology* 56 (10):4391-4403. doi: 10.1007/s13197-019-03950-z.

Dhumal, Chanda Vilas, Jasim Ahmed, Nandika Bandara, and Preetam Sarkar. 2019. "Improvement of antimicrobial activity of sago starch/guar gum bi-phasic edible films by incorporating carvacrol and citral." *Food Packaging and Shelf Life* 21:100380. doi: https://doi.org/10.1016/j.fpsl.2019.100380.

Diego Aires da Silva, Giselle Cristine Melo Aires and Rosinelson da Silva Pena. 2020. "Gums—Characteristics and Applications in the Food Industry." In *Innovation in the Food Sector Through the Valorization of Food and Agro-Food By-Products*, edited by Ana Novo de Barros and Irene Gouvinhas. IntechOpen.

Duan, J., K. Obi Reddy, B. Ashok, J. Cai, L. Zhang, and A. Varada Rajulu. 2016. "Effects of spent tea leaf powder on the properties and functions of cellulose green composite films." *Journal of Environmental Chemical Engineering* 4 (1):440-448. doi: https://doi.org/10.1016/j.jece.2015.11.029.

Dwivany, Fenny Martha, Ayesha Nilam Aprilyandi, Veinardi Suendo, and Nisrina Sukriandi. 2020. "Carrageenan Edible Coating Application Prolongs Cavendish Banana Shelf Life." *International Journal of Food Science* 2020:8861610. doi: 10.1155/2020/8861610.

Ebringerová, Anna. 2005. "Structural Diversity and Application Potential of Hemicelluloses." *Macromolecular Symposia* 232 (1):1-12. doi: https://doi.org/10.1002/masy.200551401.

Elieh-Ali-Komi, Daniel, and Michael R. Hamblin. 2016. "Chitin and Chitosan: Production and Application of Versatile Biomedical Nanomaterials." *International journal of advanced research* 4 (3):411-427.

Espitia, Paula Judith Pérez, Wen-Xian Du, Roberto de Jesús Avena-Bustillos, Nilda de Fátima Ferreira Soares, and Tara H. McHugh. 2014. "Edible films from pectin: Physical-mechanical and antimicrobial properties - A review." *Food Hydrocolloids* 35:287-296. doi: https://doi.org/10.1016/j.foodhyd.2013.06.005.

Fabra, M. J., A. López-Rubio, and J. M. Lagaron. 2014. "15 - Biopolymers for food packaging applications." In *Smart Polymers and their Applications*, edited by María Rosa Aguilar and Julio San Román, 476-509. Woodhead Publishing.

FAO. 2021a. FAO. 2021. FAOSTAT. Value of Agricultural Production. In: *FAO.* edited by FAO. Rome, Italy: FAO.

FAO. 2021b. FAO. 2021. The State of Food and Agriculture 2021. Making agrifood systems more resilient to shocks and stresses. Rome, FAO. In *The State of Food and Agriculture (SOFA)*, edited by FAO.

FAO, IFAD, UNICEF, WFP and WHO. 2020. FAO, IFAD, UNICEF, WFP & WHO. 2020. The State of Food Security and Nutrition in the World 2020. Transforming food systems for affordable healthy diets. In *The State of Food Security and Nutrition in the World (SOFI)*. Rome, Italy.

Farhan, Abdulaal, and Norziah Mohd Hani. 2017. "Characterization of edible packaging films based on semi-refined kappa-carrageenan plasticized with glycerol and sorbitol." *Food Hydrocolloids* 64:48-58. doi: https://doi.org/10.1016/j.foodhyd.2016.10.034.

Farris, Stefano, Ilke Uysal Unalan, Laura Introzzi, José Maria Fuentes-Alventosa, and Carlo A. Cozzolino. 2014. "Pullulan-based films and coatings for food packaging: Present applications, emerging opportunities, and future challenges." *Journal of Applied Polymer Science* 131 (13). doi: https://doi.org/10.1002/app.40539.

Fierens, T., K. Servaes, M. Van Holderbeke, L. Geerts, S. De Henauw, I. Sioen, and G. Vanermen. 2012. "Analysis of phthalates in food products and packaging materials sold on the Belgian market." *Food and Chemical Toxicology* 50 (7):2575-2583. doi: https://doi.org/10.1016/j.fct.2012.04.029.

García, Arantzazu Valdés, Olga B. Álvarez-Pérez, Romeo Rojas, Cristobal N. Aguilar, and María Carmen Garrigós. 2020. "Impact of Olive Extract Addition on Corn Starch-Based Active Edible Films Properties for Food Packaging Applications." *Foods* 9 (9):1339.

Geueke, Birgit, Ksenia Groh, and Jane Muncke. 2018. "Food packaging in the circular economy: Overview of chemical safety aspects for commonly used materials." *Journal of Cleaner Production* 193:491-505. doi: https://doi.org/10.1016/j.jclepro.2018.05.005.

Geyer, Roland, Jenna R. Jambeck, and Kara Lavender Law. 2017. "Production, use, and fate of all plastics ever made." *Science Advances* 3 (7):e1700782. doi: doi:10.1126/sciadv.1700782.

Go, Eun-Jeong, and Kyung Bin Song. 2019. "Antioxidant Properties of Rye Starch Films Containing Rosehip Extract and Their Application in Packaging of Chicken Breast." *Starch - Stärke* 71 (11-12):1900116. doi: https://doi.org/10.1002/star.201900116.

Goswami, Bindu. 2019. "The role of food packaging." In: *Global Challenges and Innovation in Science and Management,* ICGCISM/AP-18.

Halonen, Niina, Petra S. Pálvölgyi, Andrea Bassani, Cecilia Fiorentini, Rakesh Nair, Giorgia Spigno, and Krisztian Kordas. 2020. "Bio-Based Smart Materials for Food Packaging and Sensors – A Review." *Frontiers in Materials* 7 (82). doi: 10.3389/fmats.2020.00082.

Hashemi, Seyed Mohammad Bagher, and Dornoush Jafarpour. 2021. "Bioactive edible film based on Konjac glucomannan and probiotic Lactobacillus plantarum strains: Physicochemical properties and shelf life of fresh-cut kiwis." *Journal of Food Science* 86 (2):513-522. doi: https://doi.org/10.1111/1750-3841.15568.

Hasheminya, Seyedeh-Maryam, and Jalal Dehghannya. 2020. "Novel ultrasound-assisted extraction of kefiran biomaterial, a prebiotic exopolysaccharide, and investigation of its physicochemical, antioxidant and antimicrobial properties." *Materials Chemistry and Physics* 243:122645. doi: https://doi.org/10.1016/j.matchemphys.2020.122645.

Hassan, Bilal, Shahzad Ali Shahid Chatha, Abdullah Ijaz Hussain, Khalid Mahmood Zia, and Naseem Akhtar. 2018. "Recent advances on polysaccharides, lipids and protein based edible films and coatings: A review." *International Journal of Biological Macromolecules* 109:1095-1107. doi: https://doi.org/10.1016/j.ijbiomac.2017.11.097.

He, Xiangxiang, Wei Lu, Cuixia Sun, Hoda Khalesi, Analucia Mata, Rani Andaleeb, and Yapeng Fang. 2021. "Cellulose and cellulose derivatives: Different colloidal states and food-related applications." *Carbohydrate Polymers* 255:117334. doi: https://doi.org/10.1016/j.carbpol.2020.117334.

Hernández-Muñoz, Pilar, Josep Pasqual Cerisuelo, Irene Domínguez, Gracia López-Carballo, Ramón Catalá, and Rafael Gavara. 2019. "Chapter 8 - Nanotechnology in Food Packaging." In *Nanomaterials for Food Applications*, edited by Amparo López Rubio, Maria José Fabra Rovira, Marta martínez Sanz and Laura Gómez Gómez-Mascaraque, 205-232. Elsevier.

Hsieh, Y. S. Y., and P. J. Harris. 2019. "Xylans of Red and Green Algae: What Is Known about Their Structures and How They Are Synthesised?" *Polymers (Basel)* 11 (2). doi: 10.3390/polym11020354.

Jafarizadeh-Malmiri, Hoda, Zahra Sayyar, Navideh Anarjan, and Aydin Berenjian. 2019. "Nanobiotechnology in Food Packaging." In *Nanobiotechnology in Food: Concepts, Applications and Perspectives*, edited by Hoda Jafarizadeh-Malmiri, Zahra Sayyar, Navideh Anarjan and Aydin Berenjian, 69-79. Cham: Springer International Publishing.

Jahromi, M., M. Niakousari, M. T. Golmakani, and M. A. Mohammadifar. 2020. "Physicochemical and structural characterization of sodium caseinate based film-forming solutions and edible films as affected by high methoxyl pectin." *Int J Biol Macromol* 165 (Pt B):1949-1959. doi: 10.1016/j.ijbiomac.2020.10.057.

Jancy, S., R. Shruthy, and R. Preetha. 2020. "Fabrication of packaging film reinforced with cellulose nanoparticles synthesised from jack fruit non-edible part using response surface methodology." *Int J Biol Macromol* 142:63-72. doi: 10.1016/j.ijbiomac.2019.09.066.

Janjarasskul, Theeranun, and John Krochta. 2010a. "Edible Packaging Materials." *Annual review of food science and technology* 1:415-48. doi: 10.1146/annurev.food.080708.100836.

Janjarasskul, Theeranun, and John M. Krochta. 2010b. "Edible Packaging Materials." *Annual Review of Food Science and Technology* 1 (1):415-448. doi: 10.1146/annurev.food.080708.100836.

Jeevahan, Jeya, and M. Chandrasekaran. 2019. "Nanoedible films for food packaging: a review." *Journal of Materials Science* 54 (19):12290-12318. doi: 10.1007/s10853-019-03742-y.

Kim, Ki Myong, Jeong Hwa Son, Sung-Koo Kim, Curtis L. Weller, and Milford A. Hanna. 2006. "Properties of Chitosan Films as a Function of pH and Solvent Type." *Journal of Food Science* 71 (3):E119-E124. doi: https://doi.org/10.1111/j.1365-2621.2006.tb15624.x.

Kocira, Anna, Katarzyna Kozłowicz, Katarzyna Panasiewicz, Mariola Staniak, Ewa Szpunar-Krok, and Paulina Hortyńska. 2021. "Polysaccharides as Edible Films and Coatings: Characteristics and Influence on Fruit and Vegetable Quality—A Review." *Agronomy* 11 (5):813.

Kraśniewska, Karolina, Sabina Galus, and Małgorzata Gniewosz. 2020. "Biopolymers-Based Materials Containing Silver Nanoparticles as Active Packaging for Food Applications-A Review." *International journal of molecular sciences* 21 (3):698. doi: 10.3390/ijms21030698.

Kraśniewska, Karolina, Małgorzata Gniewosz, Alicja Synowiec, Jarosław L. Przybył, Katarzyna Bączek, and Zenon Węglarz. 2015. "The application of pullulan coating enriched with extracts from Bergenia crassifolia to control the growth of food microorganisms and improve the quality of peppers and apples." *Food and Bioproducts Processing* 94:422-433. doi: https://doi.org/10.1016/j.fbp.2014.06.001.

Kumar, Nishant, Pratibha, Neeraj, Anka Trajkovska Petkoska, Sawsan Ali AL-Hilifi, and Olaniyi Amos Fawole. 2021. "Effect of Chitosan–Pullulan Composite Edible Coating Functionalized with Pomegranate Peel Extract on the Shelf Life of Mango (Mangifera indica)." *Coatings* 11 (7):764.

Kumar, P., P. Mahajan, R. Kaur, and S. Gautam. 2020. "Nanotechnology and its challenges in the food sector: a review." *Materials Today Chemistry* 17:100332. doi: https://doi.org/10.1016/j.mtchem.2020.100332.

Kumar, Santosh, Jyotish Chandra Boro, Dharitri Ray, Avik Mukherjee, and Joydeep Dutta. 2019. "Bionanocomposite films of agar incorporated with ZnO nanoparticles as an active packaging material for shelf life extension of green grape." *Heliyon* 5 (6):e01867. doi: https://doi.org/10.1016/j.heliyon.2019.e01867.

Kumar, Vineet, Praveen Guleria, and Surinder Kumar Mehta. 2017. "Nanosensors for food quality and safety assessment." *Environmental Chemistry Letters* 15 (2):165-177. doi: 10.1007/s10311-017-0616-4.

Lazaridou A., Biliaderis C. G. 2020. "Edible Films and Coatings with Pectin." In *Pectin: Technological and Physiological Properties*, edited by Kontogiorgos V., 99-123. Springer, Cham.

Li, Hui, Hongbo Shi, Yunqing He, Xiang Fei, and Lincai Peng. 2020. "Preparation and characterization of carboxymethyl cellulose-based composite films reinforced by cellulose nanocrystals derived from pea hull waste for food packaging applications." *International Journal of Biological Macromolecules* 164:4104-4112. doi: https://doi.org/10.1016/j.ijbiomac.2020.09.010.

Li, Li, Jian Sun, Haiyan Gao, Yingbin Shen, Changbao Li, Ping Yi, Xuemei He, Dongning Ling, Jinfeng Sheng, Jiemin Li, Guoming Liu, Fengjin Zheng, Ming Xin, Zhichun Li, and Yayuan Tang. 2017. "Effects of Polysaccharide-Based Edible Coatings on Quality and Antioxidant Enzyme System of Strawberry during Cold Storage." *International Journal of Polymer Science* 2017:9746174. doi: 10.1155/2017/9746174.

Li, Li, Ping Yi, Changbao Li, Ming Xin, Jian Sun, Xuemei He, Jinfeng Sheng, Zhugui Zhou, Fengjin Zheng, Jiemin Li, Guoming Liu, Dongning Ling, Jie Tang, Zhichun Li, Ying Yang, and Yayuan Tang. 2021. "Influence of polysaccharide-based edible coatings on enzymatic browning and oxidative senescence of fresh-cut lettuce." *Food Science & Nutrition* 9 (2):888-899. doi: https://doi.org/10.1002/fsn3.2052.

Liu, Liming, Weina Wu, Luman Zheng, Jiahao Yu, Peilong Sun, and Ping Shao. 2022. "Intelligent packaging films incorporated with anthocyanins-loaded ovalbumin-carboxymethyl cellulose nanocomplexes for food freshness monitoring." *Food Chemistry* 387:132908. doi: https://doi.org/10.1016/j.foodchem.2022.132908.

Liu, Zhe, Dehui Lin, Rui Shen, Runguang Zhang, Liu Liu, and Xingbin Yang. 2021. "Konjac glucomannan-based edible films loaded with thyme essential oil: Physical properties and antioxidant-antibacterial activities." *Food Packaging and Shelf Life* 29:100700. doi: https://doi.org/10.1016/j.fpsl.2021.100700.

Liyanage, Sumedha, Noureddine Abidi, Dick Auld, and Hanna Moussa. 2015. "Chemical and physical characterization of galactomannan extracted from guar cultivars (Cyamopsis tetragonolobus L.)." *Industrial Crops and Products* 74:388-396. doi: https://doi.org/10.1016/j.indcrop.2015.05.013.

López de Dicastillo, C., F. Rodríguez, A. Guarda, and M. J. Galotto. 2016. "Antioxidant films based on cross-linked methyl cellulose and native Chilean berry for food packaging applications." *Carbohydr Polym* 136:1052-60. doi: 10.1016/j.carbpol.2015.10.013.

López de Dicastillo, Carol, Fernanda Bustos, Abel Guarda, and Maria José Galotto. 2016. "Cross-linked methyl cellulose films with murta fruit extract for antioxidant and antimicrobial active food packaging." *Food Hydrocolloids* 60:335-344. doi: https://doi.org/10.1016/j.foodhyd.2016.03.020.

Madhusha, Chamalki, Imalka Munaweera, Veranja Karunaratne, and Nilwala Kottegoda. 2020. "Facile Mechanochemical Approach to Synthesizing Edible Food Preservation Coatings Based on Alginate/Ascorbic Acid-Layered Double Hydroxide Bio-

Nanohybrids." *Journal of Agricultural and Food Chemistry* 68 (33):8962-8975. doi: 10.1021/acs.jafc.0c01879.

Makvandi, P., S. Iftekhar, F. Pizzetti, A. Zarepour, E. N. Zare, M. Ashrafizadeh, T. Agarwal, V. V. T. Padil, R. Mohammadinejad, M. Sillanpaa, T. K. Maiti, G. Perale, A. Zarrabi, and F. Rossi. 2021. "Functionalization of polymers and nanomaterials for water treatment, food packaging, textile and biomedical applications: a review." *Environmental Chemistry Letters* 19 (1):583-611. doi: 10.1007/s10311-020-01089-4.

Malm, Morgan, Andrea M. Liceaga, Fernanda San Martin-Gonzalez, Owen G. Jones, Jose M. Garcia-Bravo, and Ian Kaplan. 2021. "Development of Chitosan Films from Edible Crickets and Their Performance as a Bio-Based Food Packaging Material." *Polysaccharides* 2 (4):744-758.

Mangaraj, Satyabrata, Ranjeet Singh, Kumar Das, Naveen Kumar, and Simran Arora. 2018. "Biopolymers as packaging material in food and allied industry." *International Journal of Chemical Studies, Vol. 6, Issue 2,* 2411-2418.

Marsh, Kenneth, and Betty Bugusu. 2007. "Food Packaging—Roles, Materials, and Environmental Issues." *Journal of Food Science* 72 (3):R39-R55. doi: https://doi.org/10.1111/j.1750-3841.2007.00301.x.

Menzel, Carolin. 2020. "Improvement of starch films for food packaging through a three-principle approach: Antioxidants, cross-linking and reinforcement." *Carbohydrate Polymers* 250:116828. doi: https://doi.org/10.1016/j.carbpol.2020.116828.

Mikkonen, Kirsi S., and Maija Tenkanen. 2012. "Sustainable food-packaging materials based on future biorefinery products: Xylans and mannans." *Trends in Food Science & Technology* 28 (2):90-102. doi: https://doi.org/10.1016/j.tifs.2012.06.012.

Mohamed, Salah A. A., Mohamed El-Sakhawy, and Mohamed Abdel-Monem El-Sakhawy. 2020. "Polysaccharides, Protein and Lipid -Based Natural Edible Films in Food Packaging: A Review." *Carbohydrate Polymers* 238:116178. doi: https://doi.org/10.1016/j.carbpol.2020.116178.

Mohan, Kannan, Abirami Ramu Ganesan, Thirunavukkarasu Muralisankar, Rajarajeswaran Jayakumar, Palanivel Sathishkumar, Venkatachalam Uthayakumar, Ramachandran Chandirasekar, and Nagarajan Revathi. 2020. "Recent insights into the extraction, characterization, and bioactivities of chitin and chitosan from insects." *Trends in Food Science & Technology* 105:17-42. doi: https://doi.org/10.1016/j.tifs.2020.08.016.

Moradi, Zahra, and Nastaran Kalanpour. 2019. "Kefiran, a branched polysaccharide: Preparation, properties and applications: A review." *Carbohydrate Polymers* 223:115100. doi: https://doi.org/10.1016/j.carbpol.2019.115100.

Mostafavi, Fatemeh Sadat, and Davood Zaeim. 2020. "Agar-based edible films for food packaging applications - A review." *International Journal of Biological Macromolecules* 159:1165-1176. doi: https://doi.org/10.1016/j.ijbiomac.2020.05.123.

Motedayen, Ali Akbar, Faramarz Khodaiyan, and Esmail Atai Salehi. 2013. "Development and characterisation of composite films made of kefiran and starch." *Food Chemistry* 136 (3):1231-1238. doi: https://doi.org/10.1016/j.foodchem.2012.08.073.

Nakthong, N., Rungtiwa Wongsagonsup, and Taweechai Amornsakchai. 2017. "Characteristics and potential utilizations of starch from pineapple stem waste." *Industrial Crops and Products* 105:74-82.

Ncube, Lindani Koketso, Albert Uchenna Ude, Enoch Nifise Ogunmuyiwa, Rozli Zulkifli, and Isaac Nongwe Beas. 2020. "Environmental Impact of Food Packaging Materials: A Review of Contemporary Development from Conventional Plastics to Polylactic Acid Based Materials." *Materials (Basel, Switzerland)* 13 (21):4994. doi: 10.3390/ma13214994.

Nechita, Petronela, Roman Mirela, and Florin Ciolacu. 2021. "Xylan Hemicellulose: A Renewable Material with Potential Properties for Food Packaging Applications." *Sustainability* 13 (24):13504.

Nešić, Aleksandra, Gustavo Cabrera-Barjas, Suzana Dimitrijević-Branković, Sladjana Davidović, Neda Radovanović, and Cédric Delattre. 2020. "Prospect of Polysaccharide-Based Materials as Advanced Food Packaging." *Molecules* 25 (1):135.

Nilsen-Nygaard, Julie, Estefanía Noriega Fernández, Tanja Radusin, Bjørn Tore Rotabakk, Jawad Sarfraz, Nusrat Sharmin, Morten Sivertsvik, Izumi Sone, and Marit Kvalvåg Pettersen. 2021. "Current status of biobased and biodegradable food packaging materials: Impact on food quality and effect of innovative processing technologies." *Comprehensive Reviews in Food Science and Food Safety* 20 (2):1333-1380. doi: https://doi.org/10.1111/1541-4337.12715.

Pal, Kunal, Preetam Sarkar, Arfat Anis, Karolina Wiszumirska, and Maciej Jarzębski. 2021. "Polysaccharide-Based Nanocomposites for Food Packaging Applications." *Materials* 14 (19):5549.

Park, J. K., and T. Khan. 2009. "21 - Other microbial polysaccharides: pullulan, scleroglucan, elsinan, levan, alternant, dextran." In *Handbook of Hydrocolloids (Second Edition)*, edited by G. O. Phillips and P. A. Williams, 592-614. Woodhead Publishing.

Pavli, Foteini, Anthoula A. Argyri, Panagiotis Skandamis, George-John Nychas, Chrysoula Tassou, and Nikos Chorianopoulos. 2019. "Antimicrobial Activity of Oregano Essential Oil Incorporated in Sodium Alginate Edible Films: Control of Listeria monocytogenes and Spoilage in Ham Slices Treated with High Pressure Processing." *Materials* 12 (22):3726.

Pelissari, Franciele, Danielle Ferreira, Ludmilla Louzada, Fabiana Santos, Ana Corrêa, Francys Moreira, and Luiz Mattoso. 2019. "Starch-Based Edible Films and Coatings: An Eco-friendly Alternative for Food Packaging." In: *Starches for Food Application*, 359-420.

Perera, K. Y., S. Sharma, D. Pradhan, A. K. Jaiswal, and S. Jaiswal. 2021. "Seaweed Polysaccharide in Food Contact Materials (Active Packaging, Intelligent Packaging, Edible Films, and Coatings)." *Foods* 10 (9). doi: 10.3390/foods10092088.

Pérez-Jiménez, Jara, and Fulgencio Saura-Calixto. 2018. "Fruit peels as sources of non-extractable polyphenols or macromolecular antioxidants: Analysis and nutritional implications." *Food Research International* 111:148-152. doi: https://doi.org/10.1016/j.foodres.2018.05.023.

Piermaria, J., Mariano Canal, and Analía Abraham. 2008. "Gelling properties of kefiran, a food-grade polysaccharide obtained from kefir grain." *Food Hydrocolloids* 22:1520-1527. doi: 10.1016/j.foodhyd.2007.10.005.

Pirozzi, Annachiara, Giovanna Ferrari, and Francesco Donsì. 2021. "The Use of Nanocellulose in Edible Coatings for the Preservation of Perishable Fruits and Vegetables." *Coatings* 11 (8):990.

Pongrácz, Eva. 2007. "The Environmental Impacts of Packaging." In *Environmentally Conscious Materials and Chemicals Processing*, 237-278.

Pop, Carmen Rodica, Teodora Emilia Coldea, Liana Claudia Salanţă, Alina Lăcrămioara Nistor, Andrei Borşa, Anca Corina Fărcaş, Vasile Constantin Florian, and Ancuţa Mihaela Rotar. 2021. "The Effect of Extraction Conditions on the Barrier and Mechanical Properties of Kefiran Films." *Coatings* 11 (5):602.

Prado, Maria R., Lina Marcela Blandón, Luciana P. S. Vandenberghe, Cristine Rodrigues, Guillermo R. Castro, Vanete Thomaz-Soccol, and Carlos R. Soccol. 2015. "Milk kefir: composition, microbial cultures, biological activities, and related products." *Frontiers in Microbiology* 6 (1177). doi: 10.3389/fmicb.2015.01177.

Primožič, M., Ž. Knez, and M. Leitgeb. 2021. "(Bio)nanotechnology in Food Science-Food Packaging." *Nanomaterials (Basel)* 11 (2). doi: 10.3390/nano11020292.

Priyadarshi, Ruchir, and Jong-Whan Rhim. 2020. "Chitosan-based biodegradable functional films for food packaging applications." *Innovative Food Science & Emerging Technologies* 62:102346. doi: https://doi.org/10.1016/j.ifset.2020.102346.

Qin, Yan, Yunpeng Liu, Huimin Yong, Jing Liu, Xin Zhang, and Jun Liu. 2019. "Preparation and characterization of active and intelligent packaging films based on cassava starch and anthocyanins from Lycium ruthenicum Murr." *International Journal of Biological Macromolecules* 134:80-90. doi: https://doi.org/10.1016/j.ijbiomac.2019.05.029.

Queirós, Lúcia C. C., Sónia C. L. Sousa, Andreia F. S. Duarte, Fernanda C. Domingues, and Ana M. M. Ramos. 2017. "Development of carboxymethyl xylan films with functional properties." *Journal of food science and technology* 54 (1):9-17. doi: 10.1007/s13197-016-2389-3.

Rahmani, B., H. Hosseini, M. Khani, M. Farhoodi, Z. Honarvar, E. Feizollahi, B. Shokri, and S. Shojaee-Aliabadi. 2017. "Development and characterisation of chitosan or alginate-coated low density polyethylene films containing Satureja hortensis extract." *Int J Biol Macromol* 105 (Pt 1):121-130. doi: 10.1016/j.ijbiomac.2017.07.002.

Rodrigues, Camila, Josiane Maria Muneron de Mello, Francieli Dalcanton, Domingos Lusitâneo Pier Macuvele, Natan Padoin, Márcio Antônio Fiori, Cíntia Soares, and Humberto Gracher Riella. 2020. "Mechanical, Thermal and Antimicrobial Properties of Chitosan-Based-Nanocomposite with Potential Applications for Food Packaging." *Journal of Polymers and the Environment* 28 (4):1216-1236. doi: 10.1007/s10924-020-01678-y.

Sadeghizadeh-Yazdi, Jalal, Masoud Habibi, Ali Kamali, and Mahdi Banaei. 2019. "Application of Edible and Biodegradable Starch-Based Films in Food Packaging: A Systematic Review and Meta-Analysis." *Current Research in Nutrition and Food Science Journal* 7:624-637. doi: 10.12944/CRNFSJ.7.3.03.

Saklani, Pooja, Siddhnath Kumar, Sambit Das, and Shiv Singh. 2019. "A Review of Edible Packaging for Foods." *International Journal of Current Microbiology and Applied Sciences* 8:2885-2895. doi: 10.20546/ijcmas.2019.807.359.

Salgado, Pablo R., Luciana Di Giorgio, Yanina S. Musso, and Adriana N. Mauri. 2021. "Recent Developments in Smart Food Packaging Focused on Biobased and Biodegradable Polymers." *Frontiers in Sustainable Food Systems* 5. doi: 10.3389/fsufs.2021.630393.

Senturk Parreidt, T., K. Müller, and M. Schmid. 2018a. "Alginate-Based Edible Films and Coatings for Food Packaging Applications." *Foods* 7 (10). doi: 10.3390/foods7100170.

Senturk Parreidt, Tugce, Kajetan Müller, and Markus Schmid. 2018b. "Alginate-Based Edible Films and Coatings for Food Packaging Applications." *Foods (Basel, Switzerland)* 7 (10):170. doi: 10.3390/foods7100170.

Souza, E., L. Gottschalk, and O. Freitas-Silva. 2020. "Overview of Nanocellulose in Food Packaging." *Recent Pat Food Nutr Agric* 11 (2):154-167. doi: 10.2174/2212798410666190715153715.

Tan, Kei-Xian, Vidya N. Chamundeswari, and Say Chye Joachim Loo. 2020. "Prospects of kefiran as a food-derived biopolymer for agri-food and biomedical applications." *RSC Advances* 10 (42):25339-25351. doi: 10.1039/D0RA02810J.

Tayeb, A. H., E. Amini, S. Ghasemi, and M. Tajvidi. 2018. "Cellulose Nanomaterials-Binding Properties and Applications: A Review." *Molecules* 23 (10). doi: 10.3390/molecules23102684.

Trajkovska Petkoska, Anka, Davor Daniloski, Nathan M. D'Cunha, Nenad Naumovski, and Anita T. Broach. 2021. "Edible packaging: Sustainable solutions and novel trends in food packaging." *Food Research International* 140:109981. doi: https://doi.org/10.1016/j.foodres.2020.109981.

Trajkovska Petkoska, Anka, Davor Daniloski, Nishant Kumar, Pratibha, and Anita T. Broach. 2021. "Biobased Materials as a Sustainable Potential for Edible Packaging." In *Sustainable Packaging*, edited by Subramanian Senthilkannan Muthu, 111-135. Singapore: Springer Singapore.

Trier, X., K. Granby, and J. H. Christensen. 2011. "Polyfluorinated surfactants (PFS) in paper and board coatings for food packaging." *Environ Sci Pollut Res Int* 18 (7):1108-20. doi: 10.1007/s11356-010-0439-3.

Tumbarski, Yulian, Radosveta Nikolova, Nadezhda Petkova, Ivan Ivanov, and Anna Lante. 2019. "Biopreservation of Fresh Strawberries by Carboxymethyl Cellulose Edible Coatings Enriched with a Bacteriocin from Bacillus methylotrophicus BM47." *Food technology and biotechnology* 57 (2):230-237. doi: 10.17113/ftb.57.02.19.6128.

Umaraw, Pramila, Paulo E. S. Munekata, Akhilesh K. Verma, Francisco J. Barba, V. P. Singh, Pavan Kumar, and José M. Lorenzo. 2020. "Edible films/coating with tailored properties for active packaging of meat, fish and derived products." *Trends in Food Science & Technology* 98:10-24. doi: https://doi.org/10.1016/j.tifs.2020.01.032.

Valdés, Arantzazu, Nuria Burgos, Alfonso Jiménez, and María Carmen Garrigós. 2015. "Natural Pectin Polysaccharides as Edible Coatings." *Coatings* 5 (4):865-886.

Wang, Q., Y. Zheng, W. Zhuang, X. Lu, X. Luo, and B. Zheng. 2018. "Genome-wide transcriptional changes in type 2 diabetic mice supplemented with lotus seed resistant starch." *Food Chem* 264:427-434. doi: 10.1016/j.foodchem.2018.05.056.

Wang, W., Q. Meng, Q. Li, J. Liu, M. Zhou, Z. Jin, and K. Zhao. 2020. "Chitosan Derivatives and Their Application in Biomedicine." *Int J Mol Sci* 21 (2). doi: 10.3390/ijms21020487.

Wardana, Ata Aditya, and Tri Dewanti Widyaningsih. 2017. "Development of edible films from tapioca starch and agar, enriched with red cabbage (Brassica oleracea) as a sausage deterioration bio-indicator." *IOP Conference Series: Earth and Environmental Science* 109:012031. doi: 10.1088/1755-1315/109/1/012031.

Wu, Shengjun, Mingsheng Lu, and Shujun Wang. 2016. "Effect of oligosaccharides derived from Laminaria japonica-incorporated pullulan coatings on preservation of cherry tomatoes." *Food Chemistry* 199:296-300. doi: https://doi.org/10.1016/j.foodchem.2015.12.029.

Wu, Si, Wenqi Wang, Kun Yan, Fuyuan Ding, Xiaowen Shi, Hongbing Deng, and Yumin Du. 2018. "Electrochemical writing on edible polysaccharide films for intelligent food packaging." *Carbohydrate Polymers* 186:236-242. doi: https://doi.org/10.1016/j.carbpol.2018.01.058.

Yang, So-Young, Luyen Cao, Hyeri Kim, Song-Ee Beak, and Kyung Bin Song. 2018. "Utilization of Foxtail Millet Starch Film Incorporated with Clove Leaf Oil for the Packaging of Queso Blanco Cheese as a Model Food." *Starch - Stärke* 70 (3-4):1700171. doi: https://doi.org/10.1002/star.201700171.

Yao, Xiyu, Jing Liu, Huixia Hu, Dawei Yun, and Jun Liu. 2022. "Development and comparison of different polysaccharide/PVA-based active/intelligent packaging films containing red pitaya betacyanins." *Food Hydrocolloids* 124:107305. doi: https://doi.org/10.1016/j.foodhyd.2021.107305.

Yavari Maroufi, Leila, Marjan Ghorbani, and Mahnaz Tabibiazar. 2020. "A Gelatin-Based Film Reinforced by Covalent Interaction with Oxidized Guar Gum Containing Green Tea Extract as an Active Food Packaging System." *Food and Bioprocess Technology* 13 (9):1633-1644. doi: 10.1007/s11947-020-02509-7.

Zhao, Yuan, Bo Li, Cuicui Li, Yangfan Xu, Yi Luo, Dongwu Liang, and Chongxing Huang. 2021. "Comprehensive Review of Polysaccharide-Based Materials in Edible Packaging: A Sustainable Approach." *Foods* 10 (8):1845.

Zhu, Fan. 2020. "Underutilized and unconventional starches: Why should we care?" *Trends in Food Science & Technology* 100:363-373. doi: https://doi.org/10.1016/j.tifs.2020.04.018.

Zhu, Fan. 2021. "Polysaccharide based films and coatings for food packaging: Effect of added polyphenols." *Food Chemistry* 359:129871. doi: https://doi.org/10.1016/j.foodchem.2021.129871.

Chapter 2

Development of Nanocellulose-Based Food Packaging Films

Surya Rao and E. K. Radhakrishnan*
School of Biosciences, Mahatma Gandhi University, Kottayam, Kerala, India

Abstract

Novel food packaging techniques arise with the changing market trends. Recently there has been an increasing demand for sustainable packaging. So, the use of biodegradable polymer-based composites such as cellulose and its various derivatives are thus gaining momentum in the global market. In the previous chapter, our colleagues introduced different polysaccharides used for the development of edible food packaging films. In this chapter, we will focus our discussion to talk about the possible use of nanocellulose in food packaging. Adopting smart packaging techniques with nanocellulose composites gives an added advantage over the conventional food packaging methods. This chapter will have a snapshot of recent advances in the development and design of nanocellulose based food packaging.

Keywords: nanocellulose, smart packaging, bacterial nanocellulose, intelligent packaging

1. Introduction

Food packaging is crucial for the effective distribution and protection of food from physical damage and deterioration [1]. It enables extending the shelf life

---

* Corresponding Author's E-mail: radhakrishnanek@mgu.ac.in.

In: Food Packaging
Editor: Wing-Fu Lai
ISBN: 979-8-88697-249-8

of food products by preventing the gas and moisture penetration, light exposure and the decay caused by the microorganisms [2]. Traditionally, petroleum based plastic materials are widely used for packaging due to their versatile properties [3]. Since then, plastics production has shown tremendous increase and it reached about 299 million in the year 2013. Packaging, constructions and auto materials are the three major areas of plastic use [4]. Significantly, after its use, plastics get accumulated in the environment and the ocean, creating pollution and several other health issues [5]. Therefore, there has been a growing demand for the use of bio-based materials produced from renewable resources for food packaging [6].

Biopolymers are found to be potential candidates to replace petroleum-based packaging materials. They are produced from renewable natural sources and can be synthesized by chemical or biological methods. On the basis of their origin, generally they can be classified into natural biopolymers, synthetic biopolymers and microbial biopolymers. The synthetic biopolymers widely in use are polylactic acid (PLA) [7], polycaprolactone (PCL) [8], polyglycolic acid (PGA) [9], polyvinyl alcohol (PVA) [10] and polybutylene succinate (PBS). Some of the most commonly used natural biopolymers in the food packaging include chitosan [11], cellulose [12], soy protein and tragacanth [13].

The limiting factor in the commercial application of biopolymers is its poor thermo- mechanical and barrier properties [14]. Upon functionalization/ crosslinking, grafting or by addition of various reinforcing agents, we can easily improve the properties of biopolymers. Recent studies also show that the biopolymers properties can be enhanced by using nanocomposite materials [15]. Cellulose nanomaterials are cheap compared to other nanomaterials and also demonstrate environmental benefits like recyclability, biodegradability etc. [5]. As a result, nanocellulose based food packaging has gained tremendous attention in food packaging industries. In this chapter we will discuss nanocellulose based food packaging.

2. Cellulose

Cellulose is among the most abundantly present polysaccharide in nature. It is widely distributed among plants, algae, bacteria and some other marine chordates and protozoa [16]. Around 75 billion tons of cellulose is produced by plants each year [17]. It is a homopolysaccharide composed of D-glucopyranose units linked by **β**- 1, 4 glycosidic bonds and have free hydroxyl groups at the C-2, C-3 and C-6 atoms [18]. On the basis of chemical nature, cellulose is classified into cellulose ether and cellulose ester. They are produced by some chemical reactions or chemical modifications of cellulose [16]. Cellulose primarily exists in six different polymorphic forms, namely I, II, III_i, III_{ii}, and IV_i & IV_{ii}. Cellulose I is the native form of all these and they primarily differ from one another in their molecular orientation and hydrogen bonds [19].

Cellulose's use in various fields is increasing in the present days. The tunable aspect ratio, excellent mechanical properties, biodegradability, low density and renewability make cellulose a promising candidate in numerous fields. The large number of reactive groups present on the surface of cellulose, eases the surface modification of cellulose with various functional groups like carboxylic acid, aldehydes, amines as well as grafting by different polymers, metal nanoparticles and antibiotics [20]. The physical and chemical properties of cellulose are mainly determined by its hydrogen bond which can be chemically modified to provide better properties [5]. The hydroxyl group present is responsible for characteristic features like hydrophilicity, chirality and biodegradability [21].

The functionalization's of cellulose is mainly done with the help of functional groups like sulfate, carboxyl, phosphate, amino, aldehyde and thiol groups. The added advantage of surface modification of cellulose is that it helps in converting bulk cellulose into nanostructured forms; it also helps in inducing surface charge to cellulose and fabricating cellulose derivatives with enhanced antibacterial activity [20]. Cellulose also shows a strong tendency to self- associate and form elongated networks with the help of hydrogen bonds as a result of the strong interactions of hydroxyl groups present [22]. Cellulose moreover exhibits high stability against proteolytic enzymes, acids and different conditions like harsh temperatures [23]. Due to its enormous property, cellulose has been widely used as a base material in the development of films, hydrogels, aerogels, membranes and so on for various purposes like food packaging [24], sensing [25], optics [26], biomedical applications [27] and water treatment [28].

3. Nanocellulose

Several glucose molecules are joined by Van der Waals and hydrogen bonds and form the three-dimensional network of cellulose referred to as microfibrillated cellulose (MFC) [20]. The properties of cellulose get enhanced on conversion from its bulk material form to its nanostructured form [29]. Different top-down and bottom-up procedures are used for the synthesis of cellulosic nanomaterials having different properties. Nanocellulose is a cellulosic material that exists in the nanometer range. Mechanical or chemical treatments are used for the disintegration of cellulose. The mechanical treatment involves the use of high energy and is not so efficient whereas the latter method involves the rupture of hydrogen bonds present in the cellulose, followed by the functionalization of material [30]. The first stage in the production of cellulosic nanomaterials involves the pre-treatment of raw materials followed by converting the individual cellulosic materials either into cellulose nanofibrils or cellulose nanocrystals [31].

3.1. Classifications of Nanocellulose

Nanocellulose are broadly divided into two based on their source of origin and shape. They are cellulose nanocrystals and cellulose nanofibers. Cellulose nanocrystals (CNC) are also known as cellulose nanowhiskers (CNW), cellulose nanorods (CNR) as well as nanocrystalline cellulose (NCC). Similarly, cellulose nanofibers (CNF) are also known by another name such as nanofibrillated cellulose (NFC). The major source of both these nanocellulose are plants. Another group of nanocellulose is bacterial nanocellulose, which has been synthesized by microorganisms using the bottom-up procedure [32].

Cellulose nanocrystals (CNC) are rod-like short rigid nanocrystals within a diameter range of 3- 50 nm whereas cellulose nanofibers (CNF) are long, flexible fibrils with a diameter of about 3- 100 nm. CNC is isolated from its raw materials by purification, followed by mechanical crushing and reactions with very strong mineral acids. Further processes like centrifugation, dialysis and ultrasonication are carried out. The acids used in the process include sulfuric acid, hydrochloric acid, hydrochloric acid vapor and phosphoric acids [33]. The surface functionalization of CNC shows modified interaction with its environment. Silylation, polymer grafting, esterification, etherification, oxidation etc. are some of the common methods used for its surface functionalization [21]. CNF isolation is carried out by means of high-shear treatments with microfluidizer, grinder and ultrasonication. Among all these procedures high pressure homogenization is widely used for cellulose nanofiber production [33]. The surface modification of cellulose nanofibers is done by chemical reactions, or by covalently attaching some surfactants or polyelectrolytes. CNF based more complex nanostructures are formulated by deposition of various metals/ metal oxide nanoparticles [34].

Bacterial nanocellulose, on the other hand, is an extracellular polysaccharide synthesized by some bacteria like *Agrobacterium, Alcaligenes, Aerobacter, Pseudomonas, Rhizobium, Rhodobacter* & *Achromobacter.* They are most efficiently produced by gram negative bacteria, especially *Komagataeibacter.* Bacterial nanocellulose is familiarly known by other names such as microbial cellulose, bacterial cellulose and bionanocellulose. They are composed of **β**- D- glucopyranose units linked by **β**- 1, 4- glycosidic bonds and thus forming ultrafine web-shaped networks having high porosity. They are potential alternatives for plant derived cellulose [35]. The bacterial nanocellulose can be modified by the introduction of functional groups that results in the generation of stable composites with characteristic properties. There are two methods for functionalization of bacterial nanocellulose: in-situ technique as well as ex-situ methods. In-situ method is the most common and chiefly involves the addition of various reinforcing materials like carboxymethyl cellulose and sodium alginate during the initial stage of bacterial nanocellulose synthesis whereas the ex-situ method is carried out at the post stage of bacterial nanocellulose production process [36].

3.2. Properties of Nanocellulose

Nanocellulose shows significant mechanical and thermal properties and these characteristic features make them an interesting candidate in bionanocomposite preparation. Nanocellulose improves the mechanical properties of a bionanocomposite by increasing its own concentration up to a range of 4%. The concentration of nanocellulose above that value leads to the segregation of nanocellulose which simultaneously drops the water vapor barrier and optical properties. The addition of nanocellulose also helps in improving the strength as well as modulus of the matrix [37]. Unlike the mechanical property, the thermal stability of a film shows reduction upon increase in cellulose nanocrystal concentration [2].

CNF demonstrates excellent mechanical properties with potentially high modulus of elasticity and tensile strength but its mechanical property gets negatively affected by its immense moisture absorbing property. CNF also exhibits excellent biocompatibility characteristics [34]. The studies showed that the tensile strength of CNC occurs in the range of 7.5-7.7 GPa, which indicates a higher tensile strength than that of the steel wire [21]. Bacterial nanocellulose also has promising mechanical properties with higher crystallinity in the range of 84- 89%, in addition with higher degree of polymerization, good tensile strength and higher biodegradability. They are highly pure compared to their plant-derived counterpart, thus making their purification process cheap and easy [38].

4. Technologies for Nanocellulose Based Materials Production

The nanocellulose based materials are used for packaging and are done with the help of different techniques. Some of those promising technologies used include layer-by-layer assembly, electrospinning, and composite extrusion, casting evaporation and coating [39].

Layer-by-layer assembly is a technique used for creating a thin film coating on the surface of materials. This technique promotes multicomponent films to be formed on solid supports. It primarily works upon the electrostatic interactions that exist between polyanion and polycation, finally leading to the formation of a multi-layer structure [40]. Layer-by layer films are developed by sequential intermediate dipping and rinsing of the interested item in the media containing polyelectrolytes. For example, layer by layer multilayers have been prepared on polylactic acid substrate with CNF/poly (ethylene imine) [41]. Electrospinning is one of the other methods used for polymeric fiber formation. The technique makes use of electrical forces to create polymer fibers within a diameter range of 2 nm to several micrometers. Natural/synthetic polymers are used for this purpose. The process produces fibers with a controllable pore structure. In a study, Zheng et al. [42] developed polymeric fibers of polyethylene oxide by single needle electrospinning and by hole electrospinning setup. The study demonstrated that the hole electrospinning setup with holes generate uniform fine fibers as they exhibited a better jet behavior as well as electric field distribution.

Composite extrusion is a simpler technique and primarily involves the conversion of solid materials into liquid and vice versa with a desired thickness. Through this technique nanocellulose can be incorporated into different polymeric matrices and thus can obtain packaging materials with enhanced properties [19]. Casting from solution and evaporation is not a significant technique used in the packaging industry as the dispersibility of various components is difficult to achieve. Still many literature reports demonstrate how nanocellulose based materials have been cast successfully. The technique chiefly requires a moderate temperature and solvent to maintain the nanocellulose concentration [19]. Coating is the most accepted technique for packaging materials and in the procedure, it coats the materials with a very thin layer. The layers will have a thickness in the range of nanometer to few micrometers [43].

5. Characterization Techniques in Detection of Nanocellulose

The structural and physico- mechanical properties of nanocellulose vary based upon the bioresources from which nanocellulose had been synthesized and the production conditions like temperature, time, reagents, and liquid/solid ratio [44]. Thus, characterizing the nanocellulose formed is essential for further experiments. Electron microscopy is one best way to study the morphology of nanocellulose. The BNC pellicles, aggregated cellulose fibrils form a 3D structure with approximately 40 μm thicknesses. Its distorted fibers demonstrated a diameter of about 5 nm [45]. For TEM analysis the BNC was digested with sulfuric acid and then observed the crystals after the breakdown of fibrils in the range of 100-2000 nm length and width approximately of 5 nm. Cryo-electron microscope helps to determine the stability of nanocellulose solution as well as to determine the particle size of nanocellulose [46].

Raman spectroscopy allows the study of the crystallinity of cellulose [47]. Characteristic peaks were obtained in the study [45] for cellulose at around 381, 437, 1097, 1122,1339,1380,2900 cm^{-1}. Wood nanocellulose was kept as the standard for the study. X- ray diffraction patterns also help to determine the crystallinity of nanocellulose. Characteristic peaks are obtained in the study where high crystallinity is marked by an absorption peak having narrower peak width and higher peak height whereas a solution containing lower crystallinity, the peak tends to be wider in width and lower in height [46].

Fourier Transform Infrared Spectroscopy (FT-IR) analysis provides insight into the structure of nanocellulose and the nature of the molecular interactions involved. Signature absorption peaks are obtained [48] for nanocellulose. The thermal property of nanocellulose can be determined by differential scanning calorimetry [46]. It marks the melting point. Thus, these characterization techniques give more remarkable concepts about nanocellulose.

6. Antimicrobial Activity of Nanocellulose

Nanocellulose is marked as a potential candidate to filter several microbes according to Norrahim. The antibacterial property of nanocellulose can be accelerated by surface modifying the nanocellulose, through the addition of antibiotics or by combining nanocellulose with nanomaterials and antibacterial polymers. The glucopyranose ring of nanocellulose possesses three active hydroxyl groups on C_2, C_3 and C_6 positions and the new functional groups can easily be added to these positions by methods like oxidation, esterification or etherification [49]. Compounds like antibiotics, quaternary compounds, nanostructured materials and aldehydes are commonly incorporated into nanocellulose [50]. Good antimicrobial activity is represented by nanocellulose containing aldehyde or quaternary ammonium as functional groups. The quaternary ammonium compounds used include zwitterion betaines. Similarly, the potential metal or metal oxide nanoparticles used include copper, silver, gold, titanium and zinc oxide nanoparticles. Chitosan and its various derivatives are excellent antimicrobial polymers used as the best antibacterial agents [49].

To improve the antibacterial efficiency of nanocellulose, surface functionalization's are commonly carried out. The surface functionalization enhances the filtration efficiency even against viruses. The functionalized compounds like quaternary compounds give polyelectrolytic charge to the nanocellulose which simultaneously enhances the chemical interaction that exists between virus and nanocellulose. The electrostatic interaction between the two will entrap the virus onto the nanocellulose membrane [51]. Upon surface functionalization with certain compounds, the nanocellulose attains cationic charge and it then reacts with negative charge present in the membrane of the bacteria, which possess phosphate group in the peptidoglycan layer and the phospholipids in the outer membrane [20]. Due to electrostatic interaction, it primarily creates damage to cell membrane and then it further denatures due to fluctuation in osmotic balance of the cell [52]. In a study, upon surface functionalization with silver nanoparticle, the bacterial cellulose exhibited an excellent antimicrobial activity [53].

Antibiotic incorporation into nanocellulose showed good antimicrobial properties. For example, Jafary et al. [54], fabricated allicin conjugated nanocellulose to prepare cellulose fabrics. The result demonstrated significant antimicrobial activity against *Staphylococcus aureus*. Inorganic nanoparticles are generally incorporated with nanocellulose to enhance the antimicrobial properties. Nanoparticle incorporated nanocellulose based materials show lower minimum inhibitory concentration than metal/metal oxide alone [49]. For example, TiO_2 is an excellent metal oxide nanoparticle and its antimicrobial property is dependent on factors like its crystal structure, shape and size. Cellulose nanofibers grafted with titania nanoparticle has been successfully produced by "green chemistry" approach. Novelly synthesized nanocomposite, demonstrated antimicrobial activity against *Staphylococcus aureus* and *Escherichia coli* and the nanocomposite remained stable under UV irradiation [55]. Similarly, nanocellulose based antimicrobial materials can be produced by combining nanocellulose with polymers with immense antibacterial activities. The protonation of amino groups in polymers is mainly responsible for the antibacterial activity [49]. In a study, El- Samahy et al. [56], paper sheets have been modified by combining nanocellulose of different concentrations and the antibacterial agent, chitosan. The chitosan incorporated paper sheets showed enhanced paper resistance to various microorganisms responsible for causing food poisoning. Thus, cellulosic nanomaterials show enhanced antimicrobial activity against viruses, bacteria, fungi, algae and protozoa [50].

7. Use of Nanocellulose in Food Packaging

The food packaging system's primary aim is to protect the food product from external stimuli and prolong its shelf life. Excellent packaging systems do provide physical support to the food product [57]. An advanced version of traditional food packaging is smart packaging that overcomes the safety hazard concerns raised by traditional packaging through its interaction with both the food and the environment. Smart packaging generally consists of two types: active and intelligent packaging [58]. Nanocellulose has gained considerable attention in food packaging as an engineered nanomaterial with good mechanical strength, high surface area, biodegradability and other properties like non toxicity, sustainability and lightweight texture [33]. The composites synthesized from nanocellulose show excellent strength as they cross link with the numerous hydroxyl groups present in the nanocellulose [2].

The nanocellulose thus acts as a green option for packaging materials. In designing a food packaging material, various atmospheric conditions and some other essential properties need to be enhanced or controlled to increase the shelf life of the food. Nanocellulose based materials are able to achieve this. For example, for a longer shelf life of the food materials, a low oxygen environment is essential. It is provided by nanocellulose, as it forms a dense network with its hydrogen bonds and thus prevents various gas molecules from diffusing through its small pores. Similarly, its crystalline structure also prevents the diffusion of gas molecules [59].

Another important property is the water vapor permeability of the films. It directly affects the gas diffusion since water has the ability to plasticize the film [60]. Unmodified nanocellulose generally shows very poor water vapor barrier properties whereas modified CNF shows high crystallinity and hydrophobicity leading to reduction in pore size which prevents the water molecules diffusion [2]. Aerogels are a new subject of interest in food packaging research. They can be used as absorbers in food packages and they exhibit some characteristic features like large surface area, liquid absorption capacity etc. Recently cellulose nanocrystals are used in the fabrication of aerogels. In one of the recent works, the cellulose nanocrystals were initially prepared from rice, oats and eucalyptus cellulose and then PVA dispersion was added into it and provided the appropriate conditions for crosslinking. Chiefly designed aerogel showed excellent physico- chemical properties with rice cellulose nanocrystal showing greatest water absorption capacity (402.8%). Cellulose nanocrystals can thus be an excellent candidate as an aerogel absorber of water in food packaging industries [61].

Free radicals produced on exposure to UV radiation cause oxidation of biomolecules present in the food. Thus, leading to a reduction in quality and shelf life of food products. The free radicals also damage antioxidants. Reinforcing various biopolymers with nanocellulose produces excellent films that show better UV shielding and antioxidant properties. Li et al. [62] in a study had fabricated a cationized cellulose nanofibril-tannin extract hybrid film by vacuum filtration method. The hybrid film showed better UV shielding property and antiradical activity. Similarly, bacterial nanocellulose also acts as an excellent reinforcement agent. In a study [63], researchers have developed an antioxidant active film comprising the most abundant catechin-epigallocatechin-3- gallate (EGCG) incorporated in the chitosan matrix. The film demonstrated excellent mechanical properties. The addition of bacterial nanocellulose in the film demonstrated an increase in viscosity of film forming solution at low shear rate values. It was achieved by the interaction of macromolecular chains present in chitosan and bacterial nanocellulose. The SEM analysis suggested the presence of nanocellulose in the film that has created massive changes in the microstructure of films. The addition of bacterial nanocellulose also helped to nullify the negative effect caused by the incorporation of EGCG. Similarly, bacterial nanocellulose was used as a reinforcing agent in another film synthesized by the protein matrix- distiller's dried grains [64]. A detailed analysis was done on the composite film incorporated with bacterial nanocellulose. In the study, they observed, with an increase in bacterial nanocellulose concentration from 1.6% to 2%, the films consequently became denser. The bacterial nanocellulose containing film also showed a prominent increase in the tensile strength of the film. The optical transparency and opacity of the protein nanocomposite film were also analyzed in the study with the help of spectroscopy and it presented a decreasing trend with the increase in bacterial nanocellulose concentration. Thus, in their study they observed an enhanced property for protein-based films upon addition of bacterial nanocellulose. Some other nanocellulose based food packaging systems and their uses are given in Table 1 [65-68].

Table 1. Some nanocellulose based food packaging systems and their effects

Nanocellulose based composites	**Effects**	Reference
Cellulose nanofiber/banana starch	It showed higher mechanical properties and thermal stability than pure starch films.	[65]
Cellulose nanofiber/polydextrose/ whey protein isolate	The addition of CNF and polydextrose improved the viability of probiotic bacteria from whey protein isolate based film	[66]
TEMPO cellulose nanofiber/ chitin nanowhiskers/ polypropylene	Demonstrated high thermal recyclability over aluminium-metalized packaging	[67]
PVA/ligno nanocellulose	Exhibited improved UV barrier and oxygen barrier properties than the pure PVA films	[68]

7.1. Active Packaging

Active packaging is an alternative to preserve food and it helps in controlling the temperature and moisture level of the packaged food [69]. It is a smart packaging method that provides a sustainable and eco- friendly way to enhance the shelf life of food. Active packaging systems involve the release of various antimicrobial components to the food without any trigger or response. For example, in a diffusion based active packaging system, the antimicrobial component gets released as soon as they come in contact with the food item. Active packaging can generally be classified into two: chemical active method and biological active method. In chemo active packaging method, the active agents are certain chemicals that create changes in the chemical composition and gaseous atmosphere of the food product. Gas scavengers like oxygen scavengers and ethylene gas scavengers are widely used in food packaging. In food industries they commonly use modified atmosphere packaging methods to preserve the food items. In biological active packaging, antimicrobial agents are used. They interact with various biological molecules, which finally lead to the inhibition of microbial growth [70]. The latter method (biological active method) is the most favorable over the chemo active method as the addition of synthetic materials/chemicals create health concerns. Thus, the demand for biological components in active packaging has increased and scientists are currently researching more options for natural substances that could replace synthetic ones. Some of the active compounds used in active packaging are listed below in Table 2 [10, 69, 71-78]. The active packaging on the basis of release of antimicrobial agents into the packaged food products can again be broadly classified into responsive packaging and migratory packaging (controlled release packaging). In responsive packaging [70], the actual migration of the antimicrobial agent does not happen whereas in controlled release packaging [79] timed release of antimicrobial agent occurs.

Table 2. Some active compounds used in active packaging

Class	Active compounds	Reference
Antimicrobial Agents	• Organic acids • Peptides • Metal oxides • Essential oil • Bioactive polymers	[10, 71-74]
Antioxidant Agents	• Plant extract • Essential oil • Tocopherol	[69, 75-76]
Oxygen Scavengers	• Enzymes	[77]
Moisture Absorbers	• Egg shell	[78]

Nanocellulose incorporated with various active compounds show excellent features in food packaging. Enzymes like lysozymes have antimicrobial activity [80] and can be used in active packaging. The direct addition of these enzymes to preserve food shows some limitations, for example loss of its activity. This limitation can be resolved by immobilizing them into polymeric matrices. Bacterial nanocellulose is an excellent carrier of antimicrobial agents because of its large surface area, combined with increased porosity and ultrafine network structure. The Buruaga- Ramiro [81] and co-workers fabricated bacterial nanocellulose paper. Lysozyme was adsorbed onto it by immersing the nanocellulose paper in the lysozyme binding solution. The microstructure changes of bacterial nanocellulose were studied by X-Ray Diffraction patterns and found that no characteristic changes occurred in the crystalline structure of cellulose by the adsorption of lysozyme. The leaching of lysozyme from the bioactive paper was studied. The lysozyme migrated from the paper in a minute amount indicating the ability of bacterial nanocellulose to stabilize and sustain the release of enzymes from the composite. The incorporation of lysozyme also showed an increase in antioxidant activity of bioactive paper by 30%. Along with this, the bioactive paper showed a significant storage capacity without any additional storage conditions. Apart from this bacterial nanocellulose showed significant antimicrobial activity on incorporation with metal ions and oxides like silver, platinum, gold, ZnO, MgO and titanium dioxide. In the work of Mocanu [82] and his colleagues they prepared a bacterial nanocellulose-ZnO nanocomposite. The ethanolic propolis extract was impregnated into this nanocomposite at different concentrations.

The nanocellulose incorporation can help in another set of properties like solid particle stabilization. Due to this unique property nanocellulose is used in pickering emulsion stabilization. Pickering emulsion is a kind of emulsion which gets stabilized by addition of solid particles like proteins or polysaccharides which adsorb onto the interfaces that exist between the binary phases. The presence of nanocellulose helped in sustained release of oils from the composite system and also reduced the diffusion surface area of oil droplets [83]. Pickering emulsion of clove essential oil was stabilized with nanocellulose and added it into a binary composite containing gelatin/ agar [84]. The film was synthesized by a solvent casting method. The film showed an increased thickness due to the higher solid content obtained by stabilizing the film with nanocellulose fibers. Similarly, Liu et al. [85] developed a high internal phase pickering emulsion (HIPE) by the addition of sunflower seed onto the soy protein along with the addition of bacterial nanocellulose. Then the konjac glucomannan solutions were prepared and into it the HIPE solution was added at different ratios and the solutions were then cast to produce composite films.

Food spoilage occurs by another mechanism called oxidative degradation. It can be prevented by the addition of antioxidant additives. Essential oil, plant and fruit extracts, natural compounds like tocopherol are some examples of natural antioxidant agents [86]. Bacterial cellulose nanofibers based active membranes were fabricated by Moradian et al. [87]. The active membrane comprises green tea, rosemary and pomegranate peel extract. The addition of these various extracts had shown a significant increase in antioxidant activity of bacterial cellulose membranes. The active membrane was used to improve the shelf life of button mushrooms. Similarly in another study, Yang et al. [88] prepared bacterial cellulose nanofiber/chitin nanofiber hybridized film. The curcumin micro/ nanoparticles were incorporated into it by the antisolvent inside the film (ASIF) precipitation method. Other antimicrobial compounds that are widely used in active packaging include organic acid [89]. Sorbic acid is generally recognized as a safe additive and is commonly used in fruits, vegetables and beverages as a preservative. In a study, Jipa et al. [90], observed the release of sorbic acid from mono and multilayer bacterial nanocellulose film. The swelling degree of the film showed a characteristic decrease with increase in powdered bacterial cellulose (BCP) concentration. Sorbic acid release was prominent when BCP concentration was high but decreased as the antimicrobial concentration increased due to the formed crystal dissolution. Also, a significant decrease in the sorbic acid release was also demonstrated in the multilayer film.

Nanocellulose is known for its excellent oxygen barrier property due to its dense nanofibril network and high surface area [39]. Thus, engineered nanocellulose based films can be used to block external oxygen and UV light which simultaneously help in prolonging the freshness of food and beverages. Dou et al. [91] synthesized a self-standing thin lignin - containing cellulose nanofibril (LCNF) film. They have analyzed the oxygen and UV blocking performance of the willow bark fibril films. The hot water extracted willow bark (HWEWB) LCNF showed a very low oxygen permeability value indicating an excellent oxygen barrier property of the functionalized nanocellulose. The UV light blocking ability of willow bark LCNF was analyzed using a spectrophotometer and the WB LCNF demonstrated a significant UV protection property.

Bacterial nanocellulose is hydrophilic in nature and thus can be easily molded into three dimensional structures during its synthesis. Due to this property, bacterial nanocellulose membrane is widely used as a stabilizer in food industries. In a study, Dos Santos et al. [92], developed a bacterial nanocellulose membrane. These membranes were then immersed in different concentrations of nisin/EDTA. The study showed that the bacterial nanocellulose membrane created an effective barrier against further infection when it is loaded with antimicrobial compounds. Nisin showed better activity when it is combined with the chelating agent than alone. The spoilage of food as mentioned earlier, can occur by the presence of liquid in the food. It can be reduced with the help of absorbing materials. The nanopolymer bacterial cellulose has excellent absorption ability. A composite membrane was developed with the help of eggshell and bacterial cellulose [78]. It was noted that the increase in eggshell particles improved the absorption ability as it provided higher surface area and porosity.

7.2. Intelligent Packaging

Intelligent packaging is a new area in the food packaging sector and represents approximately 5% of the total packaging market [93]. This smart packaging system senses some of the characteristic features of the food and thus indicates the quality of food instantaneously. In contrast to the active packaging, the intelligent packaging does not prevent spoilage, but instead provides a real time monitoring of the packaged food with the help of indicators, dyes and sensors [94]. Some of the smart materials used in the food packaging are given in Figure 1. Indicators used in intelligent packaging are of two kinds: direct indicators and indirect indicators. The smart packaging system contains indicators for pH, time and temperature, freshness and sensors for detecting gas, other chemicals and for measuring even the humidity levels [93].

Freshness of the food is measured with the help of freshness indicators, that monitor the spoilage by analyzing the metabolites produced as a result of microbial growth. The indicators demonstrate a visible color change as the reaction occurs. In a recent innovation, nanocellulose smart packaging is used for observing freshness of chicken breast. In the study Lu et al. [95] fabricated a nanocellulose hydrogel system incorporated with bromothymol blue and methyl red. The mixed dyes are pH sensitive and works as a hydrophilic adsorbent for CO_2. The hydrogel system provides enough water to convert CO_2 to carbonic acid. The chicken meat upon spoilage by microorganism produces CO_2 and other volatile basic nitrogen. The mixed dye indicator shows green color in the basic solution and it turns to red in acidic solution. In the study, initially upon packaging the fresh meat it showed a green color and turned out to be bright red after storage of 2 days as the CO_2 level increased. Similarly, in another study anthocyanins were used as pH sensing freshness indicator with bacterial nanocellulose [96].

SMART MATERIALS

INDICATORS
- -Temperature/Time
- -Freshness
- -Leak
- -pH

SENSORS
- -Chemical Sensors
- -Biosensors

Figure 1. Some of the smart materials used in food packaging.

Foodstuff deterioration can also occur as a result of changes in the humidity level of the packaged food. With increase in moisture, it makes the food vulnerable to microbial attack and that can also be easily studied with the help of smart sensors incorporated in nanocomposite films. Vilela et al. [97] designed a bacterial cellulose-based nanocomposite film for sensing the humidity level in the packaged foods. The nanocomposite film apart from nanocellulose comprises radically initiated sulfobetaine methacrylate (SBMA) cross linked with poly (ethylene glycol) diacrylate. The protogenic moieties (sulfonic acid groups) present in the nanocomposite film help in monitoring the humidity level. The protonic conduction of these moieties increases when the environmental humidity level increases.

In intelligent packaging, other groups sense the gaseous or volatile compounds like CO_2, H_2S, O_2 and other gases with the help of gas sensors [98]. In a recent study [99], researchers had developed a H_2S gas sensor. H_2S is considered as spoilage gas that is generated upon the action of microorganisms and lipid oxidation. In the study they developed a bacterial cellulose nanocrystal hybrid film loaded with silver nanoparticles and alginate-molybdenum trioxide nanoparticles. The visual color change from light grayish-white to dark brown-black was observed upon generation of atomic hydrogen from the reaction of silver nanoparticles and hydrogen sulfide gas, changes in the oxidation state of molybdenum and by the further reduction of molybdenum trioxide nanoparticle. Color intensity increased with the increase in exposure time. Volatile amines generated upon microbial spoilage can be detected with the immobilization of curcumin in bacterial cellulose membranes [100]. It was designed as an on-package disposable sticker sensor that has an additional feature regeneration when treated with 0.1 HCl solution.

Recently nanocellulose films have been developed [101] that could sense electrical conductivity in response to biochemical signals. Most prominent conducting polymer, polypyrrole, is used with the bacterial nanocellulose in this study. The films thus designed had been incorporated with ZnO by the chemical synthesis method. It showed an increase in relative variation in electrical resistance with increased storage temperature and time. A noticeable change was observed with storage temperature than the latter. The changes in electrical resistance help to decide the correct expiration date of the packaged food. The conductometric nano biosensors were developed similarly by incorporation of TiO_2-Ag nanoparticles (act as semiconductor materials) into the bacterial nanocellulose and the biosensor was used to monitor the growth of pathogenic bacteria [102]. Thus, smart intelligent packaging system provides a greater boost to preserving foods.

Conclusion

The current chapter discusses nanocellulose and its use as an intelligent and smart food packaging material. Bionanocomposites is a promising alternative to the traditional plastic-based food packaging and helps in reducing the greenhouse gas effect. Nanocellulose plays a pivotal role in food packaging and is the best candidate for greener food packaging. Even though, the nanocellulose is a best candidate for packaging, certain challenges need to be resolved, especially in identifying some cost-effective method of extracting nanocellulose as well as newer methodologies for surface and end-reducing modifications. Identifying new natural active agents and smart materials to incorporate with nanocellulose would be another great initiative and might provide a promising solution to the current packaging problems. Hence, future research activities need to be directed to optimizing new innovative methodologies to produce nanocellulose based materials and to minimize the energy and time consumption required for the development of nanocellulose based materials. The unique manufacturing procedures like 3D printing and additive manufacturing should be explored further to generate technically feasible large scale nanocellulose. Further in the future investigations of nanocellulose based materials, considering and incorporating the characteristic features like artificial intelligence, machine learning based algorithms and numerical models will be a breakthrough. With improved interdisciplinary approaches, the nanocellulose based materials produced should have low environmental impact and also should open up a new research area for sustainable nanocellulose with high end performance and numerous applications.

References

[1] Wang, H., L. Wang, Developing a bio-based packaging film from soya by-products incorporated with valonea tannin, *Journal of Cleaner Production.* 143 (2017) 624–633. https://doi.org/10.1016/J.JCLEPRO.2016.12.064.

[2] Ahankari, S.S., A.R. Subhedar, S.S. Bhadauria, A. Dufresne, Nanocellulose in food packaging: A review, *Carbohydrate Polymers.* 255 (2021) 117479. https://doi.org/10.1016/J.CARBPOL.2020.117479.

[3] Kunwar, B., H.N. Cheng, S.R. Chandrashekaran, B.K. Sharma, Plastics to fuel: a review, *Renewable and Sustainable Energy Reviews.* 54 (2016) 421–428. https://doi.org/10.1016/J.RSER.2015.10.015.

[4] E.P.R.O.PlasticsEurope, Plastics—The Facts 2019. An Analysis of European Plastics Production, Demand and Waste Data, *Plast. Eur.* (2016).

[5] Vilarinho, F., A. Sanches Silva, M.F. Vaz, J.P. Farinha, Nanocellulose in green food packaging, *Critical Reviews in Food Science and Nutrition.* 58 (2018) 1526–1537. https://doi.org/10.1080/10408398.2016.1270254/SUPPL_FILE/BFSN_A_1270254_SM4398.ZIP.

[6] Lan, W., L. He, Y. Liu, Preparation and Properties of Sodium Carboxymethyl Cellulose/Sodium Alginate/Chitosan Composite Film, *Coatings* 2018, Vol. 8, Page 291. 8 (2018) 291. https://doi.org/10.3390/COATINGS8080291.

[7] Babu, S.S., N. Kalarikkal, S. Thomas, E.K. Radhakrishnan, Enhanced antimicrobial performance of cloisite 30B/poly (ε-caprolactone) over cloisite 30B/poly (l-lactic acid) as evidenced by structural features, *Applied Clay Science.* 153 (2018) 198–204. https://doi.org/10.1016/J.CLAY.2017.12.003.

[8] Thomas, R., K.R. Soumya, J. Mathew, E.K. Radhakrishnan, Electrospun Polycaprolactone Membrane Incorporated with Biosynthesized Silver Nanoparticles as Effective Wound Dressing Material, *Applied Biochemistry and Biotechnology.* 176 (2015) 2213–2224. https://doi.org/10.1007/S12010-015-1709-9/FIGURES/7.

[9] Budak, K., O. Sogut, U. Aydemir Sezer, A review on synthesis and biomedical applications of polyglycolic acid, *Journal of Polymer Research.* 27 (2020) 1–19. https://doi.org/10.1007/S10965-020-02187-1/FIGURES/8.

[10] Jayakumar, A., S. Radoor, I. C Nair, S. Siengchin, J. Parameswaranpillai, R. E.K, Polyvinyl alcohol -nanocomposite films incorporated with clay nanoparticles and lipopeptides as active food wraps against food spoilage microbes, *Food Packaging and Shelf Life.* 30 (2021) 100727. https://doi.org/10.1016/J.FPSL.2021.100727.

[11] Wang, H., J. Qian, F. Ding, Emerging Chitosan-Based Films for Food Packaging Applications, *Journal of Agricultural and Food Chemistry.* 66 (2018) 395–413. https://doi.org/10.1021/ACS.JAFC.7B04528/ASSET/IMAGES/MEDIUM/JF-2017-04528A_0013.GIF.

[12] Cazón, P., M. Vázquez, Bacterial cellulose as a biodegradable food packaging material: A review, *Food Hydrocolloids.* 113 (2021) 106530. https://doi.org/10.1016/J.FOODHYD.2020.106530.

[13] Bahrami, A., R. Rezaei Mokarram, M. Sowti Khiabani, B. Ghanbarzadeh, R. Salehi, Physico-mechanical and antimicrobial properties of tragacanth/hydroxypropyl

methylcellulose/beeswax edible films reinforced with silver nanoparticles, *International Journal of Biological Macromolecules.* 129 (2019) 1103–1112. https://doi.org/10.1016/J.IJBIOMAC.2018.09.045.

[14] Jabeen, N., I. Majid, G.A. Nayik, *Bioplastics and food packaging: A review,* Http://Www.Editorialmanager.Com/Cogentagri. 1 (2015) 1117749. https://doi.org/10.1080/23311932.2015.1117749.

[15] Honarvar, Z., Z. Hadian, M. Mashayekh, Nanocomposites in food packaging applications and their risk assessment for health, *Electronic Physician.* 8 (2016) 2531. https://doi.org/10.19082/2531.

[16] Joseph, B., V.K. Sagarika, C. Sabu, N. Kalarikkal, S. Thomas, Cellulose nanocomposites: Fabrication and biomedical applications, *Journal of Bioresources and Bioproducts.* 5 (2020) 223–237. https://doi.org/10.1016/J.JOBAB.2020.10.001.

[17] Feng, Y.H., T.Y. Cheng, W.G. Yang, P.T. Ma, H.Z. He, X.C. Yin, X.X. Yu, Characteristics and environmentally friendly extraction of cellulose nanofibrils from sugarcane bagasse, *Industrial Crops and Products.* 111 (2018) 285–291. https://doi.org/10.1016/J.INDCROP.2017.10.041.

[18] Abdul Khalil, H.P.S., Y. Davoudpour, M.N. Islam, A. Mustapha, K. Sudesh, R. Dungani, M. Jawaid, Production and modification of nanofibrillated cellulose using various mechanical processes: A review, *Carbohydrate Polymers.* 99 (2014) 649–665. https://doi.org/10.1016/J.CARBPOL.2013.08.069.

[19] Li, F., E. Mascheroni, L. Piergiovanni, The Potential of NanoCellulose in the Packaging Field: A Review, *Packaging Technology and Science.* 28 (2015) 475–508. https://doi.org/10.1002/PTS.2121.

[20] Tavakolian, M., S.M. Jafari, T.G.M. van de Ven, A Review on Surface-Functionalized Cellulosic Nanostructures as Biocompatible Antibacterial Materials, *Nano-Micro Letters* 2020 12:1. 12 (2020) 1–23. https://doi.org/10.1007/S40820-020-0408-4.

[21] George, J., S.N. Sabapathi, Cellulose nanocrystals: synthesis, functional properties, and applications, *Nanotechnology, Science and Applications.* 8 (2015) 45. https://doi.org/10.2147/NSA.S64386.

[22] Jorfi, M., E.J. Foster, Recent advances in nanocellulose for biomedical applications, *Journal of Applied Polymer Science.* 132 (2015) 41719. https://doi.org/10.1002/APP.41719.

[23] Peng, B.L., N. Dhar, H.L. Liu, K.C. Tam, Chemistry and applications of nano-crysta-line cellulose and its derivatives: A nanotechnology perspective, *The Canadian Journal of Chemical Engineering.* 89 (2011) 1191–1206. https://doi.org/10.1002/CJCE.20554.

[24] Azeredo, H.M.C., M.F. Rosa, L.H.C. Mattoso, Nanocellulose in bio-based food packaging applications, *Industrial Crops and Products.* 97 (2017) 664–671. https://doi.org/10.1016/J.INDCROP.2016.03.013.

[25] Golmohammadi, H., E. Morales-Narváez, T. Naghdi, A. Merkoçi, Nanocellulose in Sensing and Biosensing, *Chemistry of Materials.* 29 (2017) 5426–5446. https://doi.org/10.1021/ACS.CHEMMATER.7B01170/ASSET/IMAGES/MEDIUM/CM-2017-01170E_0013.GIF.

[26] Wilts, B.D., A.G. Dumanli, R. Middleton, P. Vukusic, S. Vignolini, Invited Article: Chiral optics of helicoidal cellulose nanocrystal films, *APL Photonics.* 2 (2017) 040801. https://doi.org/10.1063/1.4978387.

[27] Xue, Y., Z. Mou, H. Xiao, Nanocellulose as a sustainable biomass material: structure, properties, present status and future prospects in biomedical applications, *Nanoscale.* 9 (2017) 14758–14781. https://doi.org/10.1039/C7NR04994C.

[28] Voisin, H., L. Bergström, P. Liu, A.P. Mathew, Nanocellulose-Based Materials for Water Purification, *Nanomaterials* 2017, Vol. 7, Page 57. 7 (2017) 57. https://doi.org/10.3390/NANO7030057.

[29] Jasmania, L. Preparation of nanocellulose and its potential application, *International Journal of Nanomaterials, Nanotechnology and Nanomedicine.* 4 (2018) 014–021. https://doi.org/10.17352/2455-3492.000026.

[30] Qing, Y., R. Sabo, J.Y. Zhu, U. Agarwal, Z. Cai, Y. Wu, A comparative study of cellulose nanofibrils disintegrated via multiple processing approaches, *Carbohydrate Polymers.* 97 (2013) 226–234. https://doi.org/10.1016/J.CARBPOL.2013.04.086.

[31] Moon, R.J., A. Martini, J. Nairn, J. Simonsen, J. Youngblood, Cellulose nanomaterials review: structure, properties and nanocomposites, *Chemical Society Reviews.* 40 (2011) 3941–3994. https://doi.org/10.1039/C0CS00108B.

[32] Zhang, W., Y. Zhang, J. Cao, W. Jiang, Improving the performance of edible food packaging films by using nanocellulose as an additive, *International Journal of Biological Macromolecules.* 166 (2021) 288–296. https://doi.org/10.1016/J.IJBIOMAC.2020.10.185.

[33] Ma, T., X. Hu, S. Lu, X. Liao, Y. Song, X. Hu, *Nanocellulose: a promising green treasure from food wastes to available food materials,*Https://Doi.Org/10.1080/10408398.2020.1832440. 62 (2020) 989–1002. https://doi.org/10.1080/10408398.2020.1832440.

[34] Zhang, K., A. Barhoum, C. Xiaoqing, H. Li, P. Samyn, Cellulose Nanofibers: Fabrication and Surface Functionalization Techniques, *Handbook of Nanofibers.* (2019) 409–449. https://doi.org/10.1007/978-3-319-53655-2_58.

[35] Ludwicka, K., M. Kaczmarek, A. Białkowska, Bacterial Nanocellulose—A Biobased Polymer for Active and Intelligent Food Packaging Applications: Recent Advances and Developments, *Polymers* 2020, Vol. 12, Page 2209. 12 (2020) 2209. https://doi.org/10.3390/POLYM12102209.

[36] Cacicedo, M.L., M.C. Castro, I. Servetas, L. Bosnea, K. Boura, P. Tsafrakidou, A. Dima, A. Terpou, A. Koutinas, G.R. Castro, Progress in bacterial cellulose matrices for biotechnological applications, *Bioresource Technology.* 213 (2016) 172–180. https://doi.org/10.1016/J.BIORTECH.2016.02.071.

[37] Criado, P., C. Fraschini, S. Salmieri, D. Becher, A. Safrany, M. Lacroix, Free radical grafting of gallic acid (GA) on cellulose nanocrystals (CNCS) and evaluation of antioxidant reinforced gellan gum films, *Radiation Physics and Chemistry.* 118 (2016) 61–69. https://doi.org/10.1016/J.RADPHYSCHEM.2015.05.030.

[38] Lin, D., Z. Liu, R. Shen, S. Chen, X. Yang, Bacterial cellulose in food industry: Current research and future prospects, *International Journal of Biological*

Macromolecules. 158 (2020) 1007–1019. https://doi.org/10.1016/J.IJBIOMAC.2020.04.230.

[39] Ferrer, A., L. Pal, M. Hubbe, Nanocellulose in packaging: Advances in barrier layer technologies, *Industrial Crops and Products.* 95 (2017) 574–582. https://doi.org/10.1016/J.INDCROP.2016.11.012.

[40] Decher, G. Fuzzy Nanoassemblies: Toward Layered Polymeric Multicomposites, *Science* (1979). 277 (1997) 1232–1237. https://doi.org/10.1126/SCIENCE.277.5330.1232.

[41] Wågberg, L., J.L. Erlandsson Wågberg, J. Erlandsson, L. Wågberg, The Use of Layer-by-Layer Self-Assembly and Nanocellulose to Prepare Advanced Functional Materials, *Advanced Materials.* 33 (2021) 2001474. https://doi.org/10.1002/ADMA.202001474.

[42] Zheng, Y., S. Xie, Y. Zeng, Electric field distribution and jet motion in electrospinning process: From needle to hole, *Journal of Materials Science.* 48 (2013) 6647–6655. https://doi.org/10.1007/S10853-013-7465-8/FIGURES/12.

[43] Farris, S., L. Piergiovanni, Emerging coating technologies for food and beverage packaging materials, *Emerging Food Packaging Technologies.* (2012) 274–302. https://doi.org/10.1533/9780857095664.3.274.

[44] Kargarzadeh, H., Ahmad, I., Thomas, S., Dufresne, A. *Handbook of nanocellulose and cellulose nanocomposites,* John Wiley & Sons., 2017.

[45] Gupte, Y., A. Kulkarni, B. Raut, P. Sarkar, R. Choudhury, A. Chawande, G.R.K. Kumar, B. Bhadra, A. Satapathy, G. Das, B. Vishnupriya, S. Dasgupta, Characterization of nanocellulose production by strains of Komagataeibacter sp. isolated from organic waste and Kombucha, *Carbohydrate Polymers.* 266 (2021) 118176. https://doi.org/10.1016/J.CARBPOL.2021.118176.

[46] Qian, Y., L. Bian, K. Wang, W.Y. Chia, K.S. Khoo, C. Zhang, K.W. Chew, Preparation and characterization of curdlan/nanocellulose blended film and its application to chilled meat preservation, *Chemosphere.* 266 (2021) 128948. https://doi.org/10.1016/J.CHEMOSPHERE.2020.128948.

[47] Agarwal, U.P., R.S. Reiner, S.A. Ralph, Cellulose I crystallinity determination using FT-Raman spectroscopy: univariate and multivariate methods, *Cellulose.* Vol. 17 (2010): P. 721-733. 17 (2010) 721–733. https://doi.org/10.1007/S10570-010-9420-Z.

[48] Wu, C., D.J. McClements, M. He, L. Zheng, T. Tian, F. Teng, Y. Li, Preparation and characterization of okara nanocellulose fabricated using sonication or high-pressure homogenization treatments, *Carbohydrate Polymers.* 255 (2021) 117364. https://doi.org/10.1016/J.CARBPOL.2020.117364.

[49] Li, J., R. Cha, K. Mou, X. Zhao, K. Long, H. Luo, F. Zhou, X. Jiang, Nanocellulose-Based Antibacterial Materials, *Advanced Healthcare Materials.* 7 (2018) 1800334. https://doi.org/10.1002/ADHM.201800334.

[50] Norrahim, M.N.F., N.M. Nurazzi, M.A. Jenol, M.A.A. Farid, N. Janudin, F.A. Ujang, T.A.T. Yasim-Anuar, S.U.F. Syed Najmuddin, R.A. Ilyas, Emerging development of nanocellulose as an antimicrobial material: an overview, *Materials Advances.* 2 (2021) 3538–3551. https://doi.org/10.1039/D1MA00116G.

[51] Leung, W.W.F., Q. Sun, Electrostatic charged nanofiber filter for filtering airborne novel coronavirus (COVID-19) and nano-aerosols, *Separation and Purification Technology.* 250 (2020) 116886. https://doi.org/10.1016/J.SEPPUR.2020.116886.

[52] Tyagi, P., R. Mathew, C. Opperman, H. Jameel, R. Gonzalez, L. Lucia, M. Hubbe, L. Pal, High-Strength Antibacterial Chitosan-Cellulose Nanocrystal Composite Tissue Paper, *Langmuir.* 35 (2019) 104–112. https://doi.org/10.1021/ACS.LANGMUIR.8B02655/ASSET/IMAGES/MEDIUM/LA-2018-02655V_0013.GIF.

[53] Pal, S., R. Nisi, M. Stoppa, A. Licciulli, Silver-Functionalized Bacterial Cellulose as Antibacterial Membrane for Wound-Healing Applications, *ACS Omega.* 2 (2017) 3632–3639. https://doi.org/10.1021/ACSOMEGA.7B00442/SUPPL_FILE/AO7B00442_SI_001.PDF.

[54] Jafary, R., M. Khajeh Mehrizi, S. hossein Hekmatimoghaddam, A. Jebali, *Antibacterial property of cellulose fabric finished by allicin-conjugated nanocellulose,* Http://Dx.Doi.Org/10.1080/00405000.2014.954780. 106 (2014) 683–689. https://doi.org/10.1080/00405000.2014.954780.

[55] Galkina, O.L., K. Önneby, P. Huang, V.K. Ivanov, A.V. Agafonov, G.A. Seisenbaeva, V.G. Kessler, Antibacterial and photochemical properties of cellulose nanofiber–titania nanocomposites loaded with two different types of antibiotic medicines, *Journal of Materials Chemistry* B. 3 (2015) 7125–7134. https://doi.org/10.1039/C5TB01382H.

[56] El-Samahy, M.A., S.A.A. Mohamed, M.H. Abdel Rehim, M.E. Mohram, Synthesis of hybrid paper sheets with enhanced air barrier and antimicrobial properties for food packaging, *Carbohydrate Polymers.* 168 (2017) 212–219. https://doi.org/10.1016/J.CARBPOL.2017.03.041.

[57] Robertson, G.L. Packaging and Food and Beverage Shelf Life, *The Stability and Shelf Life of Food.* (2016) 77–106. https://doi.org/10.1016/B978-0-08-100435-7.00003-4.

[58] Wei, Ch.-I., A. Wei, P.R. Becker, D.-i. Fuchimoto, A. Takenaka, T.i Matsumoto, P. Kess, M. Mirosa, W. Xu, J. Birch, K. Lloyd, Z. Tianzhu, L. Caixia, Zhangjie(siyar), H. Qian, Y. Kaiqian, T.-R. Lee, C.-Y. Chien.*Case Studies: Insights on Agriculture Innovation 2016* (IAAS Series), Vital Wellspring Education Pte. Ltd.., 2018.

[59] Wang, J., D.J. Gardner, N.M. Stark, D.W. Bousfield, M. Tajvidi, Z. Cai, Moisture and Oxygen Barrier Properties of Cellulose Nanomaterial-Based Films, *ACS Sustainable Chemistry and Engineering.* 6 (2018) 49–70. https://doi.org/10.1021/ACSSUSCHEMENG.7B03523/ASSET/IMAGES/MEDIUM/SC-2017-03523J_0015.GIF.

[60] Cozzolino, C.A., G. Campanella, H. Türe, R.T. Olsson, S. Farris, Microfibrillated cellulose and borax as mechanical, O2-barrier, and surface-modulating agents of pullulan biocomposite coatings on BOPP, *Carbohydrate Polymers.* 143 (2016) 179–187. https://doi.org/10.1016/J.CARBPOL.2016.01.068.

[61] de Oliveira, J.P., G.P. Bruni, S.L.M. el Halal, F.C. Bertoldi, A.R.G. Dias, E. da R. Zavareze, Cellulose nanocrystals from rice and oat husks and their application in aerogels for food packaging, *International Journal of Biological Macromolecules.* 124 (2019) 175–184. https://doi.org/10.1016/J.IJBIOMAC.2018.11.205.

[62] Li, P., J.A. Sirviö, A. Haapala, A. Khakalo, H. Liimatainen, Anti-oxidative and UV-absorbing biohybrid film of cellulose nanofibrils and tannin extract, *Food Hydrocolloids.* 92 (2019) 208–217. https://doi.org/10.1016/J.FOODHYD.2019.02.002.

[63] Wang, X., Y. Xie, H. Ge, L. Chen, J. Wang, S. Zhang, Y. Guo, Z. Li, X. Feng, Physical properties and antioxidant capacity of chitosan/epigallocatechin-3-gallate films reinforced with nano-bacterial cellulose, *Carbohydrate Polymers.* 179 (2018) 207–220. https://doi.org/10.1016/J.CARBPOL.2017.09.087.

[64] Wang, X., N. Ullah, X. Sun, Y. Guo, L. Chen, Z. Li, X. Feng, Development and characterization of bacterial cellulose reinforced biocomposite films based on protein from buckwheat distiller's dried grains, *International Journal of Biological Macromolecules.* 96 (2017) 353–360. https://doi.org/10.1016/J.IJBIOMAC.2016.11.106.

[65] Pelissari, F.M., M.M. Andrade-Mahecha, P.J. do A. Sobral, F.C. Menegalli, Nanocomposites based on banana starch reinforced with cellulose nanofibers isolated from banana peels, *Journal of Colloid and Interface Science.* 505 (2017) 154–167. https://doi.org/10.1016/J.JCIS.2017.05.106.

[66] Karimi, N., A. Alizadeh, H. Almasi, S. Hanifian, Preparation and characterization of whey protein isolate/polydextrose-based nanocomposite film incorporated with cellulose nanofiber and L. plantarum: A new probiotic active packaging system, *LWT.* 121 (2020) 108978. https://doi.org/10.1016/J.LWT.2019.108978.

[67] Nguyen, H.L., T.H. Tran, L.T. Hao, H. Jeon, J.M. Koo, G. Shin, D.S. Hwang, S.Y. Hwang, J. Park, D.X. Oh, Biorenewable, transparent, and oxygen/moisture barrier nanocellulose/nanochitin-based coating on polypropylene for food packaging applications, *Carbohydrate Polymers.* 271 (2021) 118421. https://doi.org/10.1016/J.CARBPOL.2021.118421.

[68] Sánchez-Gutiérrez, M., I. Bascón-Villegas, E. Espinosa, E. Carrasco, F. Pérez-Rodríguez, A. Rodríguez, Cellulose Nanofibers from Olive Tree Pruning as Food Packaging Additive of a Biodegradable Film, *Foods* 2021, Vol. 10, Page 1584. 10 (2021) 1584. https://doi.org/10.3390/FOODS10071584.

[69] Sharma, S., S. Barkauskaite, A.K. Jaiswal, S. Jaiswal, Essential oils as additives in active food packaging, *Food Chemistry.* 343 (2021) 128403. https://doi.org/10.1016/J.FOODCHEM.2020.128403.

[70] Brockgreitens, J., A. Abbas, Responsive Food Packaging: Recent Progress and Technological Prospects, *Comprehensive Reviews in Food Science and Food Safety.* 15 (2016) 3–15. https://doi.org/10.1111/1541-4337.12174.

[71] Cruz-Romero, M.C., T. Murphy, M. Morris, E. Cummins, J.P. Kerry, Antimicrobial activity of chitosan, organic acids and nano-sized solubilisates for potential use in smart antimicrobially-active packaging for potential food applications, *Food Control.* 34 (2013) 393–397. https://doi.org/10.1016/J.FOODCONT.2013.04.042.

[72] Jayakumar, A., H. K.V., S. T.S., M. Joseph, S. Mathew, P. G., I.C. Nair, R. E.K., Starch-PVA composite films with zinc-oxide nanoparticles and phytochemicals as intelligent pH sensing wraps for food packaging application, *International Journal of Biological Macromolecules.* 136 (2019) 395–403. https://doi.org/10.1016/J.IJBIOMAC.2019.06.018.

[73] Jose, A., S. Anitha Sasidharan, C. Chacko, D. Mukkumkal Jacob, R. Edayileveettil Krishnankutty, Activity of Clove Oil and Chitosan Nanoparticles Incorporated PVA Nanocomposite Against Pythium aphanidermatum, *Applied Biochemistry and Biotechnology.* 194 (2021) 1442–1457. https://doi.org/10.1007/S12010-021-03709-3/FIGURES/7.

[74] Muxika, A., A. Etxabide, J. Uranga, P. Guerrero, K. de la Caba, Chitosan as a bioactive polymer: Processing, properties and applications, *International Journal of Biological Macromolecules.* 105 (2017) 1358–1368. https://doi.org/10.1016/J.IJBIOMAC.2017.07.087.

[75] Annu, A. Ali, S. Ahmed, Eco-friendly natural extract loaded antioxidative chitosan/polyvinyl alcohol based active films for food packaging, *Heliyon.* 7 (2021) e06550. https://doi.org/10.1016/J.HELIYON.2021.E06550.

[76] Marcos, B., C. Sárraga, M. Castellari, F. Kappen, G. Schennink, J. Arnau, Development of biodegradable films with antioxidant properties based on polyesters containing α-tocopherol and olive leaf extract for food packaging applications, *Food Packaging and Shelf Life.* 1 (2014) 140–150. https://doi.org/10.1016/J.FPSL.2014.04.002.

[77] Gaikwad, K.K., Y.S. Lee, Novel natural phenolic compound-based oxygen scavenging system for active packaging applications, *Journal of Food Measurement and Characterization.* 10 (2016) 533–538. https://doi.org/10.1007/S11694-016-9332-1/TABLES/3.

[78] Ummartyotin, S., P. Pisitsak, C. Pechyen, Eggshell and bacterial cellulose composite membrane as absorbent material in active packaging, *International Journal of Polymer Science.* 2016 (2016). https://doi.org/10.1155/2016/1047606.

[79] Lacoste, A., K.M. Schaich, D. Zumbrunnen, K.L. Yam, Advancing controlled release packaging through smart blending, *Packaging Technology and Science.* 18 (2005) 77–87. https://doi.org/10.1002/PTS.675.

[80] Saito, H., Y. Sakakibara, A. Sakata, R. Kurashige, D. Murakami, H. Kageshima, A. Saito, Y. Miyazaki, Antibacterial activity of lysozyme-chitosan oligosaccharide conjugates (LYZOX) against Pseudomonas aeruginosa, Acinetobacter baumannii and Methicillin-resistant Staphylococcus aureus, *PLOS ONE.* 14 (2019) e0217504. https://doi.org/10.1371/JOURNAL.PONE.0217504.

[81] Buruaga-Ramiro, C., S. v. Valenzuela, C. Valls, M.B. Roncero, F.I.J. Pastor, P. Díaz, J. Martinez, Development of an antimicrobial bioactive paper made from bacterial cellulose, *International Journal of Biological Macromolecules.* 158 (2020) 587–594. https://doi.org/10.1016/J.IJBIOMAC.2020.04.234.

[82] Mocanu, A., G. Isopencu, C. Busuioc, O.M. Popa, P. Dietrich, L. Socaciu-Siebert, Bacterial cellulose films with ZnO nanoparticles and propolis extracts: Synergistic antimicrobial effect, *Scientific Reports* 2019 9:1. 9 (2019) 1–10. https://doi.org/10.1038/s41598-019-54118-w.

[83] Xu, Y., Y. Chu, X. Feng, C. Gao, D. Wu, W. Cheng, L. Meng, Y. Zhang, X. Tang, Effects of zein stabilized clove essential oil Pickering emulsion on the structure and properties of chitosan-based edible films, *International Journal of Biological Macromolecules. 156* (2020) 111–119. https://doi.org/10.1016/J.IJBIOMAC.2020.04.027.

[84] Roy, S., J.W. Rhim, Gelatin/agar-based functional film integrated with Pickering emulsion of clove essential oil stabilized with nanocellulose for active packaging applications, *Colloids and Surfaces A: Physicochemical and Engineering Aspects.* 627 (2021) 127220. https://doi.org/10.1016/J.COLSURFA.2021.127220.

[85] Liu, Z., D. Lin, R. Shen, X. Yang, Characterizations of novel konjac glucomannan emulsion films incorporated with high internal phase Pickering emulsions, *Food Hydrocolloids.* 109 (2020) 106088. https://doi.org/10.1016/J.FOODHYD.2020.106088.

[86] Vilela, C., M. Kurek, Z. Hayouka, B. Röcker, S. Yildirim, M.D.C. Antunes, J. Nilsen-Nygaard, M.K. Pettersen, C.S.R. Freire, A concise guide to active agents for active food packaging, *Trends in Food Science & Technology.* 80 (2018) 212–222. https://doi.org/10.1016/J.TIFS.2018.08.006.

[87] Moradian, S., H. Almasi, S. Moini, Development of bacterial cellulose-based active membranes containing herbal extracts for shelf-life extension of button mushrooms (Agaricus bisporus), *Journal of Food Processing and Preservation.* 42 (2018) e13537. https://doi.org/10.1111/JFPP.13537.

[88] Yang, Y.N., K.Y. Lu, P. Wang, Y.C. Ho, M.L. Tsai, F.L. Mi, Development of bacterial cellulose/chitin multi-nanofibers based smart films containing natural active microspheres and nanoparticles formed in situ, *Carbohydrate Polymers.* 228 (2020) 115370. https://doi.org/10.1016/J.CARBPOL.2019.115370.

[89] Rawdkuen, S. Edible Films Incorporated with Active Compounds: Their Properties and Application, *Active Antimicrobial Food Packaging.*(2019). https://doi.org/10.5772/INTECHOPEN.80707.

[90] Jipa, I.M., A. Stoica-Guzun, M. Stroescu, Controlled release of sorbic acid from bacterial cellulose based mono and multilayer antimicrobial films, *LWT.* 47 (2012) 400–406. https://doi.org/10.1016/J.LWT.2012.01.039.

[91] Dou, T. J., Vuorinen, H. Koivula, N. Forsman, M. Sipponen, S. Hietala, Self-Standing Lignin-Containing Willow Bark Nanocellulose Films for Oxygen Blocking and UV Shielding, *ACS Applied Nano Materials.* 4 (2021) 2921–2929. https://doi.org/10.1021/ACSANM.1C00071/SUPPL_FILE/AN1C00071_SI_001.PDF.

[92] dos Santos, C.A., G.R. dos Santos, V.S. Soeiro, J.R. dos Santos, M. de A. Rebelo, M.V. Chaud, M. Gerenutti, D. Grotto, R. Pandit, M. Rai, A.F. Jozala, Bacterial nanocellulose membranes combined with nisin: a strategy to prevent microbial growth, *Cellulose.* 25 (2018) 6681–6689. https://doi.org/10.1007/S10570-018-2010-1/FIGURES/5.

[93] Balbinot-Alfaro, E., D.V. Craveiro, K.O. Lima, H.L.G. Costa, D.R. Lopes, C. Prentice, Intelligent Packaging with pH Indicator Potential, *Food Engineering Reviews.* 11 (2019) 235–244. https://doi.org/10.1007/S12393-019-09198-9/TABLES/2.

[94] Active and intelligent packaging food - research and development - a review - *Polish Journal of Food and Nutrition Sciences* - Tom 64, Number 1 (2014) - BibliotekaNauki - Yadda, (n.d.). https://yadda.icm.edu.pl/yadda/element/bwmeta1.element.agro-840f6a8c-9921-49f7-a0f3-e0406a18c77c (accessed April 19, 2022).

[95] Lu, P., Y. Yang, R. Liu, X. Liu, J. Ma, M. Wu, S. Wang, Preparation of sugarcane bagasse nanocellulose hydrogel as a colourimetric freshness indicator for intelligent food packaging, *Carbohydrate Polymers.* 249 (2020) 116831. https://doi.org/10.1016/J.CARBPOL.2020.116831.

[96] Taherkhani, E., M. Moradi, H. Tajik, R. Molaei, P. Ezati, Preparation of on-package halochromic freshness/spoilage nanocellulose label for the visual shelf-life estimation of meat, *International Journal of Biological Macromolecules.* 164 (2020) 2632–2640. https://doi.org/10.1016/J.IJBIOMAC.2020.08.177.

[97] Vilela, C., C. Moreirinha, E.M. Domingues, F.M.L. Figueiredo, A. Almeida, C.S.R. Freire, Antimicrobial and Conductive Nanocellulose-Based Films for Active and Intelligent Food Packaging, *Nanomaterials* 2019, Vol. 9, Page 980. 9 (2019) 980. https://doi.org/10.3390/NANO9070980.

[98] Kalpana, S., S.R. Priyadarshini, M. Maria Leena, J.A. Moses, C. Anandharamakrishnan, Intelligent packaging: Trends and applications in food systems, *Trends in Food Science & Technology.* 93 (2019) 145–157. https://doi.org/10.1016/J.TIFS.2019.09.008.

[99] Sukhavattanakul, P., H. Manuspiya, Fabrication of hybrid thin film based on bacterial cellulose nanocrystals and metal nanoparticles with hydrogen sulfide gas sensor ability, *Carbohydrate Polymers.* 230 (2020) 115566. https://doi.org/10.1016/J.CARBPOL.2019.115566.

[100] Kuswandi, B., Jayus, T.S. Larasati, A. Abdullah, L.Y. Heng, Real-Time Monitoring of Shrimp Spoilage Using On-Package Sticker Sensor Based on Natural Dye of Curcumin, *Food Analytical Methods.* 5 (2012) 881–889. https://doi.org/10.1007/S12161-011-9326-X/FIGURES/9.

[101] Pirsa, S., T. Shamusi, Intelligent and active packaging of chicken thigh meat by conducting nano structure cellulose-polypyrrole-ZnO film, *Materials Science and Engineering:* C. 102 (2019) 798–809. https://doi.org/10.1016/J.MSEC.2019.02.021.

[102] Ghasemi, S., M.R. Bari, S. Pirsa, S. Amiri, Use of bacterial cellulose film modified by polypyrrole/TiO2-Ag nanocomposite for detecting and measuring the growth of pathogenic bacteria, *Carbohydrate Polymers.* 232 (2020) 115801. https://doi.org/10.1016/J.CARBPOL.2019.115801.

Chapter 3

Food Packaging Films Generated from 2D Biopolymer-Based Nanocomposites: An Overview

**Nahed A. Abd El-Ghany*
and Mahmoud H. Abu Elella**
Chemistry Department, Faculty of Science, Cairo University, Giza, Egypt

Abstract

The designing of high-quality food packaging materials has gained more interest in modern societies for food safety. In the preceding two chapters, our colleagues have presented different types of polysaccharides for use in food packaging. In fact, besides polysaccharides, many other materials show high potential for use in food packaging applications. A good example is the two-dimensional nanosheet materials (2DM)-based nanocomposites, especially in the presence of biocompatible polymers, which have been widely employed for food packaging. The as-prepared 2DM nanocomposites have fascinating features, including cost-effectiveness, good mechanical and thermal stability, eco-friendliness, good O_2 barrier performance, high electrocatalytic performance, and large surface area. In this chapter, we will summarise recent food packaging applications of 2DM nanosheets (such as graphene, layered double hydroxides, hexagonal boron nitride, and nanoclays (e.g., Montmorillonite)-based nanocomposites in the presence of different biocompatible polymers such as natural polysaccharides (chitosan and cellulose) and synthetic polymers (polyvinyl alcohol and polyurethane). It shows their usage in active and intelligent food packaging. This chapter will demonstrate the increased

* Corresponding Author's Email: dr_nahed_055@yahoo.com.

In: Food Packaging
Editor: Wing-Fu Lai
ISBN: 979-8-88697-249-8

number of reported studies that confirm the versatility of 2DM and polymer nanocomposites in food packaging.

Keywords: food packaging, two-dimensional nanosheet materials, nanocomposites, biocompatible polymers

1. Introduction

Packaging has assumed increasing importance in maintaining the quality of food products from manufacturing to marketing and final consumption, due to its protection from environmental challenges [1, 2]. In fact, the role of food packaging has continued to evolve in response to the changes in human demands. Consumer preferences for high quality as well as safe, rapid food are on the rise with the globalisation and civilization. Researchers are working to build multi-functional food packaging that will preserve the goodness and freshness of food while also promoting the spread and advertising of food products [3]. Thus, studies on food packaging generally aim at enhancing, integrating, or broadening the four principal functions of conventional food packaging: containment, protection, communication, and convenience [4]. In functional food packaging systems, active packaging is defined as "packaging" in which auxiliary constituents have been included in or on either the packaging material or the package surface to enhance the performance of the package system [5]. Besides, intelligent packaging possesses advanced functions with respect to identification and communication, such as providing information about the real-time quality of food products to consumers [4].

Food packaging materials now on the market are petroleum-based polymers that are non-biodegradable. Because of their non-degradability and toxicity, traditional plastics hurt the ecosystem and exacerbate the problem of trash disposal, posing a threat to the marine environment and physical security [6]. Furthermore, without aseptic processing, the majority of them fail to maintain food freshness, which is damaging to human health [7]. As a result, developing biodegradable, antimicrobial, and low-cost polymer films for food packaging is critical [8]. Nanotechnology has opened up new ways of developing active and intelligent food packaging materials with enhanced properties. Nanotechnology is defined as the fabrication and modification of materials with a size of not more than 100 nm. These ultra-small materials show unique physical, chemical, and biological effectiveness. In the food industry, nanomaterials are substantially utilised in food processing and food packaging [9]. Various organic and inorganic nanomaterials have unique properties such as high surface adhesion, light hinder, anti-oxidation and biological activities and have been used in preparing functional food package materials. A biocompatible polymer has been widely studied in food packaging due to its potential in many fields such as material science [10] and pharmaceutical application [11], because it can be obtained at a low cost, is nontoxic, and biodegradable [12]. They do not present all the necessary requirements intended for food packaging, namely in terms of mechanical resistance and water/gas barrier properties. Furthermore, there is an ongoing trend towards active and intelligent packaging, i.e., packaging that is able to interact with food to actively extend its shelf life and to inform the consumer about the safety state of the food products, respectively [13]. Barrier properties can be impregnated into the food package by adding a layer of impermeable material. The permeability of the polymeric material can be improved by the addition of lamellar fillers such as clay, boron nitride (BN), and graphene. Whereas the filler in this case is considered impermeable and acts as a barrier to the diffusion of gases and vapors.

In this respect, two-dimensional materials may play a function inimparting activity and intelligence to the packaging, in addition to other properties [14]. Two-dimensional materials (2DMs) are ultrathin nanomaterials with unique characteristics such as high surface-to-volume ratio, surface charge, shape, high degree of anisotropy, and adjustable chemical functionality, they only have single or a few atoms in thickness (often less than 5 nm) [15]. Members of the 2DM family, such as graphene, graphitic, layered double hydroxide, Montmorillonite clay, and hexagonal boron nitrides, have been extensively studied in many fields, includingbiomedicine [16] and energy [17, 18]. However, in the field of food packaging, the emergence of 2DMs provides more opportunities to develop new active and intelligent food packaging technologies [19].

On the other hand, biocompatible polymers have been widely used in different applications such as drug delivery, tissue engineering, biomedical applications, wastewater treatment, and food packaging applications. They have some fascinating physicochemical properties, such as non-toxic; mechanical and thermal stability; antioxidant activity; and biocompatible [20-22]. They can be divided into natural polysaccharides and synthetic polymers. Natural polysaccharides are carbohydrate polymeric oligosaccharides which are connected by glycosidic bonds. They can be categorised based on the composition of monosaccharide repeating units into homopolymers such as cellulose and heteropolymers such as chitosan. Synthetic polymers, on the other hand, such as poly (vinyl alcohol) and poly urethane (Figure 1) [23-28]. In this chapter, we focus on 2DMs/ biocompatible polymer-based nanocomposites and their recent advancement in food packaging applications.

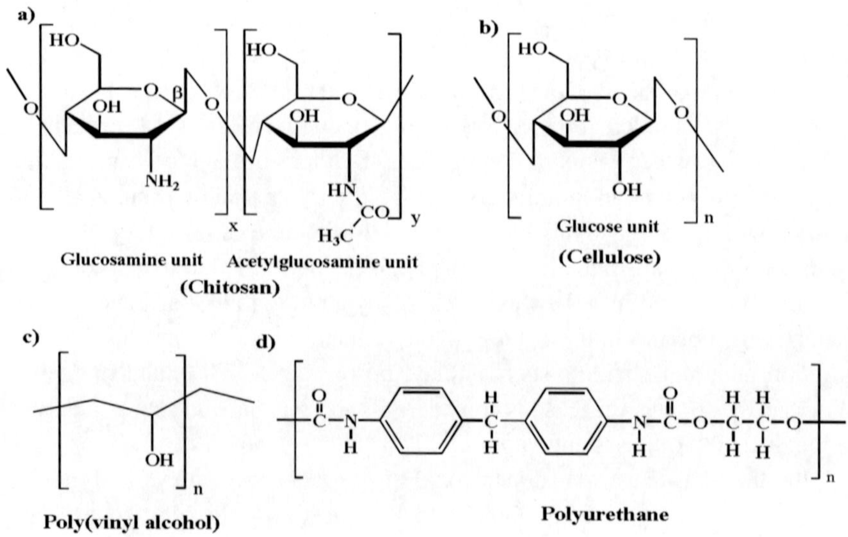

Figure 1. The chemical structure of different biocompatible polymers a) chitosan, b) cellulose, c) poly(vinyl alcohol, and d) polyurethane.

2. 2DM/ Biocompatible Polymers - Nanocomposites Films

2.1. Graphene Nanocomposite Films

Graphene is a monolayer of sp2-hybridized carbon atoms that are arranged into hexagonal lattice units (Figure 1a). Graphene and its derivatives have many intriguing properties, such as superior electron mobility, large specific surface area, ultra-strong mechanical stiffness, and a fracture strength of approx. 125 GPa [29]. Graphene and its derivatives have been applied in many fields [30]. Graphene has many advantages such as It has flexible and transparent carbon monolayered atoms, it has better thermal and electrical properties and it is the finest and hardest material known [15, 31-34]. Polymers can be used with graphene in food packaging to improve barrier, thermal, and mechanical properties. The interaction between graphene and the polymer matrix includes Van der Waal electrostatic interaction between the polymer function groups and the graphene double bonds, causing a helical conformation formed within the main chain of the polymer [35]. Furthermore, graphene exhibits a strong antibacterial activity because it destroys the permeability of the cell membrane, resulting in leakage of the cell components like phospholipids and proteins [36, 37]. Therefore, some polymer-graphene nanocomposites show good antibacterial activities. Due to the large surface area and good electrocatalytic activity of graphene, it is effectively applied to construct preservative new devices for intelligent packaging [38]. Graphene sheets with lateral dimensions of less than 100 nm, have very low toxicity and exhibit high stability, excellent solubility, stable photoluminescence, and good biocompatibility [39]. Moreover, several advantages and some drawbacks of the prepared graphene nanocoposites films have been reported in Table 1.

Table 1. The advantages (pros) and disadvantages (cons) of 2DM nanocomposites films

2DM nanocomposites	Advantage (Pros)	Disadvantage (Cons)	Ref.
Graphene nanocomposites	• High thermal and mechanical stability • High specific surface area • Good electrical conductivity • High Young's modulus • Good antibacterial activity • Good electrocatalytic activity	• Strong agglomeration of graphene • Low cytotoxicity	[36, 37, 97-99]
Boron nitride nanocomposites	• High chemical stability • Good mechanical strength • Good thermal conductivity • Resistance to oxidation • They are electrical insulators	Although remarkable progress in the development of BN nanocomposites has been reported, their practical applications are still challenging, and there are many obstacles that must be overcome, including their durability and recovery.	[100]
Layered double hydroxides nanocomposites	• Exceptional mechanical properties • Good gas and moisture barrier	By increasing LDHs content, the tensile strength decreased and there was agglomeration of LDHs on the nanocomposite surface, which leads to decrease its efficiency.	[84]
Montmorillonite nanocomposites	• Good mechanical and thermal stability • Good barrier property • Good antimicrobial activity • They are used as biosensors and oxygen scavengers	Using MMT nanofiller in food packaging showed some toxicological effects due to its migration from food packaging into food products. So, its nanocomposite preparation needs proper considerations in terms of safety for end use.	[101, 102]

The oxidative processes, along with microbiological growth, are one of the mechanisms of food quality deterioration, being responsible for many important changes such as loss of nutritional values, texture modifications, and the development of undesirable compounds such as off flavors, coloured substances, and even toxic substances to humans [40]. The antioxidant, antifungal, and antibacterial activities of chitosan have been attracting special attention in the food packaging domain [4, 41, 42]. Chitosan is a promising alternative as an ecologically packaged material, which can provide suitable atmospheric conditions for fruits and vegetables, reducing gas exchange and delaying ripening [43]. For example, Vilvert et al. [43] reported the preparation of an environmentally friendly graphene oxide/chitosan nanocomposite bag to keep the 'Palmer' mango's postharvest quality within cold storage (Figure 2). They found that after the coverage of the mango, the ripening is delayed, preserving its external appearance and skin and flesh colors. In addition, they found that firmness and soluble solids as well as chlorophyll and β-carotene constituents are preserved.

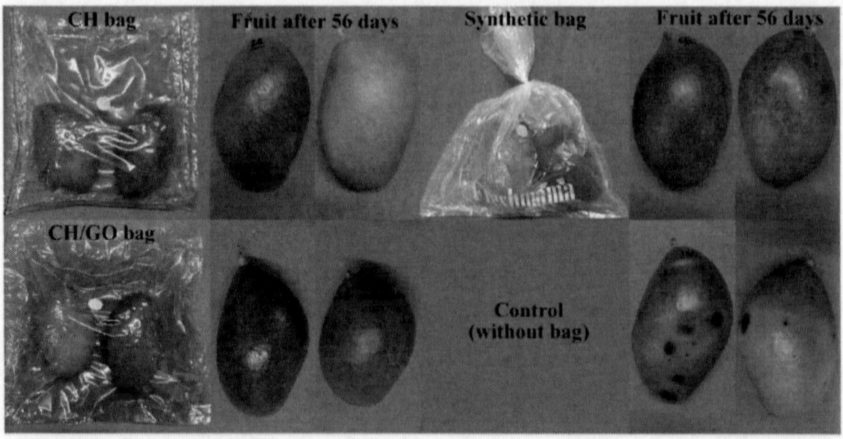

Figure 2. The packaging of 'Palmer' mangoes at different conditions (12°C for 56 days) using graphene oxide/chitosan nanocomposite bags [43].

It has been investigated as an edible coating for fruit packaging applications because it delays the rate of respiration, inhibits postharvest diseases, improves safety and increases the antioxidant process, allowing extension the of the food product shelf life while maintaining the fruits' overall quality. When chitosan is used as a film coating, it gives rise to transparent, flexible, and good oxygen barrier properties [44, 45]. The low water vapour barrier properties and low mechanical strength of CS limit its usage in food packaging applications. To overcome these fragilities for food packaging applications [46, 47]. The antimicrobial activity of chitosan, the mechanical properties as well as the water vapour barrier properties can be enhanced by the addition of bioactive ingredients with antioxidant and/or antimicrobial activity [47]. Among these active materials, nanomaterials and two-dimensional nano-materials can significantly improve the physical and biological properties of chitosan. Chitosan loaded with small-sized two-dimensional materials such as nano- graphene oxide, double layered hydroxide, nitrides, and nano-clay exhibited higher antioxidant and antimicrobial activity that prevented the food oxidation and growth of spoilage microorganisms [48]. Recently, a study showed that graphene oxide cross-linked chitosan nanocomposite films significantly affect the physical and biological properties of the CSGO film. The thermal and mechanical stability as well as the antimicrobial activity against bacteria and fungi of chitosan were improved upon loading of GO. Moreover, the brown colour of GO sheets imparts a dark colour to the CSGO film, which could prevent the damage of the food by ultra violet light, thus maintaining food packaging quality [49].

Cellulose is a linear polymer of β-(1-4) linked D-glucose units. It is the most abundant natural biopolymer. It is biodegradable, and it is derived from wood pulp by a delignification process [50]. Neat cellulose does not have a thermoplastic nature and is unable to form films due to high crystallinity and long chains. Nanocrystalline cellulose (CNC) is a material derived from cellulose by acid hydrolysis. It has better properties such as elasticity, optical transparency, low thermal expansion coefficient, potential gas barrier ability, and film forming ability as well as biocompatibility, biodegradability, and nontoxicity [51]. CNC has been used as a gas barrier and filler of nanocomposites in food packaging applications [51-54]. Nanocrystalline cellulose/graphene composites exhibit several more advantageous properties than materials containing only cellulose nanoparticles or only graphene. Adding carbon nanoparticles like graphene or graphene oxide to CNC can improve their mechanical strength, thermal stability, and nonporous nature [55, 56]. Recently, cellulose nanocrystals (CNCs) and reduced graphene oxide (rGO) have been used as reinforcing nanofiller incorporated into poly lactic acid (PLA) via the solution casting method. The CNC/rGO/PLA nanocomposite exhibited improved mechanical, barrier, thermal, antimicrobial, and biomedical properties, making it promising in food packaging and biomedical applications [57].

Polyvinyl alcohol (PVA) is an ethynyl polymerthat exhibits the following features: excellent film-forming and adhesion tendencies; high chemical polarity and solubility in water; degradability; nontoxicity; and biocompatibility. It has distinguishable chemical resistance, mechanical and physical properties. In addition to the benefits of PVA in many applications such asdrug delivery [58], paper lamination [59], and textile spinning and weaving [60], it more useful as a packaging material. Mixture of PVA with starch provides a stable supporting medium, with improved physical and mechanical properties. PVA has been widely used in the packaging industry due to its low cost and biodegradability [61, 62].

The physical and mechanical as well as the antimicrobial properties were improved by the incorporation of 2D nanomaterials between PVA chains as reinforcing materials, which significantly improved the physical and mechanical properties of PVA. Graphene and reduced GO are not compatible with all polymer matrices and tend to form aggregates [63]. GO and its derivatives lose their tendency to aggregation when dispersed in hydrophilic polymers due to the strong hydrogen bonding formed between oxygen-containing groups of each GO sheet and the hydroxyl groups present in polymer chains [64]. PVA treated with GO exhibited improvements in the mechanical properties, thermal stability, and higher electrical conductivity performance of the PVA/GO nanocomposite, allowing it to be used in many industrial applications, especially in the food package domain [65]. The PVA/GO/ZnO nanocomposite film with a mass ratio of 3:7 showed great promise as an environmentally friendly, antibacterial packaging material with UV light barrier qualities to extend the shelf life of packaged foods [66].

Many different adhesive substances are used in food packaging due to the different materials that must be bonded and to the production process requirements. Flexible laminates make up the largest component of the packaging industry in many applications, such as: packages that go from the freezer to the oven; pouches for food and beverages;containers for medical and pharmaceutical products [67]. In most of these applications, the packaged product should be protected from not only the environmental conditions (e.g., moisture, oxygen, UV), but also from the adhesive and packaging chemicals that may be present in the package itself or that result from processing. Polyurethanes in the form of solvent solutions, aqueous dispersion, hot melt, and solventless for lamination of flexible films is the most common form of adhesive material used in the fabrication of flexible laminates for packaging products. Polyurethane is a kind of block copolymer with oligomer polyols as the ssoft segment and isocyanates as the hard segment. They contain many polar bonds in their structure, such as carbamate bonds, ester bonds, urea bonds [68], which result in excellent adhesion properties and film forming properties. Polyurethane adhesives are widely used in soft packaging of food materials due to their excellent mechanical properties, serviceability, heatresistance, and high hygienic integrity [68, 69]. The incorporation of BN into polyurethane (PU) adhesives is a promising method to decrease polymer permeability and improve the practical adhesion of aluminium foil in food packaging. The improvement of the adhesive barrier properties of the PU by boron nitride incorporation resulted from the reduction of the thickness of deposited aluminium layers or even their elimination. PU impregnated with BN and h-BN displayed better moisture barrier properties and high adhesive peeling resistance that is required for flexible laminated food packaging [52].

2.2. Boron Nitrides-Based Nanocomposites

BN is a synthetic binary discovered as a commercial material in the latter half of the 20th century [70, 71]. It shows physical properties similar to graphite. It has good lubricating properties according to its layered structure. It displays very high thermal conductivity, better thermal stability, a good resistance to corrosion and oxidation, and also exhibits a strong UV emission [72-74]. According to its shape and synthesis process, it can offer super hydrophobicity [75]. BN has several remarkable advantages, such as nontoxic and good resistance to chemical attack. It also has high environmental compatibility and good mechanical behaviour properties [19]. Due to the lamellar structure of boron nitride, it displays gas barrier properties. Its high aspect ratio contributes to promoting longer pathways for gas molecules and increasing gas barrier properties. Hexagonal-BN (h-BN), also known as "white graphite," has stacked graphene-like 2D layer structures. Equal amounts of nitrogen and boron atoms are placed in a ring-shaped hexagon in each layer, and they are bound together with strong covalent bonds. The surface area of h-BN is large, and the in-plane elastic modulus is high. h-BN has been used as a filler in polymers to increase the mechanical properties of polymers [76, 77]. Moreover, BN nanocomposite films have several advantages and some disadvantages have been mentioned in Table 1.

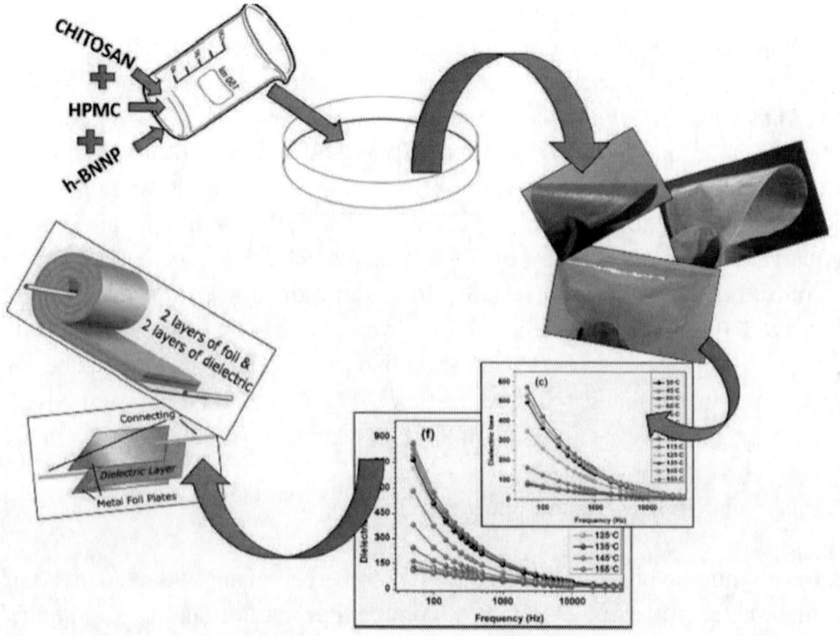

Figure 3. Schematic preparation of hexagonal boron nitride nanoparticles/ chitosan/ hydroxypropyl methylcellulose (HPMC) ternary nanocomposites with casting solution technique and enhancement of their dielectric properties [78].

Previously, Shareef et al. [78] mentioned hexagonal boron nitride nanoparticles/ biopolymer ternary nanocomposites in the presence of chitosan and Hydroxypropyl methylcellulose (HPMC) with casting solution technique for food packaging application. They investigated their dielectric properties as a function of frequency and temperature. The results showed that the maximum value of dielectric constant was reported to be 10200 at 100 Hz and 35 °C. In addition, the dielectric loss of 12476 at 100 Hz and 35 °C was achieved, using 5% (w/w) of nanoparticles within polymer chains. Also, the findings revealed good interaction among boron nitride nanoparticles and polymer chains. The TGA analysis showed good thermal stability of the obtained nanocomposites with a minimum weight loss of around 5.6% at 100 °C. The authors exhibited the enhancement of the mechanical properties by increasing Young's modulus up to 3.715 GPa, which indicated excellent stiffness (Figure 3). In addition, Behera et al. [79] also reported nanobiocomposite food packaging films based on chitosan/boron nitride nanoplatelets. The results exhibited good dispersion of nanoplatelets inside chitosan chains; however, after their dispersion, they found that the water solubility of nanocomposite films declined from 41.2% to 27.8% (for 7% (w/w) of nanoplatelets) and their moisture absorption as well.Moreover, water vapour permeation of the as-prepared films decreased to reach up to 2.9 x 10^{-11} g/m.s.Pa after loading the same boron nitride concentration. In addition, the decreased O2 permeability of nanocomposite films was reported to be 84%. On the other hand, compared with unmodified chitosan, the obtained nanocomposite films illustrated better NaOH resistance and excellent thermal stability. However, the addition of 5% (w/w) of nanoplatelet enhanced the mechanical properties (Young's modulus) up to 45%. The authors examined the cytotoxicity of nanocomposites and found that their cytotoxicity declined after the addition of boron nitrides.

Sustainable boron nitride nanosheets/cellulose nanofiber nanocomposite films with good O2 barrier with cytotoxicity and change in their colour were prepared by Nguyen et al. [80] as active food packaging materials. Notably, cellulose nanofibers show an ideal O2 transmission rate of less than 1 cc/m^2 per day. So, after the addition of 5% (w/w) of boron sheets, the findings exhibited an improvement in O2 transmission rate to 4.7 cc/m^2/day, which is sufficient to preserve the cheese and meat. The authors reported that boron nitride nanosheets reduce pinhole formation during the nanocomposite fabrication. Also, the addition of nanosheets improved the mechanical properties of the as-prepared nanocomposite films owing to the good miscibility of boron nanosheets with cellulose nanofibers in the aqueous solution. Moreover, 2D nanosheets produced negligible colour change and cytotoxicity in in-vitro cell lines. Years ago, Nacas et al. [68] reported the synthesis of a polyurethane/boron nitride nanocomposite and used it as an adhesive food packaging with a good O2 barrier layer. They used this prepared nanocomposite as an alternative pack to aluminium foil in food packaging applications. They prepared the PU nanocomposites in the presence of two different boron sizes: micro- and nano-structure. The as-prepared nanocomposites achieved a decreased in the water vapour permeance of up to 50% using 1% and 2% of micro-boron and nano-boron, respectively. Inaddition, they increased in mechanical adhesion properties by 37% compared to the native PU adhesive. The adhesive property of the BN nanocomposite was higher peeling resistance than virgin PU adhesive, at 37.63% compared to 7.14% (native PU). Moreover, the findings demonstrated that the nano-BN reduced the permeability of PU adhesives and could be used for flexible food packaging.

Recently, Ge et al. [81] have reported the preparation of thermal conductive and flexible ternary composite films based on h-BN/polyvinyl alcohol/microfibril cellulose for food packaging applications. In this approach, h-BN was modified with polydopamine and PVA was employed as an environmentally friendly polymeric matrix in the preparation of nanocomposites. The mechanical findings illustrated that PDA/h-BN/microfibril cellulose improved the tensile strength and the break of elongation of PVA films by up to 2.4% and 12.5%, respectively. The contact angle of nanocomposite films was measured of 52° at 30% (w/w) of hybrid powder, which increased by 49.8% compared with the pure PVA films. Moreover, the thermal conductivity of the as-prepared films was investigated, and the findings showed the increased thermal conductivity after the addition of modified h-BN/microfibril cellulose and cellulose acted as guides to mitigate the PDA/h-BN agglomerate.

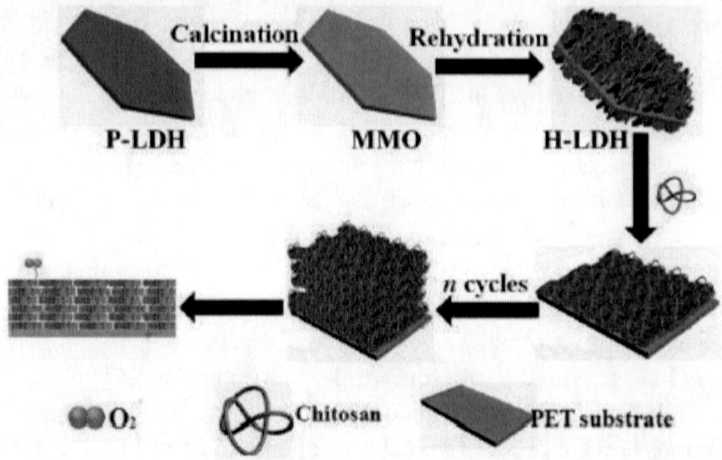

Figure 4. Schematic preparation of hierarchical LDHs/chitosan nanocomposite films with remarkable performance of O2 barrier property using the spin-coating method [88].

3. Layered Double Hydroxides-Based Nanocomposites

Layered double hydroxides (LDHs) belong to twodimensional materials consisted of positively charged sheets of hydroxides intercalated by negatively charged anions. LDHs are called anionic clays. Their structure can be represented by the general formula: $[M_{1-x}^{2+}M_x^{3+}(OH)_2]^{x+}[A_{x/n}]^{n-}mH2\ O$

where M^{2+} is a divalent metal cation such as Mg, Fe, Co, Cu, Zn... and M^{3+} is trivalent such as Al, Mn, Fe, Ga, and $A^{n-}$is the inorganic or organic anion. Recent findings prove that A^{n-}can exchange rapidly with polymers to produce intercalation compounds [82, 83]. LDHs dispersed in a polymer matrix can result in the formation of LDHs–polymer nanocomposites with potential mechanical and barrier properties as a result of the large aspect ratio of LDHs [84]. Many LDHs-polymer composites further display antimicrobial activities since the LDHs contain metal ions with antibacterial potency as Zn^{++} [85-87]. All these features make LDHs promising in preparing food packaging materials. Additionally, LDHs nanocomposite films have several advantages and some disadvantages have been mentioned in Table 1.

In an earlier study, Pan et al. [88] reported the fabrication of a high-performance O2 barrier nanocomposite film based on CS and hierarchical LDHs using the spin-coating method (Figure 5) to be employed in the food packaging field. They prepared hierarchical LDHs via a calcination–rehydration approach, which formed nanowalls aligned vertically on LDH platelets, which they incorporated the CS matrix and created a hybrid nanocomposite films. The hierarchical LDH films exhibited significantly enhanced O2 barrier properties with an O2 transmission rate of 0.005 cm^3/ m^2/day atm. The remarkable O2 barrier performance of nanocomposite films is related to the tortuous diffusion of hierarchical LDHs. Additionally, the authors measured the adhesive force between O2 gas and hierarchical LDHs, an adsorption energy which decreases the O2 diffusion rate and promotes O2 barrier property, at -2.46 eV.

Kim et al. [89] recently reported the effect of LDHs on the relatively simple preparation of PVA/poly(acrylic acid), PAA, and composite films as green food packaging materials. . The findings showed improvement inthe gas barrier and mechanical properties of nanocomposite films after incorporation of different concentrations of LDHs (0%, 1%, 2%, 3%, and 5%) within polymer matrix chains. The scanning electron microscopy illustrated that above 3% of LDHs showed aggregation, so both the gas barrier and mechanical properties raised sharply up to 2% of LDHs and then declined at >3%. Moreover, the O2 permeability decreased by 50% after the addition of LDHs, while the thermal stability and melting point of the nanocomposite increased. On the other hand, the ultraviolet blocking property was enhanced, while the transparency in the visible light region was maintained.

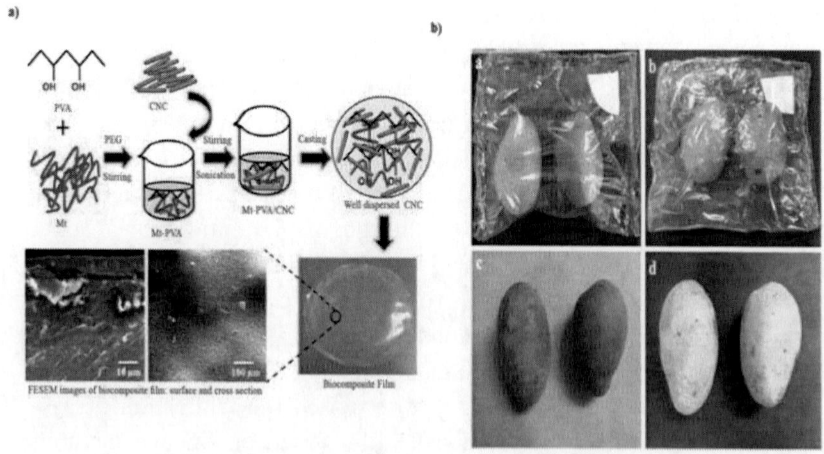

Figure 5. Schematic diagram of a) biocomposite packaging films (MMT/PVA/CNC) preparation for b) postharvest mango fruit [94].

4. Montmorillonite Clay (MMT)-Based Nanocomposites

MMT clay belongs to the structural family known as phyllosilicates, the physical structure of montmorillonite particles is generally recognisable in sheets and layers. Each layer is made up of two types of constitutional sheets: octahedral and tetrahedral [90]. They have exchangeable cations, hydroxyl groups, and adsorbed water molecules in their structure. Montmorillonite is used for enhancing the performance of a variety of materials and products in the fields of food additives, antibacterial applications, polymers, and sorbents. Several layers may be joined in a clay structure by interlayer cations, Van der Waals force, electrostatic force, and hydrogen bonding. The important functional utilisation of montmorillonite can be observed in food packaging applications [91]. Table1 shows several advantages and some disadvantages of MMT nanocomposite films.

Mixing MMT clay with CS is one of the most promising ways to improve the mechanical properties of the as-prepared food- packaging-based nanocomposite films. Basically, the enhancement depends on well-ordered MMT intercalated structure with stiff interfacial adhesion bonds, and ionic bonds as well, which form high-rigid MMT/CS nanocomposite films [92]. For example, Kusmono and Abdurrahim reported the preparation of CS/MMT nanocomposite films with sustainable mechanical properties, which are endowed with increased tensile strength after the addition of 5% MMT [93]. In this study, the highest tensile strength was recorded at 5% of clay. Interestingly, at higher MMT content, the tensile strength declined sharply owing to the agglomeration of MMT clay throughout the CS matrix. Perumal et al. [94] also studied the effect of multiscale rice straw (Oryza sativa) as biofiller throughout the preparation of MMT clay and PVA nanocomposite food packaging films for improving the storability of postharvest mango fruit, as shown in Figure 5a &b. They isolated cellulose nanocrystals from the biodegradable filler of rice straws (RS) via sulfuric acid hydrolysis and found that their length and diameter were 831 ± 45.0 nm and 44 ± 12 nm, respectively. On the other hand, the scanning electron microscope exhibited the needle-like structures of CNC and the XRD technique illustrated the increased crystallinity of CNC after many treatments. Authors exhibited that the thermal stability and mechanical properties of bicomposite films were significantly enhanced after the addition of 6% (w/w) of CNC. Moreover, they sealed the postharvest mango fruit with biocomposite packaging films (Figure 5b) and the findings showed that the shelf-life extended up to 19 days with excellent quality parameters such as phenol and DPPH antioxidant activity, in addition decreasing the total mass and O2 level, while the level of CO2 increased. Different examples for food packaging materials based on 2DMs/biocompatible polymers to preserve foods are shown in Table 2.

Table 2. Different 2DM biocomposite formulations based on biocompatible polymers to preserve food products

Type of package	Biopolymer/ 2DM formulation	Main results	Ref.
Nanocomposite bags	Chitosan/ graphene oxide bags	Bags package of mango resulted in: • Delayed ripening, preserves external appearance, skin and flesh colors, firmness, soluble solids as well as chlorophyll and β-carotene constituents. • Reduced weight loss, and respiration rate and anthracnose incidence and seriousness during storage.	[43]
Nanocomposite film	Chitosan-Graphene oxide/ Zn(II)	Nanocomposite film significantly: • Enhanced water resistance, oxygen barrier, and thermomechanical properties. • Decreased water vapor and oxygen permeability. • Enhanced antioxidant and antibacterial activities. • Decreased UV transmittance.	[92]

Table 2. (Continued)

Type of package	Biopolymer/2DM formulation	Main results	Ref.
Nanocomposite film	PVA/starch/GO	Nanocomposite film showed: • Enhanced mechanical properties, transmittance and water vapour permeability of package. • T increased tensile strength with an increase of GO concentration.	[103]
Nanocomposite film	PVA/GO	Nanocomposite film: • Increased the thermal and mechanical properties. • Decreased water intake. • Reduced the dimensional change. • Improved the electrical conductivity.	[104]
Nanocomposite film	PVA/GO	• The tensile strength of the nanocomposite film increased with an increase in the GO content. • Optimal mechanical properties were obtained at 20% GO concentration. • The optimal tensile strength was higher than pure PVA due to the formation of multi-hydrogen bonding between the oxygen-containing functional groups of GO and -OH groups of PVA.	[105]
Nanofilm	Chitosan/MMT	Incorporation of clay into chitosan matrix significantly: • Imparted unique physical and chemical properties to the pristine chitosan films and coatings. • Improved the mechanical and barrier properties. • Improved the antioxidant and antimicrobial properties as well as increased the thermal stability.	[92]
Nanocomposite film	Chitosan (CS)/boron nitride (BNNP)	Loading of BNNP on chitosan: • Improved the water and UV barrier properties of CS film and reduced the oxygen permeability. • Decreases the biodegradation rate compared with pure CS. The composite films were resistant to NaOH. • Improved the thermal stability of CS film. • Exhibited superior mechanical properties compared with pure CS film.	[79]
Nanocomposite adhesive	Polyurethane/boron nitride	• The water vapor permeance reduced by up to 50% • Increased the mechanical adhesion properties by 37%. • Increased the peeling resistance by up to 37.63%.	[68]
Biocomposite film	MMT/PVA/CNC	• Improved the thermal stability and mechanical property. • The shelf-life of mango fruits is extended by up to 19 days.	[94]

5. Future Perspective on 2DMs Application

Because packaging reduces food waste, experts are focusing their efforts on inventing novel food packaging technologies. With the introduction of 2DMs, more options for creating more cost-effective, environmentally friendly food packaging materials with improved performance and numerous uses are now available. First, in the food packaging field, 2DMs' large aspect ratios, hydrophobicity, and high surface-to-volume ratio dictate the performance of their composite packaging materials, which include barrier, mechanical, and antibacterial qualities, among other things. Second, in terms of intelligent packaging, 2DMs' physiochemical features, such as strong conductivity and catalytic activity, allowed them to be used in difficult domains such as indicators, sensors, and smart labels, among others [95].

Despite various papers demonstrating the adaptability and great potential of 2DMs in the food packaging industry, there are still certain gaps that need to be filled: i) Only a few well-known members of the 2DM family, such as graphene and LDHs, have been extensively researched for use in food packaging. Many more 2DMs have yet to be fully exploited. These yet-to-be-discovered 2DMs could lead to the development of next-generation materials or technologies for food packaging. ii) Because the migration behaviours of 2DMs in food packaging materials have been little studied, the toxicity and eco-toxicity of these 2DMs-based food packaging materials remain unknown. Given that many 2DMs have some toxicity [96], it will be necessary to examine the risks of 2DM-based food packaging materials. iii) Despite the fact that 2DMs showed great promise and potential in the lab, the lack of regulations, as well as consumers' unwillingness to pay a premium for products with better packaging, creates uncertainty and concerns, preventing the widespread commercialization of nanomaterials, including 2DMs in food packaging. The market penetration of 2DMs-based food packaging is influenced by industrial production, legislation, and customer acceptance. Overall, it is expected that 2DMs will provide additional opportunities to fundamentally update present food packaging technology, assuming that academics, businesses, and governments give it the attention it deserves.

Conclusion

Food packaging applications are one of the most interesting fields for the government and industrial sectors. So many academic and industrial researchers do their best to develop new food packaging material to prevent food waste. Due to their remarkable properties such as high thermal and mechanical properties and improving the gas barrier properties of materials, as well as their large surface area, the recent emergence of 2DM provides a new alternative route for designing high-cost-effective and eco-friendly materials for food packaging such as graphene, layered double hydroxides, hexagonal boron nitride, and Montmorillonite nanoclay-based nanocomposites. Particularly, in the presence of biocompatible polymers such as natural polysaccharides like chitosan and cellulose and synthetic polymers including PVA and poly urethanes, which have excellent features such as beingeco-friendly, biocompatible, low-cost materials. In this chapter, we have shown different preparations of 2DM-based nanocomposites as effective food packaging materials and improvedtheir O2 barrier, mechanical and thermal stability to improve the shelf-life of the food products. In the next chapter, our colleagues will focus on graphene oxide nanocomposites and discuss the use of those nanocomposites in the development of food packaging films.

References

[1] Kalpana, S., S. Priyadarshini, M.M. Leena, J. Moses, C. Anandharamakrishnan, Intelligent packaging: Trends and applications in food systems, *Trends in Food Science & Technology* 93 (2019) 145-157.

[2] Risch, S.J., Food packaging history and innovations, *Journal of agricultural and food chemistry* 57(18) (2009) 8089-8092.

[3] Ketelsen, M., M. Janssen, U. Hamm, Consumers' response to environmentally-friendly food packaging-A systematic review, *Journal of Cleaner Production* 254 (2020) 120123.

[4] Ansorena, M.R., N.E. Marcovich, M. Pereda, Food biopackaging based on chitosan, *Handbook of Ecomaterials;* Springer: Berlin, Germany (2019) 2057-2083.

[5] Yam, K.L., P.T. Takhistov, J. Miltz, Intelligent packaging: concepts and applications, *Journal of food science* 70(1) (2005) R1-R10.

[6] Kaur, K., R. Jindal, M. Maiti, S. Mahajan, Studies on the properties and biodegradability of PVA/Trapa natans starch (N-st) composite films and PVA/N-st-g-poly (EMA) composite films, *International journal of biological macromolecules* 123 (2019) 826-836.

[7] Yu, Z., B. Li, J. Chu, P. Zhang, Silica in situ enhanced PVA/chitosan biodegradable films for food packages, *Carbohydrate polymers* 184 (2018) 214-220.

[8] Akhavan, A., F. Khoylou, E. Ataeivarjovi, Preparation and characterization of gamma irradiated Starch/PVA/ZnO nanocomposite films, *Radiation Physics and Chemistry* 138 (2017) 49-53.

[9] Duncan, T.V. Applications of nanotechnology in food packaging and food safety: barrier materials, antimicrobials and sensors, *Journal of colloid and interface science* 363(1) (2011) 1-24.

[10] Castillejo, M., E. Rebollar, M. Oujja, M. Sanz, A. Selimis, M. Sigletou, S. Psycharakis, A. Ranella, C. Fotakis, Fabrication of porous biopolymer substrates for cell growth by UV laser: The role of pulse duration, *Applied surface science* 258(22) (2012) 8919-8927.

[11] Chan, S.W., H. Mirhosseini, F.S. Taip, T.C. Ling, C.P. Tan, Comparative study on the physicochemical properties of κ-carrageenan extracted from Kappaphycus alvarezii (doty) doty ex Silva in Tawau, Sabah, Malaysia and commercial κ-carrageenans, *Food Hydrocolloids* 30(2) (2013) 581-588.

[12] Mano, J.F. Viscoelastic properties of chitosan with different hydration degrees as studied by dynamic mechanical analysis, *Macromolecular bioscience* 8(1) (2008) 69-76.

[13] Vilela, C., M. Kurek, Z. Hayouka, B. Röcker, S. Yildirim, M.D.C. Antunes, J. Nilsen-Nygaard, M.K. Pettersen, C.S. Freire, A concise guide to active agents for active food packaging, *Trends in Food Science & Technology* 80 (2018) 212-222.

[14] Barra, A., J.D. Santos, M.R. Silva, C. Nunes, E. Ruiz-Hitzky, I. Gonçalves, S. Yildirim, P. Ferreira, P.A. Marques, Graphene derivatives in biopolymer-based composites for food packaging applications, *Nanomaterials* 10(10) (2020) 2077.

[15] Goda, E.S., M.H. Abu Elella, H. Gamal, S.E. Hong, K.R. Yoon, *Two-Dimensional Nanomaterials as Smart Flame Retardants for Polyurethane, Materials and Chemistry of Flame-Retardant Polyurethanes* Volume 1: A Fundamental Approach, ACS Publications2021, pp. 189-219.

[16] Russier, J., V. León, M. Orecchioni, E. Hirata, P. Virdis, C. Fozza, F. Sgarrella, G. Cuniberti, M. Prato, E. Vázquez, Few-Layer Graphene Kills Selectively Tumor Cells from Myelomonocytic Leukemia Patients, *Angewandte Chemie* 129(11) (2017) 3060-3065.

[17] Nasti, G., A. Abate, Tin halide perovskite (ASnX3) solar cells: a comprehensive guide toward the highest power conversion efficiency, *Advanced Energy Materials* 10(13) (2020) 1902467.

[18] Elella, M.H.A., E.S. Goda, H.M. Abdallah, S. Elyamny, H. Gamal, E.S.A. Serea, A.E. Shalan, Electroactive Polymers for Electrochromic Applications, *Electroactive Polymeric Materials,* CRC Press, pp. 163-201.

[19] Yu, Y., J. Zheng, J. Li, L. Lu, J. Yan, L. Zhang, L. Wang, Applications of two-dimensional materials in food packaging, *Trends in Food Science & Technology* 110 (2021) 443-457.

[20] M.H.A.J.J.o.C.E.R.U. Elella, *Journal of Chemical Engineering Research Updates,* 73 (2021) 97.

[21] El-Ghany, A., A. Nahed, Z.M.J.P.B. Mahmoud, *Synthesis, characterization and swelling behavior of high-performance antimicrobial amphoteric hydrogels from corn starch,* 78(11) (2021) 6161-6182.

[22] Abd El-Ghany, N.A., M.S.A. Aziz, M.M. Abdel-Aziz, Z.J.I.j.o.b.m. Mahmoud, *Antimicrobial and swelling behaviors of novel biodegradable corn starch grafted/poly (4-acrylamidobenzoic acid) copolymers,* 134 (2019) 912-920.

[23] Ngwuluka, N.C., N.Y. Abu-Thabit, O.J. Uwaezuoke, J.O. Erebor, M.O. Ilomuanya, R.R. Mohamed, S.M. Soliman, M.H.A. Elella, N.A. Ebrahim, *Natural Polymers in Micro-and Nanoencapsulation for Therapeutic and Diagnostic Applications:* Part I: Lipids and Fabrication Techniques, Nano-and Micro-Encapsulation-Techniques and Applications, IntechOpen2020.

[24] Elella, M.H.A., E.S. Goda, M.A. Gab-Allah, S.E. Hong, B. Pandit, S. Lee, H. Gamal, A. ur Rehman, K.R. Yoon, Xanthan gum-derived materials for applications in environment and eco-friendly materials: a review, *Journal of Environmental Chemical Engineering* (2020) 104702.

[25] Elella, M.H.A., E.S. Goda, H.M. Abdallah, A.E. Shalan, H. Gamal, K.R. Yoon, Innovative bactericidal adsorbents containing modified xanthan gum/montmorillonite nanocomposites for wastewater treatment, *International Journal of Biological Macromolecules* 167 (2021) 1113-1125.

[26] Mohamed, R.R., M.H.A. Elella, M.W. Sabaa, Synthesis, characterization and applications of N-quaternized chitosan/poly (vinyl alcohol) hydrogels, *International journal of biological macromolecules* 80 (2015) 149-161.

[27] Mohamed, R.R., M.H.A. Elella, M.W. Sabaa, Cytotoxicity and metal ions removal using antibacterial biodegradable hydrogels based on N-quaternized chitosan/poly (acrylic acid), *International journal of biological macromolecules* 98 (2017) 302-313.

[28] Elella, M.H.A., E. Goda, M. Gab-Allah, S.E. Hong, Y.G. Lijalem, K.R. Yoon, *Biodegradable Polymeric Nanocomposites for Wastewater Treatment, Advances in Nanocomposite Materials for Environmental and Energy Harvesting Applications*, Springer 2022, pp. 245-298.

[29] Lee, C., W. Xiaoding, W. Jeffrey, K.; James, H, Meas. Elastic Prop. Intrinsic Strength Monolayer Graphene. *Sci* 321 (2008) 385-388.

[30] Zhu, Y., S. Murali, W. Cai, X. Li, J.W. Suk, J.R. Potts, R.S. Ruoff, Graphene and graphene oxide: synthesis, properties, and applications, *Advanced materials* 22(35) (2010) 3906-3924.

[31] Hossain, K., M. Rafatullah, S.Z. Abbas, A. Ahmad, N. Ismail, A.Y. Maruthi, *Antimicrobial activity of graphene-based nanomaterials: current development and challenges, Graphene-Based Nanotechnologies for Energy and Environmental Applications,* Elsevier 2019, pp. 293-314.

[32] Goda, E.S., M.H.A. Elella, S.E. Hong, B. Pandit, K.R. Yoon, H. Gamal, Smart flame retardant coating containing carboxymethyl chitosan nanoparticles decorated graphene for obtaining multifunctional textiles, *Cellulose* 28(8) (2021) 5087-5105.

[33] Elella, M.H.A., E.S. Goda, K.R. Yoon, S.E. Hong, M.S. Morsy, R.A. Sadak, H. Gamal, Novel v apor polymerization for integrating flame retardant textile with multifunctional properties, *Composites Communications* 24 (2021) 100614.

[34] Goda, E.S., M.H.A. Elella, M. Sohail, B.S. Singu, B. Pandit, A. El Shafey, A.M. Aboraia, H. Gamal, S.E. Hong, K.R. Yoon, N-methylene phosphonic acid chitosan/graphene sheets decorated with silver nanoparticles as green antimicrobial agents, *International Journal of Biological Macromolecules* 182 (2021) 680-688.

[35] Tiwari, S.K., S. Sahoo, N. Wang, A. Huczko, Graphene research and their outputs: Status and prospect, *Journal of Science: Advanced Materials and Devices* 5(1) (2020) 10-29.

[36] Valapa, R.B., G. Pugazhenthi, V. Katiyar, Effect of graphene content on the properties of poly (lactic acid) nanocomposites, *Rsc Advances* 5(36) (2015) 28410-28423.

[37] Son, E.J., J.S. Lee, M. Lee, C.H.T. Vu, H. Lee, K. Won, C.B. Park, Self-adhesive graphene oxide-wrapped TiO2 nanoparticles for UV-activated colorimetric oxygen detection, *Sensors and Actuators B: Chemical* 213 (2015) 322-328.

[38] Li, X., M. Guo, P. Bandyopadhyay, Q. Lan, H. Xie, G. Liu, X. Liu, X. Cheng, N.H. Kim, J.H. Lee, Two-dimensional materials modified layered double hydroxides: A series of fillers for improving gas barrier and permselectivity of poly (vinyl alcohol), *Composites Part B: Engineering* 207 (2021) 108568.

[39] Sundramoorthy, A.K., S. Gunasekaran, Applications of graphene in quality assurance and safety of food, *TrAC Trends in Analytical Chemistry* 60 (2014) 36-53.

[40] Amit, S.K., M. Uddin, R. Rahman, S. Islam, M.S. Khan, A review on mechanisms and commercial aspects of food preservation and processing, *Agriculture & Food Security* 6(1) (2017) 1-22.

[41] Elmehbad, N.Y., N.A. Mohamed, N.A.J.I.J.o.B.M. Abd El-Ghany, *Evaluation of the antimicrobial and anti-biofilm activity of novel salicylhydrazido chitosan derivatives impregnated with titanium dioxide nanoparticles,* (2022).

[42] Mohamed, N.A., N.A. Abd El-Ghany, M.M.J.I.J.o.B.M. Abdel-Aziz, *Synthesis, characterization, anti-inflammatory and anti-Helicobacter pylori activities of novel benzophenone tetracarboxylimide benzoyl thiourea cross-linked chitosan hydrogels,* 181 (2021) 956-965.

[43] Vilvert, J.C., S.T. de Freitas, M.A.R. Ferreira, R.H. de Lima Leite, F.K.G. dos Santos, C.d.S.R. Costa, E.M.M. Aroucha, Chitosan and graphene oxide-based biodegradable bags: An eco-friendly and effective packaging alternative to maintain postharvest quality of 'Palmer'mango, *LWT* 154 (2022) 112741.

[44] Tang, X., P. Kumar, S. Alavi, K. Sandeep, Recent advances in biopolymers and biopolymer-based nanocomposites for food packaging materials, *Critical reviews in food science and nutrition* 52(5) (2012) 426-442.

[45] Ouattara, B., R. Simard, G. Piette, A. Begin, R. Holley, Diffusion of acetic and propionic acids from chitosan-based antimicrobial packaging films, *Journal of food science* 65(5) (2000) 768-773.

[46] Jost, V., K. Kobsik, M. Schmid, K. Noller, Influence of plasticiser on the barrier, mechanical and grease resistance properties of alginate cast films, *Carbohydrate polymers* 110 (2014) 309-319.

[47] Lim, H., N. Huang, C. Loo, Facile preparation of graphene-based chitosan films: Enhanced thermal, mechanical and antibacterial properties, *Journal of Non-Crystalline Solids* 358(3) (2012) 525-530.

[48] Carson, L., C. Kelly-Brown, M. Stewart, A. Oki, G. Regisford, Z. Luo, V.I. Bakhmutov, Synthesis and characterization of chitosan–carbon nanotube composites, *Materials letters* 63(6-7) (2009) 617-620.

[49] Grande, C.D., J. Mangadlao, J. Fan, A. De Leon, J. Delgado-Ospina, J.G. Rojas, D.F. Rodrigues, R. Advincula, Chitosan Cross-Linked Graphene Oxide Nanocomposite Films with Antimicrobial Activity for Application in Food Industry, *Macromolecular symposia,* Wiley Online Library, 2017, p. 1600114.

[50] Rajinipriya, M., M. Nagalakshmaiah, M. Robert, S. Elkoun, Importance of agricultural and industrial waste in the field of nanocellulose and recent industrial developments of wood based nanocellulose: a review, *ACS Sustainable Chemistry & Engineering* 6(3) (2018) 2807-2828.

[51] Bergamonti, L., M. Potenza, A.H. Poshtiri, A. Lorenzi, A.M. Sanangelantoni, L. Lazzarini, P.P. Lottici, C. Graiff, Ag-functionalized nanocrystalline cellulose for paper preservation and strengthening, *Carbohydrate Polymers* 231 (2020) 115773.

[52] Jiang, S.-j., T. Zhang, Y. Song, F. Qian, Y. Tuo, G. Mu, Mechanical properties of whey protein concentrate based film improved by the coexistence of nanocrystalline cellulose and transglutaminase, *International journal of biological macromolecules* 126 (2019) 1266-1272.

[53] Fortunati, E., I. Armentano, Q. Zhou, A. Iannoni, E. Saino, L. Visai, L.A. Berglund, J. Kenny, Multifunctional bionanocomposite films of poly (lactic acid), cellulose nanocrystals and silver nanoparticles, *Carbohydrate polymers* 87(2) (2012) 1596-1605.

[54] Jiang, G., M. Zhang, J. Feng, S. Zhang, X. Wang, High oxygen barrier property of poly (propylene carbonate)/polyethylene glycol nanocomposites with low loading of cellulose nanocrytals, *ACS Sustainable Chemistry & Engineering* 5(12) (2017) 11246-11254.

[55] Trache, D., V.K. Thakur, R. Boukherroub, Cellulose nanocrystals/graphene hybrids—a promising new class of materials for advanced applications, *Nanomaterials* 10(8) (2020) 1523.

[56] Li, F., H.-Y. Yu, Y.-Y. Wang, Y. Zhou, H. Zhang, J.-M. Yao, S.Y.H. Abdalkarim, K.C. Tam, Natural biodegradable poly (3-hydroxybutyrate-co-3-hydroxyvalerate) nanocomposites with multifunctional cellulose nanocrystals/graphene oxide hybrids for high-performance food packaging, *Journal of agricultural and food chemistry* 67(39) (2019) 10954-10967.

[57] Pal, N., P. Dubey, P. Gopinath, K. Pal, Combined effect of cellulose nanocrystal and reduced graphene oxide into poly-lactic acid matrix nanocomposite as a scaffold and its anti-bacterial activity, *International journal of biological macromolecules* 95 (2017) 94-105.

[58] Yang, E., X. Qin, S. Wang, Electrospun crosslinked polyvinyl alcohol membrane, *Materials Letters* 62(20) (2008) 3555-3557.

[59] Gong, X., C.Y. Tang, L. Pan, Z. Hao, C.P. Tsui, Characterization of poly (vinyl alcohol)(PVA)/ZnO nanocomposites prepared by a one-pot method, *Composites Part B: Engineering* 60 (2014) 144-149.

[60] Asran, A.S., S. Henning, G.H. Michler, Polyvinyl alcohol–collagen–hydroxyapatite biocomposite nanofibrous scaffold: Mimicking the key features of natural bone at the nanoscale level, *Polymer* 51(4) (2010) 868-876.

[61] Wang, J., L. Ye, Structure and properties of polyvinyl alcohol/polyurethane blends, *Composites Part B: Engineering* 69 (2015) 389-396.

[62] Young, R.J., M. Liu, I.A. Kinloch, S. Li, X. Zhao, C. Vallés, D.G. Papageorgiou, The mechanics of reinforcement of polymers by graphene nanoplatelets, *Composites Science and Technology* 154 (2018) 110-116.

[63] J., Zhang, Y. He, P. Zhu, S. Lin, S. Ju, D. Jiang, In situ reduction of graphene oxide in the poly (vinyl alcohol) matrix via microwave irradiation, *Polymer Composites* 40(1) (2019) 170-178.

[64] Salavagione, H.J., G. Martínez, M.A. Gómez, Synthesis of poly (vinyl alcohol)/reduced graphite oxide nanocomposites with improved thermal and electrical properties, *Journal of Materials Chemistry* 19(28) (2009) 5027-5032.

[65] Hurayra–Lizu, K.A., M.W. Bari, F. Gulshan, M.R. Islam, GO based PVA nanocomposites: tailoring of optical and structural properties of PVA with low percentage of GO nanofillers, *Heliyon* 7(5) (2021) e06983.

[66] Zhang, R., Y. Wang, D. Ma, S. Ahmed, W. Qin, Y. Liu, Effects of ultrasonication duration and graphene oxide and nano-zinc oxide contents on the properties of polyvinyl alcohol nanocomposites, *Ultrasonics sonochemistry* 59 (2019) 104731.

[67] Otoukesh, M., C. Nerín, M. Aznar, A. Kabir, K.G. Furton, Z. Es' haghi, Determination of adhesive acrylates in recycled polyethylene terephthalate by fabric phase sorptive extraction coupled to ultra performance liquid chromatography-mass spectrometry, *Journal of Chromatography* A 1602 (2019) 56-63.

[68] Nacas, A., L. Antonino, A. Chinellato, D. dos Santos, Nano boron nitride/polyurethane adhesives in flexible laminated food packaging: peeling resistance and permeability properties, *International Journal of Adhesion and Adhesives* 93 (2019) 102326.

[69] Qiao, J. Jiang, H. Li, X. Gui, J. Zhang, Y. Zhang, Polyurethane Adhesives Used in Food Packaging Materials, *Advance Journal of Food Science and Technology* 12(12) (2016) 705-708.

[70] Paine, R.T., C.K. Narula, Synthetic routes to boron nitride, *Chemical Reviews* 90(1) (1990) 73-91.

[71] Balmain, W. Bemerkungen über die Bildung von Verbindungen des Bors und Siliciums mit Stickstoff und gewissen Metallen, *Journal für Praktische Chemie* 27(1) (1842) 422-430.

[72] Han, W.-Q. Anisotropic hexagonal boron nitride nanomaterials: synthesis and applications, *Nanotechnologies for the Life Sciences* (2009).

[73] Yu, J., L. Qin, Y. Hao, S. Kuang, X. Bai, Y.-M. Chong, W. Zhang, E. Wang, Vertically aligned boron nitride nanosheets: chemical vapor synthesis, ultraviolet light emission, and superhydrophobicity, *ACS nano* 4(1) (2010) 414-422.

[74] Zhang, S., G. Lian, H. Si, J. Wang, X. Zhang, Q. Wang, D. Cui, Ultrathin BN nanosheets with zigzag edge: one-step chemical synthesis, applications in wastewater treatment and preparation of highly thermal-conductive BN–polymer composites, *Journal of Materials Chemistry* A 1(16) (2013) 5105-5112.

[75] Yu, Y., H. Chen, Y. Liu, V. Craig, L.H. Li, Y. Chen, Superhydrophobic and superoleophilic boron nitride nanotube-coated stainless steel meshes for oil and water separation, *Advanced Materials Interfaces* 1(1) (2014) 1300002.

[76] Kim, T.Y., E.H. Song, B.H. Kang, S.J. Kim, Y.-H. Lee, B.-K. Ju, Hydrolyzed hexagonal boron nitride/polymer nanocomposites for transparent gas barrier film, *Nanotechnology* 28(12) (2017) 12LT01.

[77] Cai, W., D. Zhang, B. Wang, Y. Shi, Y. Pan, J. Wang, W. Hu, Y. Hu, Scalable one-step synthesis of hydroxylated boron nitride nanosheets for obtaining multifunctional polyvinyl alcohol nanocomposite films: Multi-azimuth properties improvement, *Composites Science and Technology* 168 (2018) 74-80.

[78] Shareef, S., K. Chidambaram, S.K. Pasha, Structure, morphology and dielectric properties of hexagonal boron nitride nanoparticles reinforced biopolymer nanocomposites, *Polymer-Plastics Technology and Materials* 58(11) (2019) 1210-1225.

[79] Behera, K., M. Kumari, Y.-H. Chang, F.-C. Chiu, Chitosan/boron nitride nanobiocomposite films with improved properties for active food packaging applications, *International Journal of Biological Macromolecules* 186 (2021) 135-144.

[80] Nguyen, H.-L., Z. Hanif, S.-A. Park, B.G. Choi, T.H. Tran, D.S. Hwang, J. Park, S.Y. Hwang, D.X. Oh, Sustainable boron nitride nanosheet-reinforced cellulose nanofiber composite film with oxygen barrier without the cost of color and cytotoxicity, *Polymers* 10(5) (2018) 501.

[81] Ge, X., W.-J. Liang, J.-F. Ge, X.-J. Chen, J.-Y. Ji, X.-Y. Pang, M. He, X.-M. Feng, Hexagonal boron nitride/microfibril cellulose/poly (vinyl alcohol) ternary composite film with thermal conductivity and flexibility, *Materials* 12(1) (2019) 104.

[82] Markland, C., G.R. Williams, D. O'Hare, The intercalation of flavouring compounds into layered double hydroxides, *Journal of Materials Chemistry* 21(44) (2011) 17896-17903.

[83] Lobo-Sánchez, M., G. Nájera-Meléndez, G. Luna, V. Segura-Pérez, J.A. Rivera, G. Fetter, ZnAl layered double hydroxides impregnated with eucalyptus oil as efficient hybrid materials against multi-resistant bacteria, *Applied Clay Science* 153 (2018) 61-69.

[84] Barik, S., S.K. Badamali, Layer double hydroxide reinforced polymer bionanocomposites for packaging applications, *Bionanocomposites for packaging applications,* Springer 2018, pp. 269-290.

[85] Cheng, H.-M., X.-W. Gao, K. Zhang, X.-R. Wang, W. Zhou, S.-J. Li, X.-L. Cao, D.-P. Yan, A novel antimicrobial composite: ZnAl-hydrotalcite with p-hydroxybenzoic acid intercalation and its possible application as a food packaging material, *New Journal of Chemistry* 43(48) (2019) 19408-19414.

[86] Saikia, P., A. Borah, R.L. Goswamee, Hybrid nanocomposites of layered double hydroxides: an update of their biological applications and future prospects, *Colloid and Polymer Science* 295(5) (2017) 725-747.

[87] Kuthati, Y., R.K. Kankala, C.-H. Lee, Layered double hydroxide nanoparticles for biomedical applications: Current status and recent prospects, *Applied Clay Science* 112 (2015) 100-116.

[88] Pan, T., S. Xu, Y. Dou, X. Liu, Z. Li, J. Han, H. Yan, M. Wei, Remarkable oxygen barrier films based on a layered double hydroxide/chitosan hierarchical structure, *Journal of Materials Chemistry* A 3(23) (2015) 12350-12356.

[89] Kim, H.J., K. Charoensri, J.A. Ko, H.J. Park, Effects of layered double hydroxides on poly (vinyl alcohol)/poly (acrylic acid) films for green food packaging applications, *Progress in Organic Coatings* 163 (2022) 106634.

[90] Uddin, F., Clays, nanoclays, and montmorillonite minerals, *Metallurgical and Materials Transactions* A 39(12) (2008) 2804-2814.

[91] Hu, C., L. Gu, Z. Luan, J. Song, K. Zhu, Effects of montmorillonite–zinc oxide hybrid on performance, diarrhea, intestinal permeability and morphology of weanling pigs, *Animal Feed Science and Technology* 177(1-2) (2012) 108-115.

[92] Qu, B., Y. Luo, A review on the preparation and characterization of chitosan-clay nanocomposite films and coatings for food packaging applications, C*arbohydrate Polymer Technologies and Applications* 2 (2021) 100102.

[93] Abdurrahim, I. Water sorption, antimicrobial activity, and thermal and mechanical properties of chitosan/clay/glycerol nanocomposite films, *Heliyon* 5(8) (2019) e02342.

[94] Perumal, A.B., P.S. Sellamuthu, R.B. Nambiar, E.R. Sadiku, G. Phiri, J. Jayaramudu, Effects of multiscale rice straw (Oryza sativa) as reinforcing filler in montmorillonite-polyvinyl alcohol biocomposite packaging film for enhancing the storability of postharvest mango fruit (Mangifera indica L.), *Applied Clay Science* 158 (2018) 1-10.

[95] Qadir, A., T.K. Le, M. Malik, K.A.A. Min-Dianey, I. Saeed, Y. Yu, J.R. Choi, P.V.J.R.A. Pham, *Representative 2D-material-based nanocomposites and their emerging applications: a review,* 11(39) (2021) 23860-23880.

[96] Guiney, L.M., X. Wang, T. Xia, A.E. Nel, M.C.J.A.n. Hersam, *Assessing and mitigating the hazard potential of two-dimensional materials,* 12(7) (2018) 6360-6377.

[97] Tiwari, S.K., S. Sahoo, N. Wang, A.J.J.o.S.A.M. Huczko, *Devices, Graphene research and their outputs: Status and prospect,* 5(1) (2020) 10-29.

[98] Sundramoorthy, A.K., S.J.T.T.i.A.C. Gunasekaran, *Applications of graphene in quality assurance and safety of food,* 60 (2014) 36-53.

[99] Díez-Pascual, A.M., J.A.J.P. Luceño-Sánchez, *Antibacterial activity of polymer nanocomposites incorporating graphene and its derivatives: A state of art,* 13(13) (2021) 2105.

[100] Gonzalez-Ortiz, D., C. Salameh, M. Bechelany, P.J.N.b.n.b.m.s. Miele, applications, *Materials Today Advances,* (2020) 100-107.

[101] Othman, S.H.J.A., A.S. Procedia, *Bio-nanocomposite materials for food packaging applications: types of biopolymer and nano-sized filler,* 2 (2014) 296-303.

[102] Vilarinho, F., M.F. Vaz, A.S.J.R.P.o.F. Silva, Nutrition, Agriculture, *The use of montmorillonite (MMT) in food nanocomposites: Methods of incorporation, characterization of MMT/polymer nanocomposites and main consequences in the properties,* 11(1) (2020) 13-26.

[103] Wu, Z., Y. Huang, L. Xiao, D. Lin, Y. Yang, H. Wang, Y. Yang, D. Wu, H. Chen, Q. Zhang, Physical properties and structural characterization of starch/polyvinyl alcohol/graphene oxide composite films, *International journal of biological macromolecules* 123 (2019) 569-575.

[104] Gahlot, S., V. Kulshrestha, G. Agarwal, P.K. Jha, Synthesis and characterization of PVA/GO nanocomposite films, *Macromolecular Symposia,* Wiley Online Library, 2015, pp. 173-177.

[105] Cheng-an, T., Z. Hao, W. Fang, Z. Hui, Z. Xiaorong, W. Jianfang, Mechanical properties of graphene oxide/polyvinyl alcohol composite film, *Polymers and Polymer Composites* 25(1) (2017) 11-16.

Chapter 4

Development of Graphene Oxide/Biopolymer-Based Nanocomposite Food Packaging Films

Mahmoud H. Abu Elella*
Chemistry Department, Faculty of Science, Cairo University, Giza, Egypt

Abstract

Food packaging has been considered one of the most important applications in recent years. However, microbial contamination using pathogenic microorganisms, which resist a broad range of antibiotics, is considered one of the serious challenges facing food packaging material development over a century. The quality, safety, and microbial control of the food packed has gained more interest from industrial and academic researchers around the world since the packaging section is very important to food protection. Therefore, various attempts have been made to design active food packaging through advanced strategies utilising smart and multifunctional materials. In the previous chapter, my colleagues have introduced how 2D biopolymer nanocomposites are used in film development for food packaging. Based on the foundation laid in the previous chapter, I will discuss in more depth the potential and applications of graphene oxide/ biopolymer-based nanocomposites in food packaging. These nanocomposites have received a great deal of attention in food packaging applications owing to their outstanding properties such as good antimicrobial activity, high specific surface area, good stability, biodegradable, biocompatible, and cost-effectiveness. This chapter will serve as a good basis for industrial and academic researchers to explore new avenues in food packaging development.

---

* Corresponding Author's Email: mahmoudhussien3766@yahoo.com.

In: Food Packaging
Editor: Wing-Fu Lai
ISBN: 979-8-88697-249-8

Keywords: food packaging, multifunctional materials, biopolymer, graphene oxide, nanocomposites

1. Introduction

Food packaging is considered as one of the most important stages in the food industry since the active packaging function is to preserve the fresh food product from industry to customers against different environmental deterioration factors including biological, chemical, and physical factors [1]. Furthermore, they can extend the shelf-life of food products to improve their quality, absorbing and scavenging properties, microbial and temperature control, as well as prolong their shelf-life and releasing, emitting, and sensory properties [2, 3]. On the other hand, different plastic materials have been widely used in food packaging applications because they have the different valuable features such as easy molding, cost-effectiveness, excellent mechanical properties, and good barrier qualities. However, they have main environmental drawback of low degradability, which creates very dangerous chemical waste [4]. To solve this issue, hybrid polymer nanocomposites, especially biopolymer-based nanocomposites, are receiving great attention from global food researchers to fabricate high-performance active food packaging materials. Because they have outstanding features, including low cost, high abundance, easy availability, biodegradability, biocompatibility, and non-toxicity [5]. Generally, hybrid polymer nanocomposites are fundamentally fabricated by incorporating different nanofillers: nanocarbons, 2D layered nanoclay, metal, and metal oxide nanoparticles [6-11].

Biopolymers can be divided into two categories: polysaccharides and proteins. Polysaccharides are long chains of carbohydrate polymeric oligosaccharides connected by glycosidic bonds between monosaccharide repeating units [12]. Polysaccharide extraction is dependent on the source of the polysaccharide: plant, animal, microbial, or algal origin. Also, they can be categorised based on the composition of monosaccharide repeating units into homopolymers such as starch and heteropolymers such as chitosan [7, 13-19]. Moreover, proteins are defined as natural polymers with a polyamide structure. They are mainly formed of sequenced amino acids through peptide bonds and are considered one of the main components in the human body due to their essential roles in controlling cell fates, catalysing reactions, and building cellular structures. These days, different proteins such as collagen and gelatin have been widely used in the food packaging industry thanks to their outstanding features, including non-toxic, low cost, biocompatibility, and biodegradability [20-23].

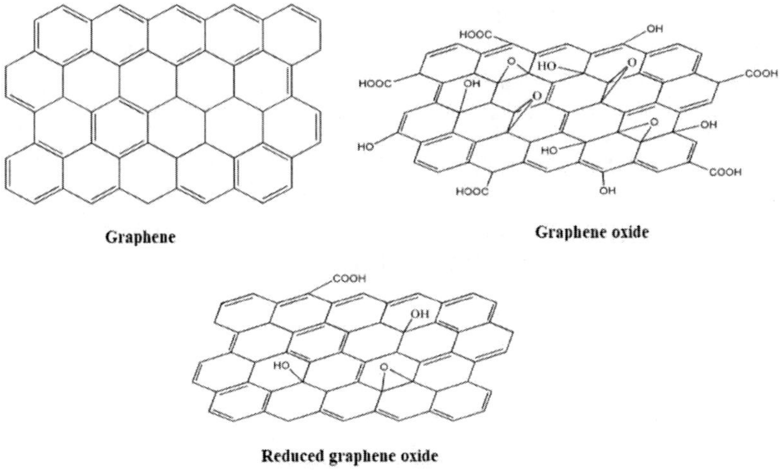

Figure 1. Chemically illustrated of the chemical structure of 2D carbon layered nanosheets: graphene, graphene oxide, and reduced graphene oxide.

Usually, polymer/2D carbon layered nanofiller-based nanocomposites contain platelet-shaped sheets that have a thickness in the range of a few nanometers to several hundred nanometers. They have more attractions in the recent food packaging development application. 2D carbon layers come in a variety of chemical forms, including graphene, graphene oxide (GO), and reduced graphene oxide (rGO) nanosheets (Figure 1) [24]. Graphene is a 2D carbon layered nanosheet that is the vastly employed in form of polymer nanocomposites in multifunctional applications owing to its fabulous physicochemical properties including high thermal and mechanical stability, good diffraction strength, high Young's modulus, good electrical conductivity, and large specific surface area [9, 22, 25-27]. Up to now, graphene nanosheets can be commonly prepared using different techniques such as layer-by-layer (exfoliation), chemical vapour deposition, chemical and electrochemical methods, and ice-crystal-template [27-30]. However, it has main drawbacks such as its hydrophobicity, which is an obstacle to its dispersion in polar solvents and a strong agglomeration tendency. In addition, it has poor solubility because of strong van der Waals force interactions between its sheets that hinder its applications. Therefore, to solve its disadvantage, several graphene derivatives have been prepared, from which graphene oxide (GO) is considered one of the most popular graphene derivatives that is widely used in wide applications such as food packaging, biomedical, and wastewater treatment applications, as a high-performance nanofiller inside polymer chain nanocomposites [31].

Graphene oxide (GO) is the oxidised form of graphene and is considered one of the most popular carbon nanofillers that has been employed to design hierarchical nano-architectures owing to the presence of abundant highly reactive groups such as alkenyl, hydroxyl, epoxide, carbonyl, and carboxyl groups on its surface (Figure 1) [32, 33]. GO contains SP^3 hybridization carbon atoms, which boost its interlayer spacing and enhance its ability to maintain compounds. The attached functional groups modify the electronic structure of graphene and serve as massive scattering centres which affect its electrical transport. Therefore, GO shows insulating properties and low carbon mobility (0.1 $cm^2/V.s$) and sheet resistance as well, at around 1012 W/sq [34]. However, it has the ability to improve the physicochemical properties of biopolymer nanocomposite films. GO demonstrates sustainable features including aqueous processability, versatility, surface functionalization capability, and amphiphilicity. Additionally, GO has a good hydrophilic nature and has the ability to fabricate stable aqueous colloids to boost the macroscopic structure assembly, which is very important for large-scale applications [35]. For example, xanthan gum, which is an anionic polysaccharide and composed of three repeating unites: glucose, mannose, and glucuronic acid with the molar ratio of 2:2:1, and graphene oxide nanocomposites films have been mentioned by Liu et al. with a green and high-friendly coordination method with ferric ions [36]. The authors illustrated the improvement of the mechanical properties of the prepared nanocomposites by reporting a high strength and toughness of 93.6 MPa and 0.91 MJ/m^3, respectively.

GO can be prepared according to "bottom-up" and "top-down" techniques. The former includes simple carbon molecules like graphene but has disadvantages, for instance, time consuming and scalable problems. As a result, another technique "top-down" is widely applied and includes four approaches: Hummers′, Hofmann's, Staudenmaier's, and Brodie's methods [34]. The Hummers′ method is the most employed technique for the preparation of GO nanosheets. It contains the addition of potassium permanganate ($KMnO_4$) to graphite solution, and also both $NaNO_3$, and H_2SO_4 are added. Many factors are changed in this method to improve its preparation and to achieve a low-cost process [37]. In another route, GO can also be prepared from graphite oxide using sonication and stirring. The sonication factor completely exfoliates graphite oxide in a short time and declines the size of GO from a few microns to nanometer sheets. While mechanical stirring is another approach, it consumes a long time and is less heavy-handed [38].

2. Antibacterial Action of Graphene Oxide-Based Biopolymer Nanocomposites

Bacterial contamination is a growing issue in our lives everywhere and causes noteworthy harm in different industries, such as food packaging, biomedical, and water treatment applications. Pathogenic microorganisms have recently developed the ability to resist several antibiotics, causing academic researchers to focus more on developing new treatments for controlling bacterial growth [38]. For example, some of Gram-positive bacteria like Staphylococcus aureus and Gram-negative bacteria like Escherichia coli are considered multi-drug resistant (MDR) microorganisms and have the capability to cause many hospital infections. Consequently, new high-effective and cheap antimicrobial food packaging materials to maintain bacterial growth are urgently needed and are considered a very promising approach nowadays [38, 39]. According to Figure 1, GO is the oxidised form of the pristine graphene and includes different high-reactive functional groups such as epoxides, carbonyl, carboxylic, and hydroxyl groups. Because of its acquired fantastic properties such as amphiphilicity, aqueous processability, and biocompatibility, these functional groups award its ability to interact with biological cells and to employ in wide application [33]. Hu et al. mentioned the first antimicrobial activity of GO nanosheets and the results showed sustainable antimicrobial activity [40]. After that, a lot of antimicrobial studies have been reported for GO [41-43]. Moreover, the degree of hydrophilicity is very useful for antibacterial inhibition. Thus, graphene nanosheets are hydrophobic in nature, so they tend to aggregate thanks to bi-bi interaction, but it plays a crucial role in favouring strong interaction with bacteria's cell wall. Conversely, GO offers outstanding features: hydrophilicity and amphiphilicity, which improve its adhesion to bacteria's cell walls. Additionally, there are other features such as its nanomaterial surface, size, and shape that display negative charges on their surface, owing to deprotonation of its hydroxyl and carboxylic moieties. Its nanosheet size is in the range of a few nm to πm, which has great influence on its bacterial adhesion and increases its antibacterial action by 4-fold when its nanosheet size decreases between 0.65 πm^2 and 0.01 πm^2 [44].

Due to their excellent antimicrobial activity, several studies based on the incorporation of GO inside biopolymer (polysaccharide and protein) chains have recently been cited. In this regard, different polysaccharides have been used, such as chitosan, alginate, cellulose, and starch, in the form of GO nanocomposites. For example, chitosan (Ch) is one of the most abundant polysaccharides in food packaging applications due to its great potential antimicrobial activity. Its features are affected by different parameters, including pH, molecular weight, and degree of acetylation. Notably, low molecular weight and high deacetylation degree have been demonstrated high antimicrobial activity against several bacterial strains [45]. For instance, Díez-Pascual et al. [45] reported the preparation of Ch/ poly (lactide-co-glycolide) electrospun fiber. And then, it was mixed with GO nanosheets in the presence of Ag nanoparticles via chemical electrostatic interactions between carboxylic acid in GO nanosheets and amino groups in the Ch fibre structure. The authors found that the prepared GO nanocomposite had high effectiveness against different bacterial growth, such as Gram-negative bacteria including E. coli and P. aeruginosa, and also against Gram-positive bacteria like S. aureus. Moreover, nanocomposite films based on Ch and GO in the presence of TiO_2 nanoparticles were prepared by Xu et al. [46]. They studied their antibacterial inhibition against two bacterial strains such as *A. niger* and *B. subtilis*, and the findings exhibited great antibacterial activity that reached 99% against two bacteria using the ratio of 20:1:4 between the three reactants. The antibacterial mechanism was elucidated via the fluorescence technique and examination of the morphological structure of the examined bacterial cell wall. So, the results showed the damage to their cell wall and also their cell membrane owing to the nanocomposite exposure.

In another study, chitosan was incorporated with polyvinylpyrrolidone and GO nanosheets to design a nanocomposite in which GO can interact with Ch in two different ways (covalent and non-covalent bonds) (Figure 2). Consequently, in Figure 2a, the epoxy moieties on the GO structure can interact with the abundant amino groups on the Ch structure via nucleophilic covalent bonds. Nonetheless, amino groups of Ch can interact through H-bonding interactions with hydroxyl and carboxylic groups on GO nanosheets (Figure 2b). Notably, the low content of GO can be interacted with Ch in the single filler form with a low crosslinking degree. Furthermore, the sheet and plane structures of GO give the ability to make the Ch chain segment easier than a slide. However, increasing the GO content led to aggregate and decreased the efficiency of the physicochemical properties of its nanocomposites [47].

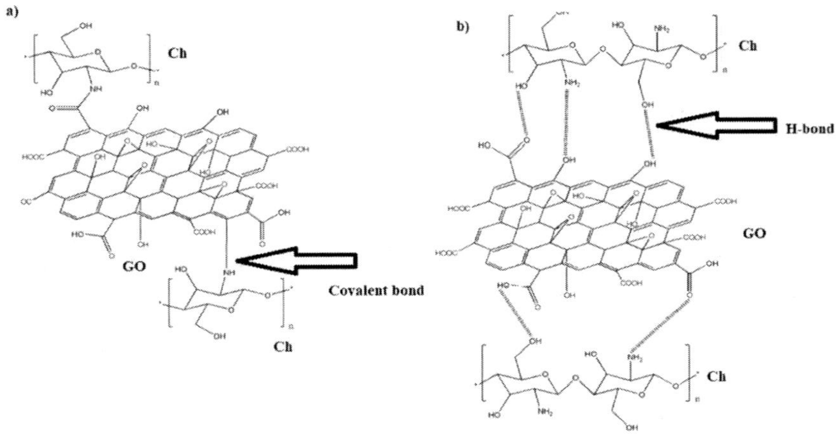

Figure 2. Chemical illustration of the interaction between Ch and GO via a) chemical covalent bond and b) physical (H-bond) interaction bonds.

3. Food Packaging Based on Graphene Oxide/Biopolymer Nanocomposites

3.1. Graphene Oxide/Chitosan Nanocomposites

Chitosan (Ch) is a partially deacetylated cationic natural biopolymer which is extracted from chitin through partial alkaline hydrolysis for N-acetylated units (Figure 3a). Ch is one of the top listed antibacterial biopolymer-based hydrogels over the last two decades. Ch has fabulous features, including its antimicrobial, biocompatible, biodegradable, non-toxic, antioxidant, and anti-inflammatory properties [16, 48, 49]. For example, Xie et al. [46] reported the synthesis of low-cost nanocomposite film for food packaging applications consisting of GO and chitosan in the presence of titanium dioxide (TiO_2) nanoparticles by a self-assembly method (Figure 4a & b). The use of Ch and GO contributed to preparing uniform and stable TiO_2 NPs. Hence, Ch/GO/TiO_2 nanocomposite films were endowed with excellent antibacterial activity assays against *Aspergillus niger* and *Bacillus subtilis.* The as-prepared nanocomposites exhibited fabulous features such as strong antibacterial activity, an easy processing method, and the production of graphene oxide in large amounts. Therefore, they were applied as an antibacterial food packaging material.

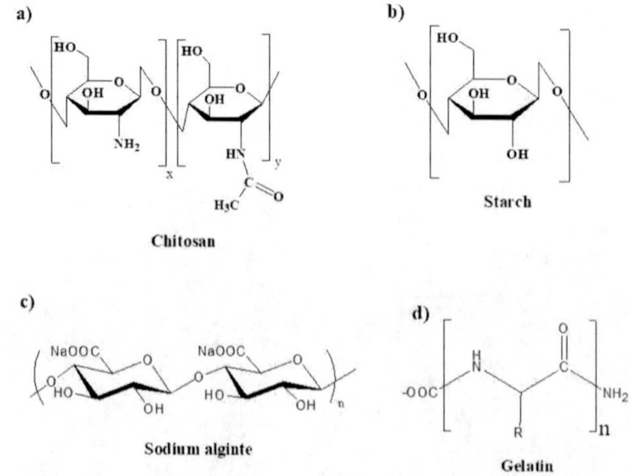

Figure 3. The chemical structure of different biopolymers, such as a) chitosan, b) starch, c) alginate, and d) gelatin.

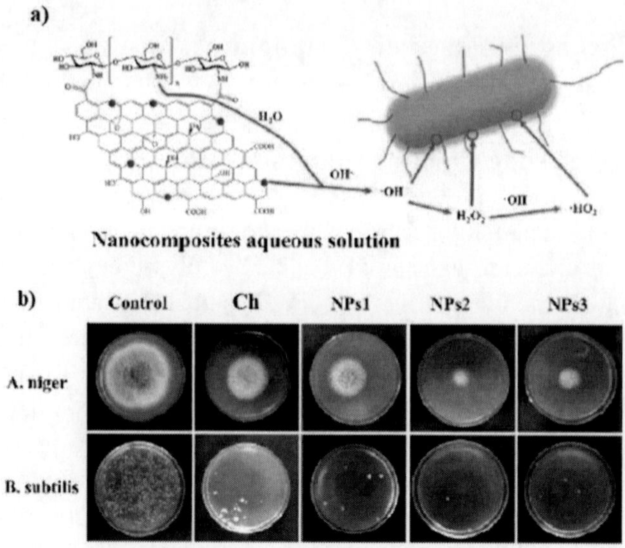

Figure 4. A) Schematic synthesis of antibacterial nanocomposites (Ch/GO/TiO2) and B) images of antibacterial activity against *A. niger* and *B. subtilis* using different concentration ratios (GO:CS:TiO2 ratios of 1:20:2 (NPs1), 1:20:4 (NPs2), and 1:20:8 (NPs3), respectively. Reproduced from [46] with permission from Elsevier B.V.

To improve the thermal and mechanical stabilities of food packaging films, Ch nanocomposite films were fabricated in the presence of different dosages of GO nanosheets: 0.5, 1, and 2 w/w%. The findings demonstrated that the addition of nanosheets did not improve the mechanical rigidity of nanocomposites in solution; however, their dielectric properties were enhanced. In addition, compared with neat films in the absence of GO nanosheets, the mechanical, thermal, and barrier properties of the as-prepared nanocomposites improved by enhancing the tensile strength and glass transition temperature of the nanocomposite. This, in turn, significantly decreased the lightness value as well as UV light transmittance. Furthermore, GO nanosheets increased the diffusive path tortuosity, allowing oxygen and water vapours to easily diffuse within the Ch/GO nanocomposite films [50].

Recently, Mahmoudi et al. [51] reported the preparation of antibacterial GO nanocomposite films in the presence of polyvinylpyrrolidone (PVP) and chitosan using the solvent casting method. The thickness of the prepared nanocomposite films was measured at 1 nm, with lateral dimensions of a few μm. The findings showed that the addition of PVP to GO/Ch films significantly enhanced their water vapour permeability and optical transmittance. Additionally, the thermal analysis study revealed that the superior thermal stability of the fabricated films may be owing to the strong H-bonding interaction between Ch and GO. The antibacterial assay of the GO nanocomposite was investigated against two bacterial strains such as S. aureus and E.coli, and the results confirmed that GO improved the antibacterial activity of Ch through a synergistic effect for bacterial inhibition growth through damaging of the cell membranes and the antibacterial of Ch/GO nanocomposite was more vigorous against S. aureus than E.coli. In the same framework, Lyn et al. [52] are interested in preparing GO/Ch nanocomposite films using the solvent casting method. They prepared GO via various oxidation degrees of graphite using potassium permanganate ($KMnO_4$) at different concentrations: 1:2 (GO1), 1:4 (GO2), 1:6 (GO3), and 1:8 (GO4). The physicochemical properties of the fabricated films showed that increasing the oxidation degree in GO up to 1:8 (GO4) led to improved the mechanical strength and decreased UV light transmittance compared to native Ch. Also, the highest degree of temperature improved the hydrophobicity of the GO/Ch nanocomposite, and all prepared nanocomposite films were completely degraded in soil within 28 days.

3.2. Graphene Oxide/Starch Nanocomposites

Starch is another example of the most common biopolymers, which are widely applied in different applications such as food packaging. Based on its origin source, starch is composed of linear amylose and highly branched amylopectin in proportions of 20-30% and 70-80%, respectively (Figure 3b). However, while starch has outstanding properties such as non-toxicity, biocompatibility, and low cost, it also has drawbacks such as poor mechanical and thermal stability, moisture sensitivity, and bacterial attack, which limit its practical application [53]. For instance, Usman et al. [54] reported the preparation of antibacterial food packaging nanocomposite films based on starch and GO nanosheets in the presence of poly(vinyl alcohol) and silver nanoparticles by the solution casting method. The mentioned results exhibited good thermal stability with an increase in the maximum temperature (Tmax) up to 295°C owing to the strong physical bonding interactions among GO nanolayers and polymer chains. Tmax of neat PVA was 266°C, while Tmax of PVA/GO and PVA/GO/Ag nanocomposites were 290°C. Moreover, the authors reported the improvement of the mechanical stability and their antimicrobial activity. On the other hand, the physical properties of the starch/GO/PVA nanocomposite films, which were prepared by the casting technique, were studied by Wu et al. [55]. They found that the increase of GO nanosheets concentration inside the film structure resulted in an increase in both thickness and tensile strength (25.28 MPa) using 2 mg of GO. In addition, the improvement of the film's moisture resistance. The authors found that the addition of GO led to a change in the phenomenon of nanocomposite agglomeration and boosted their thermal stability. Additionally, they noticed a decrease in water vapour permeability.

Recently, Krystyjan et al. [56] demonstrated the fabrication of bacteriostatic bio-nanocomposite films consisting of starch and Ch in the presence of GO nanosheets. They discovered that the as-prepared nanocomposite has excellent tolerability by human cells as well as antibacterial activity. The analysis techniques confirmed that GO nanosheets were homogenous distributed within polymer chains through strong physical interaction bonds. The GO changed their thermal features and declined their brightness. The cytotoxicity assay showed that bio-nanocomposite films revealed non-toxicity to the human cells (HaCat keratinocytes and HepG2 hepatoma cells). Moreover, the mechanical study exhibited that nanocomposite films had better resistance against rupture and showed excellent elasticity/expandability. The addition of GO significantly enhanced the nanocomposite film elongation at break. The antibacterial assay illustrated excellent bacteriostatic activity for the as-prepared nanocomposites against several pathogenic bacterial strains such as E. coli, Proteus mirabilis, and S. aureus.

In another study, Narayanan et al. [57] reported the usage of starch as a reducing agent to prepare reduced GO by hydrothermal technique with 0.2 wt.% of the soluble potato starch and 0.5 mg of GO. After that, the amylose structure of starch was linked to polyiodide to form a reduced GO/starch nanocomposite (Figure 5a). Additionally, the UV-Vis spectroscopy and XRD analysis as well confirmed the nanocomposite preparation by the presence of polyiodide in nanocomposites (Figure 5b & c). The antibacterial activity of nanocomposite was investigated against S. aurous and E. coli, and the findings showed good antibacterial activity against two pathogens with inhibition diameter zones of 20.2 nm and 22.2 nm, respectively, with low bactericidal concentration values (2.5 and 5 mg/mL). The antibacterial activity of the as-prepared nanocomposite is considered as a potential vehicle for the sustained release of polyiodide that destroys bacterial cell membranes, meaning that they can be efficient materials for food packaging applications.

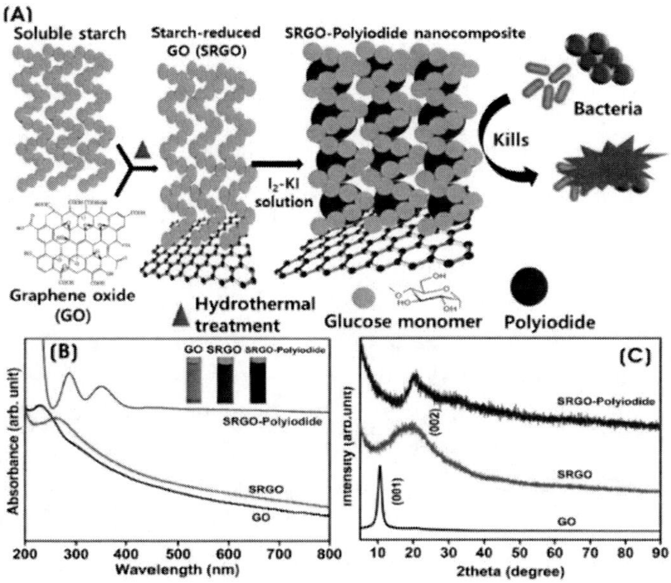

Figure 5. A) Schematic preparation of a Starch/r-GO nanocomposite and its antibacterial activity, b) UV-Vis spectroscopy, and c) XRD analysis of nanocomposites. Reproduced from [57] with permission from Elsevier B. V.

3.3. Graphene Oxide/Alginate Nanocomposites

Among biopolymer polysaccharides, alginate has attracted attention in the development of biopolymers for food packaging applications owing to its natural origin in the brown algae (Phaeophyceae) cell walls [58]. Alginate consists of a linear anionic structure composed of β-D-mannuronate (M) and α-L-guluronate (G) attached by 1-4-glycosidic bonds (Figure 3c) [59]. The physico-chemical properties of alginate are affected by the ratio between D-mannuronate and α-L-guluronate and their distribution through the polymer chains. Alginates have more interest in the fabrication of nanocomposite-based on GO for food packaging because they have excellent features including biocompatibility, non-toxicity, hydrophilicity, low-cost, biodegradability, and easy availability [60]. For example, Marti et al. [61] reported the preparation of crosslinked alginate/GO nanocomposites as antibacterial food packaging materials by solvent casting in the presence of a biocompatible crosslinker such as $ZnCl_2$ aqueous. The findings showed that the incorporation of a low concentration of GO (1.0 wt.%) into nanocomposite films enhanced their antibacterial activity against S. aureus and S. epidermidis. Additionally, they found that the as-prepared nanofilms were non-toxic to human cells. A similar study has been reported about the fabrication of alginate /GO nanocomposite with higher antibacterial inhibition activity against S. aureus and S. epidermidis using the same previously mentioned GO concentration.

In another approach, Weng et al. [62] reported the preparation of bioinspired assembly of ultralarge graphene oxide nanosheets among alginate chains using a one-step casting technique (Figure 6). In aqueous alginate suspension, the GO self-assembled into lyotropic nematic liquid crystal at a concentration above 2 wt.% and, after evaporation, the liquid crystal nanocomposite yielded a highly ordered and dense nanocomposite layered structure, which was confirmed by SEM technique. The addition of GO induced a remarkable enhancement in elastic modulus, reaching up to 230% (12.59 GPa) at 15.0 wt.% of GO content. Also, the TGA technique showed high thermal stability for the incorporation of ultralarge GO inside polymer chains.

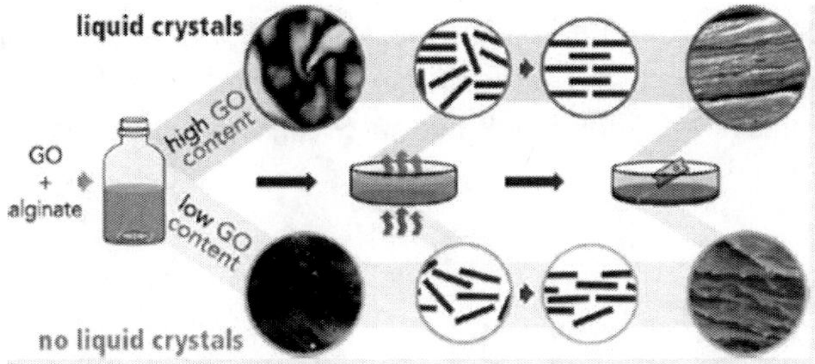

Figure 6. Schematic preparation of ultralarge graphene oxide/alginate nanocomposite films. Reprinted with permission from the American Chemical Society from [62].

3.4. Graphene Oxide/Gelatin Nanocomposites

Gelatin is a low-cost proteinic biopolymer that is derived from partial hydrolysis of collagen, which has different functional groups along its backbone such as carboxyl, amino, and hydroxyl groups (Figure 3d) [63, 64]. It has potential properties including low-cost, non-toxicity, biocompatibility, non-immunogenicity, and easy availability [65], so it is widely applied in different applications, especially food packaging applications in the presence of GO nanosheets and different natural biopolymers to design high-performance food packaging materials. For example, Afshar et al. [66] reported the fabrication of biodegradable gelatin nanocomposite films by using GO nanosheets and starch for potential food packaging applications using Design Expert software via RSM in the presence of different GO content. The SEM and XRD techniques proved the intercalation of GO layers inside starch/gelatin chains with proper dispersion of GO nanofillers via the solution casting method. The results exhibited improvement in the mechanical strength, surface hydrophilicity, thermal stability, and water vapour permeability in the presence of GO related to the reinforcing effect and good barrier properties of graphene oxide. Moreover, the degradation study of nanocomposite films in soil was investigated within 6 weeks and the findings showed that at least 30% of the total weight of films was lost. In another study, Li et al. [67] reported the fabrication of gelatin/cellulose nanofibril nanocomposite films using GO nanosheets modified by dopamine and immobilised with Ag NPs. The data showed that both the tensile strength and toughness of the as-prepared nanocomposite films increased to a maximum of 35.99 MPa and 13.49 MJ/m3, respectively (Figure 7a & b). Additionally, the nanocomposite films exhibited good thermal stability, water resistance, and UV-shielding capacity. In addition, they exhibited excellent antibacterial efficiency against two pathogens including *S. aureus* and *E. coli.*

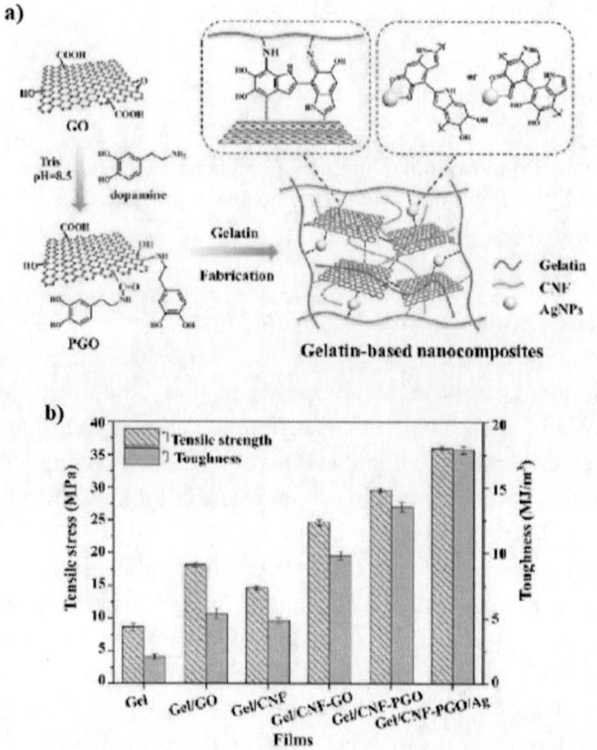

Figure 7. A) the schematic preparation of gelatin-based nanocomposites in the presence of modified GO nanosheets with dopamine and cellulose nanofibrils and Ag NPs, and b) the mechanical properties (tensile strength and toughness) of the as-prepared nanocomposite films. Reproduced from [67] with permission from Elsevier B. V.

Recently, Li et al. [68] mentioned the fabrication of gelatin/GO nanocomposite films in the presence of sodium alginate and rosemary extract (RE) with antioxidant and UV-shielding properties for the food packaging field. The authors added Polyvinylpyrrolidone (PVP) to overcome the mechanical properties due to the usage of RE. The findings demonstrated an enhancement in the tensile strength by 13.6 MPa without a reduction in the elongation at break. Moreover, they also showed excellent ultraviolet-shielding capacity by 52.5% in the range of 300-400 nm thanks to RE. Additionally, the as-prepared nanocomposite films exhibited superior 1,1-diphenyl-2-picrylhydrazyl radical scavenging activity in aqueous ethanol and water. The peroxide value of the soybean sealed with films was 0.51 times that of the ternary nanocomposite film after 1 month of storage, endowing it with fabulous antioxidant activity.

4. Prospects and Challenges

Approaches to prepare graphene nanocomposite films on a large scale and at a relatively low cost are in high demand. Although great efforts have been made in this direction, current strategies are seriously limited by their low efficiencies, which should be addressed for commercial applications. A reliable method for meeting the strong demand for graphene and its derivatives via an eco- friendly approach with high yield is still lacking. Another problem is that the real specific surface area of graphene materials is significantly less than those predictions owing to the strong agglomeration of graphene sheets and their derivatives within hydrophobic interactions, and their limited antibacterial activity. In this regard, novel green methods to fabricate and exfoliate graphene sheets are being pursued. Another limitation for the practical application of graphene-based nanocomposites is the lack of in-vivo antibacterial activity studies on animal models. While graphene nanocomposite may be valuable for wound healing and other biomedical applications, future uses in vivo cannot be ruled out. Studying the selectively antibacterial activity of graphene against pathogenic microorganisms without affecting normal mammalian cells or non-pathogenic bacteria is very scarce currently, because their toxicity is not clear yet. Despite great efforts in studying the effects of these graphene nanocomposites on human health and the environment, results are often inconsistent. They might provoke cytotoxicity of the graphene-based nanomaterials in humans, and this issue should be clarified. It is important to note that the effects of graphene and its derivatives may differ depending on their intrinsic properties.

Overall, the lack of techniques for producing simple, reproducible, and cost-effective nanomaterials on a large scale is impeding the development and application of antimicrobial graphene nanocomposites as a high-efficiency food packaging system. Research in this field is still in its infancy. The combination of graphene structure and biopolymers has opened new properties and applications due to their synergistic effects. Nonetheless, more work is needed to ensure that the developed nanocomposites are easy to manufacture and have no toxicity. It can be envisaged that after wide-ranging research in the field and continuous innovative efforts, nanocomposites incorporating polymers and graphene materials could provide a new outlook for the development of food packaging materials with excellent antimicrobial activity and physicochemical properties as well.

Conclusion

Scholar researchers are paying close attention to food packaging materials to reduce food waste. So they do their best to design new food packaging materials with excellent features such as good mechanical and thermal stability, good antibacterial activity, gas barrier, water, water resistance, and UV-shielding capacity. This chapter focuses on the GO nanocomposite films based on biopolymers. GO nanosheets have some fascinating features including large surface area, good antimicrobial activity, excellent mechanical and thermal properties, and unique electrocatalytic activity. Hence, they have been widely employed for different applications such as food packaging. Additionally, biopolymer materials such as chitosan, starch, alginate, and gelatin are also widely used in food packaging due to their excellent properties including easily availability, low-cost materials, biodegradable, and biocompatible polymers. Therefore, the addition of GO nanolayers among biopolymer chains showed excellent promise and potential food packaging nanocomposite materials with excellent mechanical and thermal stability, excellent antimicrobial activity, good water resistance, and UV-shielding capacity. Overall, it is anticipated that GO/biopolymer-based nanocomposites will provide a significant opportunity for academic and industrial researchers, as well as governments, to fundamentally update food packaging fabrication.

References

[1] Soleimani-Rambod, A., S. Zomorodi, S. Naghizadeh Raeisi, A. Khosrowshahi Asl, S. A. Shahidi, The effect of xanthan gum and flaxseed mucilage as edible coatings in cheddar cheese during ripening, *Coatings* 8(2) (2018) 80.

[2] Nile, S., V. Baskar, D. Selvaraj, A. Nile, J. Xiao, G. Kai, Nanotechnologies in food science: applications, recent trends, and future perspectives. *Nano-Micro Lett* 12: 45–51, 2020.

[3] Prasad, P., A. Kochhar, Active packaging in food industry: a review, *Journal of Environmental Science, Toxicology and Food Technology* 8(5) (2014) 1-7.

[4] Elella, M. H. A., E. S. Goda, M. A. Gab-Allah, S. E. Hong, B. Pandit, S. Lee, H. Gamal, A. ur Rehman, K. R. Yoon, Xanthan gum-derived materials for applications in environment and eco-friendly materials: A review, *Journal of Environmental Chemical Engineering* 9(1) (2021) 104702.

[5] Bratovčić, A., A. Odobašić, S. Ćatić, I. Šestan, Application of polymer nanocomposite materials in food packaging, *Croatian journal of food science and technology* 7(2) (2015) 86-94.

[6] Janaki, V., K. Vijayaraghavan, B.-T. Oh, K.-J. Lee, K. Muthuchelian, A. Ramasamy, S. Kamala-Kannan, Starch/polyaniline nanocomposite for enhanced removal of reactive dyes from synthetic effluent, *Carbohydrate polymers* 90(4) (2012) 1437-1444.

[7] Elella, M. H. A., E. S. Goda, H. M. Abdallah, A. E. Shalan, H. Gamal, K. R. Yoon, Innovative bactericidal adsorbents containing modified xanthan gum/montmorillonite nanocomposites for wastewater treatment, International *Journal of Biological Macromolecules* 167 (2021) 1113-1125.

[8] Elella, M. H. A., E. S. Goda, H. Gamal, S. M. El-Bahy, M. A. Nour, K. R. Yoon, Green antimicrobial adsorbent containing grafted xanthan gum/SiO2 nanocomposites for malachite green dye, *International journal of biological macromolecules* 191 (2021) 385-395.

[9] Goda, E. S., M. H. Abu Elella, H. Gamal, S. E. Hong, K. R. Yoon, Two-Dimensional Nanomaterials as Smart Flame Retardants for Polyurethane, *Materials and Chemistry of Flame-Retardant Polyurethanes Volume 1: A Fundamental Approach*, ACS Publications2021, pp. 189-219.

[10] Pandit, B., E. S. Goda, M. H. A. Elella, A. ur Rehman, S. E. Hong, S. R. Rondiya, P. Barkataki, S. F. Shaikh, A. M. Al-Enizi, S. M. El-Bahy, One-pot hydrothermal preparation of hierarchical manganese oxide nanorods for high-performance symmetric supercapacitors, *Journal of Energy Chemistry* 65 (2022) 116-126.

[11] Elella, M. H. A. J. J. o. C. E. R. U., *Journal of Chemical Engineering Research Updates*, 73 (2021) 97.

[12] Elella, M. H. A., E. S. Goda, H. M. Abdallah, S. Elyamny, H. Gamal, E. S. A. Serea, A. E. Shalan, Electroactive Polymers for Electrochromic Applications, Electroactive Polymeric Materials, CRC Press, pp. 163-201.

[13] Ngwuluka, N. C., N. Y. Abu-Thabit, O. J. Uwaezuoke, J. O. Erebor, M.O. Ilomuanya, R. R. Mohamed, S. M. Soliman, M. H. A. Elella, N. A. Ebrahim, *Natural Polymers in Micro-and Nanoencapsulation for Therapeutic and Diagnostic Applications: Part I: Lipids and Fabrication Techniques, Nano-and Micro-Encapsulation-Techniques and Applications*, IntechOpen2020.

[14] Elella, M. H. A., E. S. Goda, M. A. Gab-Allah, S. E. Hong, B. Pandit, S. Lee, H. Gamal, A. ur Rehman, K. R. Yoon, Xanthan gum-derived materials for applications in environment and eco-friendly materials: a review, *Journal of Environmental Chemical Engineering* (2020) 104702.

[15] Mohamed, R. R., M. H. A. Elella, M. W. Sabaa, Synthesis, characterization and applications of N-quaternized chitosan/poly (vinyl alcohol) hydrogels, *International journal of biological macromolecules* 80 (2015) 149-161.

[16] Mohamed, R. R., M. H. A. Elella, M. W. Sabaa, Cytotoxicity and metal ions removal using antibacterial biodegradable hydrogels based on N-quaternized chitosan/poly (acrylic acid), *International journal of biological macromolecules* 98 (2017) 302-313.

[17] Abu Elella, M. H., H. M. Abdallah, H. Gamal, E. B. Moustafa, E. S. Goda, Rational design of biocompatible IPNs hydrogels containing carboxymethyl starch and trimethyl chitosan chloride with high antibacterial activity, *Cellulose* (2022) 1-14.

[18] Elella, M. H. A., N. Aamer, Y. Mohamed, H. A. El Nazer, R. R. Mohamed, Innovation of high-performance adsorbent based on modified gelatin for wastewater treatment, *Polymer Bulletin* (2022) 1-17.

[19] Elella, M. H. A., A. E. Shalan, M. W. Sabaa, R. R. Mohamed, One-pot green synthesis of antimicrobial chitosan derivative nanocomposites to control foodborne pathogens, *RSC advances* 12(2) (2022) 1095-1104.

[20] Abu Elella, M. H., E. A. ElHafeez, E. S. Goda, S. Lee, K. R. Yoon, Smart bactericidal filter containing biodegradable polymers for crystal violet dye adsorption, *Cellulose* 26(17) (2019) 9179-9206.

[21] Sabaa, M. W., D. H. Hanna, M. H. Abu Elella, R. R. Mohamed, Encapsulation of bovine serum albumin within novel xanthan gum based hydrogel for protein delivery, *Materials Science and Engineering: C* 94 (2019) 1044-1055.

[22] Elella, M. H. A., E. S. Goda, K. R. Yoon, S. E. Hong, M. S. Morsy, R. A. Sadak, H. Gamal, Novel vapor polymerization for integrating flame retardant textile with multifunctional properties, *Composites Communications* 24 (2021) 100614.

[23] Elella, M. H. A., E. Goda, M. Gab-Allah, S. E. Hong, Y. G. Lijalem, K. R. Yoon, Biodegradable Polymeric Nanocomposites for Wastewater Treatment, *Advances in Nanocomposite Materials for Environmental and Energy Harvesting Applications,* Springer2022, pp. 245-298.

[24] de Carvalho, A. P. A., C. A. C. Junior, Green strategies for active food packagings: A systematic review on active properties of graphene-based nanomaterials and biodegradable polymers, *Trends in Food Science & Technology* (2020).

[25] Hossain, K., M. Rafatullah, S. Z. Abbas, A. Ahmad, N. Ismail, A. Y. Maruthi, *Antimicrobial activity of graphene-based nanomaterials: current development and challenges, Graphene-Based Nanotechnologies for Energy and Environmental Applications*, Elsevier2019, pp. 293-314.

[26] Goda, E. S., M. H. A. Elella, S. E. Hong, B. Pandit, K. R. Yoon, H. Gamal, Smart flame retardant coating containing carboxymethyl chitosan nanoparticles decorated graphene for obtaining multifunctional textiles, *Cellulose* 28(8) (2021) 5087-5105.

[27] Goda, E. S., M. H. A. Elella, M. Sohail, B. S. Singu, B. Pandit, A. El Shafey, A. M. Aboraia, H. Gamal, S. E. Hong, K. R. Yoon, N-methylene phosphonic acid chitosan/graphene sheets decorated with silver nanoparticles as green antimicrobial agents, *International Journal of Biological Macromolecules* 182 (2021) 680-688.

[28] Chen, Z., W. Ren, L. Gao, B. Liu, S. Pei, H.-M. Cheng, Three-dimensional flexible and conductive interconnected graphene networks grown by chemical vapour deposition, *Nature materials* 10(6) (2011) 424-428.

[29] Liu, F., T. S. Seo, A controllable self-assembly method for large-scale synthesis of graphene sponges and free-standing graphene films, *Advanced Functional Materials* 20(12) (2010) 1930-1936.

[30] Kuilla, T., S. Bhadra, D. Yao, N. H. Kim, S. Bose, J. H. Lee, Recent advances in graphene based polymer composites, *Progress in polymer science* 35(11) (2010) 1350-1375.

[31] Díez-Pascual, A. M., J. A. Luceño-Sánchez, Antibacterial Activity of Polymer Nanocomposites Incorporating Graphene and Its Derivatives: A State of Art, *Polymers* 13(13) (2021) 2105.

[32] Yao, Q., B. Fan, Y. Xiong, C. Jin, Q. Sun, C. Sheng, 3D assembly based on 2D structure of cellulose nanofibril/graphene oxide hybrid aerogel for adsorptive removal of antibiotics in water, *Scientific reports* 7(1) (2017) 1-13.

[33] Dreyer, D. R., S. Park, C. W. Bielawski, R. S. Ruoff, The chemistry of graphene oxide, *Chemical society reviews* 39(1) (2010) 228-240.

[34] Smith, A. T., A. M. LaChance, S. Zeng, B. Liu, L. Sun, Synthesis, properties, and applications of graphene oxide/reduced graphene oxide and their nanocomposites, *Nano Materials Science* 1(1) (2019) 31-47.

[35] Diez-Pascual, A. M., A. L. Diez-Vicente, Poly (propylene fumarate)/polyethylene glycol-modified graphene oxide nanocomposites for tissue engineering, *ACS applied materials & interfaces* 8(28) (2016) 17902-17914.

[36] Liu S., F. Yao, M. Kang, S. Zhao, Q. Huang, G. Fu, Hierarchical xanthan gum/graphene oxide nanocomposite film induced by ferric ions coordination, *Materials & Design* 113 (2017) 232-239.

[37] Chen, J., B. Yao, C. Li, G. Shi, An improved Hummers method for eco-friendly synthesis of graphene oxide, *Carbon* 64 (2013) 225-229.

[38] Díez-Pascual, A. M., Antibacterial Action of Nanoparticle Loaded nanocomposites based on graphene and its derivatives: A mini-review, *International Journal of Molecular Sciences* 21(10) (2020) 3563.

[39] Díez-Pascual, A. M., *Antibacterial activity of nanomaterials*, Multidisciplinary Digital Publishing Institute, 2018, p. 359.

[40] Hu, W., C. Peng, W. Luo, M. Lv, X. Li, D. Li, Q. Huang, C. Fan, Graphene-based antibacterial paper, *ACS nano* 4(7) (2010) 4317-4323.

[41] Liu, S., T. H. Zeng, M. Hofmann, E. Burcombe, J. Wei, R. Jiang, J. Kong, Y. Chen, Antibacterial activity of graphite, graphite oxide, graphene oxide, and reduced graphene oxide: membrane and oxidative stress, *ACS nano* 5(9) (2011) 6971-6980.

[42] Hui, L., J. G. Piao, J. Auletta, K. Hu, Y. Zhu, T. Meyer, H. Liu, L. Yang, Availability of the basal planes of graphene oxide determines whether it is antibacterial, *ACS applied materials & interfaces* 6(15) (2014) 13183-13190.

[43] Chen, J., H. Peng, X. Wang, F. Shao, Z. Yuan, H. Han, Graphene oxide exhibits broad-spectrum antimicrobial activity against bacterial phytopathogens and fungal conidia by intertwining and membrane perturbation, *Nanoscale* 6(3) (2014) 1879-1889.

[44] Perreault, F., A. F. De Faria, S. Nejati, M. Elimelech, Antimicrobial properties of graphene oxide nanosheets: why size matters, *ACS nano* 9(7) (2015) 7226-7236.

[45] Díez-Pascual, A. M., A. L. Díez-Vicente, Electrospun fibers of chitosan-grafted polycaprolactone/poly (3-hydroxybutyrate-co-3-hydroxyhexanoate) blends, *Journal of Materials Chemistry* B 4(4) (2016) 600-612.

[46] Xu, W., W. Xie, X. Huang, X. Chen, N. Huang, X. Wang, J. Liu, The graphene oxide and chitosan biopolymer loads TiO2 for antibacterial and preservative research, *Food chemistry* 221 (2017) 267-277.

[47] Sivashankari, P., M. Prabaharan, *Chitosan/carbon-based nanomaterials as scaffolds for tissue engineering, Biopolymer-Based Composites*, Elsevier2017, pp. 381-397.

[48] Eliyahu, S., A. Galitsky, E. Ritov, H. Bianco-Peled, Hybrid Acrylated Chitosan and Thiolated Pectin Cross-Linked Hydrogels with Tunable Properties, *Polymers* 13(2) (2021) 266.

[49] Maciel, V. B. V., C. M. Yoshida, T. T. Franco, Chitosan/pectin polyelectrolyte complex as a pH indicator, *Carbohydrate polymers* 132 (2015) 537-545.

[50] Ahmed, J., M. Mulla, Y. A. Arfat, Mechanical, thermal, structural and barrier properties of crab shell chitosan/graphene oxide composite films, *Food Hydrocolloids* 71 (2017) 141-148.

[51] Mahmoudi, N., F. Ostadhossein, A. Simchi, Physicochemical and antibacterial properties of chitosan-polyvinylpyrrolidone films containing self-organized graphene oxide nanolayers, *Journal of Applied Polymer Science* 133(11) (2016).

[52] Tan, C. P., R. Zawawi, N. H. ZA, Effect of sonication time and heat treatment on the structural and physical properties of chitosan/graphene oxide nanocomposite films, *Food Packaging and Shelf Life* 28 (2021) 100663.

[53] Ogunsona, E., E. Ojogbo, T. Mekonnen, Advanced material applications of starch and its derivatives, *European Polymer Journal* 108 (2018) 570-581.

[54] Usman, A., Z. Hussain, A. Riaz, A. N. Khan, Enhanced mechanical, thermal and antimicrobial properties of poly (vinyl alcohol)/graphene oxide/starch/silver nanocomposites films, *Carbohydrate polymers* 153 (2016) 592-599.

[55] Wu, Z., Y. Huang, L. Xiao, D. Lin, Y. Yang, H. Wang, Y. Yang, D. Wu, H. Chen, Q. Zhang, Physical properties and structural characterization of starch/polyvinyl alcohol/graphene oxide composite films, *International journal of biological macromolecules* 123 (2019) 569-575.

[56] Krystyjan, M., G. Khachatryan, M. Grabacka, M. Krzan, M. Witczak, J. Grzyb, L. Woszczak, Physicochemical, bacteriostatic, and biological properties of starch/chitosan polymer composites modified by graphene oxide, designed as new bionanomaterials, *Polymers* 13(14) (2021) 2327.

[57] Narayanan, K. B., G. T. Park, S. S. Han, Antibacterial properties of starch-reduced graphene oxide–polyiodide nanocomposite, *Food Chemistry* 342 (2021) 128385.

[58] Jeon, O., D. S. Alt, S. M. Ahmed, E. Alsberg, The effect of oxidation on the degradation of photocrosslinkable alginate hydrogels, *Biomaterials* 33(13) (2012) 3503-3514.

[59] Abd El-Ghany, N. A., M. S. A. Aziz, M. M. Abdel-Aziz, Z. Mahmoud, Antimicrobial and swelling behaviors of novel biodegradable corn starch

grafted/poly (4-acrylamidobenzoic acid) copolymers, *International journal of biological macromolecules* 134 (2019) 912-920.

[60] Shelke, N. B., R. James, C. T. Laurencin, S. G. Kumbar, Polysaccharide biomaterials for drug delivery and regenerative engineering, *Polymers for Advanced Technologies* 25(5) (2014) 448-460.

[61] Martí, M., B. Frigols, B. Salesa, Á. Serrano-Aroca, Calcium alginate/graphene oxide films: Reinforced composites able to prevent Staphylococcus aureus and methicillin-resistant Staphylococcus epidermidis infections with no cytotoxicity for human keratinocyte HaCaT cells, *European Polymer Journal* 110 (2019) 14-21.

[62] Weng, S. X., N. Yousefi, N. Tufenkji, Self-assembly of ultralarge graphene oxide nanosheets and alginate into layered nanocomposites for robust packaging materials, *ACS Applied Nano Materials* 2(3) (2019) 1431-1444.

[63] Kozlov, P., G. J. P. Burdygina, *The structure and properties of solid gelatin and the principles of their modification*, 24(6) (1983) 651-666.

[64] Hanani, Z. N., Y. H. Roos, J. P. Kerry, Use and application of gelatin as potential biodegradable packaging materials for food products, *International journal of biological macromolecules* 71 (2014) 94-102.

[65] Bello, A. B., D. Kim, D. Kim, H. Park, S. H. J. T. E. P. B. R. Lee, *Engineering and functionalization of gelatin biomaterials: From cell culture to medical applications,* 26(2) (2020) 164-180.

[66] Afshar, S., H. Baniasadi, Investigation the effect of graphene oxide and gelatin/starch weight ratio on the properties of starch/gelatin/GO nanocomposite films: The RSM study, *International journal of biological macromolecules* 109 (2018) 1019-1028.

[67] Li, K., S. Jin, J. Li, H. Chen, Improvement in antibacterial and functional properties of mussel-inspired cellulose nanofibrils/gelatin nanocomposites incorporated with graphene oxide for active packaging, *Industrial Crops and Products* 132 (2019) 197-212.

[68] Li, J., J. Yang, J. Zhong, F. Zeng, L. Yang, X. Qin, Development of sodium alginate-gelatin-graphene oxide complex film for enhancing antioxidant and ultraviolet-shielding properties, *Food Packaging and Shelf Life* 28 (2021) 100672.

Chapter 5

Development of Food Packaging Films from Microorganism-Generated Polyhydroxyalkanoates

Özcan Bulantekin[1], PhD and Duygu Alp[2,*], PhD

[1]Department of Food Engineering, Faculty of Engineering, Süleyman Demirel University, Isparta, Turkey

[2]Department of Gastronomy and Culinary Arts, School of Tourism and Hospitality Management, Ardahan University, Ardahan, Turkey

Abstract

Petroleum-based packaging materials cause environmental pollution and toxic substance accumulation because they cannot decompose in nature for a long time. To prevent these problems, a wide variety of food packaging materials emerge as alternatives to petroleum-based packaging materials. So far our colleagues have already discussed how polyscahharides and biopolymer-based nanocomposites are used in the development of food packaging films. As the last chapter of this Section, we will introduce how the microorganism-generated biopolymer, polyhydroxyalkanoates (PHAs) to be specific, is used in food packaging. PHAs, also known as green plastics, have positive social and environmental impact when compared to traditional plastics in terms of production and recycling. The food industry produces tons of waste every year, but these wastes have a wide range of applications such as chemistry, medicine, cosmetics or animal husbandry thanks to their biocompatibility and biodegradability. Considering that industrial wastes contain high quality polysaccharides, essential oils and proteins, using them in the production of biodegradable packaging will both reduce

* Corresponding Author's Email: duygualp1@gmail.com.

In: Food Packaging
Editor: Wing-Fu Lai
ISBN: 979-8-88697-249-8

environmental problems and provide economic gain by reprocessing the wastes into products with higher added value. However, it has some disadvantages in competition with synthetic plastics and applications as biomaterials due to some properties such as poor mechanical properties, high production costs, limited functionality, incompatibility with conventional heat treatment techniques and susceptibility to thermal degradation. In this chapter, we will discuss the future and potential difficulties that may be experienced in the production or dissemination of PHA as a packaging material.

1. Introduction

The packaging of foods plays an important role in the preservation of the product throughout the storage and distribution chain. The main purpose of packaging is to protect food from contamination and extend its shelf life (Popović et al., 2018). To meet this purpose, packages are produced from different raw materials. Plastic packages are frequently used in food packaging due to their high quality, easy processing and combination properties. However, today, the harmful effects of plastic waste on the environment have led the food industry to use biodegradable packaging materials (Ahmadzadeh and Khaneghah, 2020). Biodegradable packaging materials are obtained from renewable bio-based resources and are often referred to as "biopolymers" (Bajpai, 2019). Biopolymers are strong alternatives to non-biodegradable and non-renewable plastic packaging materials due to their environmentally friendly nature (Shankar and Rhim, 2018). Synthetic films used as packaging materials negatively affect the environment since they are not biodegradable, while biological-based packaging materials obtained from renewable materials draw attention with their biological compatibility, biodegradability, environmental protection and reliability (Bajpai, 2019). Biodegradable polymers are defined as polymers that can be broken down into simple molecules such as carbon dioxide, methane, water and inorganic compounds in a certain process in nature as a result of the enzymatic activity of microorganisms (Peelman et al., 2013; Siracusa and Lotti, 2018). Bioplastics are generally materials that can be produced from renewable resources such as sugar, corn and potatoes, containing proteins, lipids and polysaccharides or from certain microorganisms, algae and fungi (Sarasa et al., 2009; Karana, 2012). Environmentally friendly, biodegradable polymers have good mechanical and barrier properties and are thus seen as potential packaging materials to replace traditional plastic packaging materials (Shankar and Rhim, 2018; Siracusa and Lotti, 2018).

2. Bioplastics and Their Sources

Bioplastics which biodegradable materials made from renewable resources are the new materials of the 21st century and are of great importance (Mohanty et al., 2002; Boonniteewanich et al., 2014). Starches from corn, potatoes, wheat, rice, barley, oats; fibres obtained from pineapple; jute, hemp, banana stems, cassava, newspaper pulp, waste paper, citrus waste are renewable sources for bioplastics. The use of new techniques for bioplastic production that promotes sustainable solution and reduces plastic waste has been greatly promoted in recent years (Shah et al., 2021). Bioplastics can be produced from the inedible parts of food. Food wastes such as orange peel, pomegranate peel, banana peel, potato peel are used in Bioplastic production. In recent years, cellulose, hemicellulose, starch, pectin, these lignocellulosic raw materials are made useful for bioplastic film production as a trend (Shah et al., 2021). Biobased (biodegradable) packaging materials are a potential alternative to replace petroleum-based (synthetic) polymers. One of the reasons for this situation is the decrease in the demand for petroleum-based products and the movement towards renewable resources to produce plastics and the reduction of the amount of gases released into the atmosphere. Another reason is to reduce the solid waste problem by turning biodegradable materials into compostable organic residues after use (Väisänen et al., 2016; Berthet et al., 2016).

Starch is the most widely used source in the production of biopolymer packaging materials due to its edibility and ease of raw material supply. The main source of starch used for these purposes is usually corn the fact remains that the mechanical properties such as strain rate, tensile and flexibility strength of films produced from starch are not sufficient. Therefore, starch can be chemically modified or mixed with other substances. Plasticizers such as glycerol, polyether and urea are used to reduce the fracturability of starch. (Liu, 2006). Since starch shows hydrophilic properties, it is not suitable for liquid food products with high moisture content, however starch-based films have good oxygen barrier properties (Cha and Chinnan, 2004), as well they used as an alternative to petroleum-derived materials because it is inexpensive and biodegradable (Xie et al., 2006; Lu et al., 2009).The ability of starch to be hydrolyzed by microorganisms, and used as a carbon source and whether they have the ability to produce α-amylase enzyme is strain specific feature. For this reason, an external source of this enzyme is needed in order for starch to be hydrolyzed by microorganisms and used as a carbon source (Raza, 2018). Figure 1 shows bioplastics and their sources.

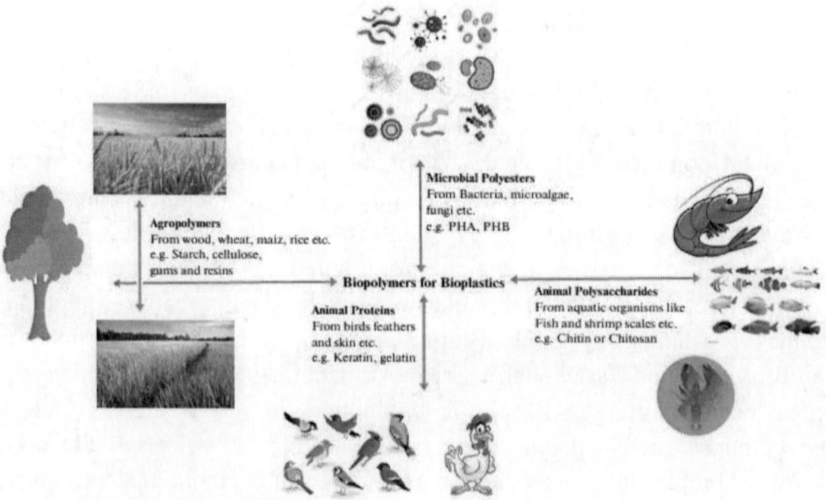

Figure 1. Bioplastics and their sources.

Cellulose consists of glucose monomer units linked by glycosidic bonds and is an inexpensive source because it is found in all plants. However, its hydrophilic properties, low solubility and high crystal structure create difficulties in its use in packaging, and due to the successive hydroxyl side chains, it causes low moisture barrier properties in cellulose-based packages. Also, the packaging material formed due to the high crystalline structure is brittle and has poor flexibility and tensile strength (Liu, 2006, Venkatachalam and Palaniswamy, 2020). For these reasons, research now focuses on cellulose derivatives for packaging purposes. Cellulose-based biopolymer cellophane films, known as candy wrappers, are highly transparent and colorful. Cellophane treated by lamination, injection and extrusion molding exhibits good film-forming properties (Zepnik et al., 2010). Because it is insoluble and has excellent dimensional stability, it is also used in packaging products ranging from laminate, flower wrapping, cheese to coffee and chocolate (Venkatachalam and Palaniswamy, 2020). Resrearchers stated that starch/PVA which is composite biodegradable films reinforced with cellulosic fiber are suitable for use in food packaging (Priya et al., 2014). Cellulose derivatives are obtained by the reaction of cellulose in the presence of an aqueous solution of sodium hydroxide and an esterifying reagent. Cellulose derivatives such as Hydroxypropyl cellulose, Hydroxypropyl methylcellulose, Carboxymethylcellulose and Methylcellulose are used for edible films/coating. Suspensions of these substances have thermogelation properties that form gels when heated and, on the contrary, regain their original consistency when cooled. Such films are poor water barriers and show poor mechanical properties due to the hydrophilic nature of the molecules (Phillips and Williams, 2009). The quality of the moisture barrier can be improved by adding hydrophobic compounds such as fatty acids to the cellulose matrix (Morillon et al., 2002). Due to the high production cost, the use of cellulose-based plastic is limited in the market.

Chitosan, the second most abundant natural polymer after cellulose, is obtained by partial deacetylation of the natural polysaccharide chitin (Casetteria et al., 2012). Chitosan is an important waste of the fishing industry. It has many functions such as antimicrobial effect against bacteria, molds and yeasts, moisture adsorbing, precipitation, film formation, enzyme immobilization (Bostan et al., 2007). It also exhibits good oxygen & carbon dioxide permeability, but major drawback is that it has poor solubility in neutral solutions. Sun et al., (2017) determined that chitosan films combined with apple polyphenols can be used as bioactive packaging material to increase the shelf life of foods. Pectin, which is a heteropolysaccharide found in the cell walls of plants, is frequently used to thicken jams and jellies (Müller-Maatsch et al., 2016). In the industry, mostly apple pulp and citrus peels are used as a source. These types of edible films are used for certain food-related functions, such as food packaging materials, anti-browning, flavor enhancer and antimicrobial functions, their production usually using casting method or extrusion (Espitia et al., 2014, Janjarasskul and Krochta, 2010). Pectins, which are frequently used to produce biodegradable films, can be supported with nano-structured fillers to compete with commercial polymers because they show poor physical and barrier properties. The nanocrystals obtained from cellulose, a natural component, impart rigidity and strength to the films (Karakuş and Ayhan, 2019).

Whey, which is a by-product of the cheese and casein industry and makes up 90% of the processed milk volume, is used for humans and animals, while some is thrown into the environment. Whey, which is an inexpensive substrate and carbon source for bacterial growth, is also preferred in the production of PHA (Koller et al., 2008). Wheat bran contains high protein, carbohydrate, and mineral, and they suitable for use as the waste. Various studies have been carried out to evaluate wheat bran waste. One of them is the use by Van-Thuoc et al., (2008) as a carbon source for bacterial growth and PHA production. They determined that by growing Halomonas boliviensis LC1 resulting in biomass production of 3.19 g/l and PHB production of 1.08 g/l. Soy proteins are generally a by-product of the soybean oil industry. Generally produced by wet casting, soybeans are preferred in edible films and coatings because of their good film forming properties. Although their water permeability is high, their oxygen permeability is good like other protein films (Coltelli et al., 2016, Khwaldia et al., 2010, Zink et al., 2016). Gelatin is a naturally occurring hydrocolloid polymer derived from animal skin, bones and related tissues. Gelatin-based bioplastic films are widely used as packaging material in the food industry (Hanani et al., 2014). Many studies showed that the physical and mechanical properties of biofilms made of gelatin can be improved with biocomposite films such as gelatin-starch or gelatin-starch-glycerol (Podshivalov et al., 2017). Lipids are used as protective coatings against transfer, but besides lipids, wax and other resin materials have several disadvantages as packaging materials, such as a waxy taste and texture, oily surface and bitterness. Among its advantages in terms of packaging, the lipid component reduces water transmission, while the hydrocolloid component acts as a gas barrier and even contributes to strength and structural integrity (Aulin and Ström, 2013, Janjarasskul and Krochta, 2010).

In the food industry, large amounts of both solid and liquid waste occur due to the production, preparation and consumption of food. These wastes cause problems while being destroyed, but they are valuable products as biomass, and nutritional components (Deniz et al., 2015; Kanmaz and Saral, 2017). Especially fruit and vegetable industry is an area that generates a lot of waste. Fruits and vegetables are consumed fresh as well as processed into juice and jam, and vegetables into canned products, and as a result of all these processes, wastes such as shell, seed and pulp with high polysaccharide, protein and lipid content are formed. These wastes are reusable for different processes. For this reason, a "Zero Waste Approach" is being tried to be adopted. The zero waste approach is based on the use of organic waste generated after processing, in a different field such as chemistry, medicine, cosmetics, or as a raw material again in food production by subjecting it to various processes. Especially fruit and vegetable industry wastes contain plenty of pectin and different essential oils. These organic substances in the structure of waste are used not only to enhance the mechanical and barrier feature of packaging materials, but also for the production of biodegradable films (Wikiera et al., 2016; Pereira et al., 2016; Oliveira et al., 2016). In recent years, the production of biodegradable packaging materials obtained from fruit and vegetable wastes has gained importance. These biodegradable films obtained from these are unfortunately not at a level to compete with commercial polymers in terms of mechanical and barrier properties. However, it is possible to develop these properties with nanotechnology applications (Da Silva et al., 2017). Considering the mechanical and barrier properties of commercial polymers, although they are suitable for use as packaging materials, interest in natural polymers is increasing because they are not sustainable and biodegradable. However, films obtained from natural polymers show poor barrier and mechanical properties. In this respect, biodegradable polymers are not yet competitive with commercial polymers. For this reason, developed biopolymers are obtained by supporting many biopolymers with organic or inorganic additives (Romero-Bastida et al., 2016; Grigoriadi et al., 2015). By integrating nanoparticles into biopolymers, the negative mechanical and barrier properties of biodegradable films can be eliminated, and new materials with completely different properties can be developed. Polymer/clay composites improve barrier properties of thin films (Trifol et al., 2016; Staroszczyk et al., 2017). It is stated in the literature that bionanocomposites and nanoparticles have an inhibitory effect on the growth of some bacteria (Ahmed et al., 2017), act as a carrier of antimicrobial substances (Siripatrawan and Vitchyakitti, 2016), and directly form an antimicrobial film (Reesha et al., 2015). Antimicrobial bionanocomposite films are formed by using fillers such as chitosan (Al-Naamani et al., 2016), nano-silver (Li et al., 2017), zinc oxide (Sanuja et al., 2015) and titanium dioxide (Long et al., 2014). The use of food industry wastes in the production of biodegradable films has recently been one of the topics of interest in terms of environmentalist approach. The components contained in the waste can add different properties to the packaging material such as elasticity, strength, biodegradability, transparency and antimicrobial activity.

Orange peels are some waste rich in pectin content, and when the films obtained by using the powder form of this waste were dried, seen that the films obtained from orange peel powder, values close to the tensile strength of commercial polymers such as low density polyethylene, high density polyethylene, polytetrafluorethylene, polypropylene and polystyrene (16-32 MPa) were obtained (Bátori et al., 2017). Another industrial output evaluated for use in bioplastic production is soybean waste. Soybean is a raw material that is processed in large quantities and produces excessive waste. Therefore, it is obtained economically at a lower cost than other bioplastic produced materials (Kowalczyk and Piorkowska, 2012). In a study on the evaluation of food industry waste, a film was obtained by mixing lemon peels and potato pulp at different rates. The amount of potato pulp released during the production of potato chips is between 12-20% of the total production. Potato pulp is rich in starch, cellulose, hemicellulose and fermentable sugar content and is a potential waste for biopolymer film production (Rogols et al., 2003). Lemon, on the other hand, is a product that is usually consumed fresh or processed into fruit juice, and its peel is very rich in flavonoids, pectin and essential oils.

In addition, antioxidant and antimicrobial properties of lemon pulp are also mentioned in the literature (Chutia et al., 2009). Pomegranate is a raw material that is generally processed into products such as pomegranate juice or jam, after processing, approximately 55% pulp is produced. In addition to its antioxidant and antibacterial properties, it has a pulp rich in pectin, tannin and moisture (Canbolat et al., 2014). Studies are carried out to obtain a film from pomegranate peels (Oliveria et al., 2016). After the banana processing process, up to 30% of the fruit is exposed. Banana peel is a waste not only rich in moisture, protein, pectin and potassium, but also rich in dietary fiber (cellulose), antioxidant and phenolic compounds (González-Montelongo et al., 2010). In a study conducted by adding pectin and cellulose nanocrystals obtained from banana peels to the film formulation, it was desired to improve physical and barrier properties. This study shows that nanofillers can improve the mechanical and barrier properties of biodegradable films (Oliveira et al., 2017). Pectins, which are frequently used to produce biodegradable films, can be supported with nano-structured fillers to compete with commercial polymers because they show poor physical and barrier properties. The nanocrystals obtained from cellulose, a natural component, impart rigidity and strength to the films.In the production of biodegradable films, not only fruit and vegetable wastes, but also the shells of nuts are used. It has been stated in the literature that shells containing a high percentage of starch are a suitable material for the production of biothermoplastics (Yuliana et al., 2012). The same is true for walnut shells. In biodegradable films, walnut shells act as absorbent (Ding et al., 2013) and reinforcing material (Sarsari et al., 2016). In a study in which starch obtained from cashew shells was used as a film matrix, cellulose obtained from walnut shells was also used as a filler. This study proved that the cellulose obtained from walnut shell shows very good barrier properties against oxygen (Harini et al., 2018).

3. Polyhydroxyalkanoate (PHA)

3.1. Structure and Classification of PHA

Recently, Zhang et al., (2006) have reported about 150 different type of PHAs. They have various types and structures, this diversity is due to the number of carbon atoms, molecular structure and chain lengths they have (Pötter et al., 2006; Muneer, 2020). However, when the carbon chain length is taken into account, three types of PHAs emerge. These; short chain (scl-PHAs), medium chain (mcl PHAs), and long chain (lcl-PHAs) (Sharma et al., 2017; Raza et al., 2018; Muneer et al., 2020). The most common and known member of the polyhydroxyalkanoate family is poly-3-hydroxybutyrate (P3HB), it polymerizes to give a polymeric chain and consists of (R)-3HB repeating unit (monomer) (Sudesh et al., 2000; Amache et al., 2009; Koller et al., 2010; Sharma et al., 2017; Muneer, 2020). Poly(3-hydroxybutyrate), poly(4-hydroxybutyrate), poly(3-hydroxyvalerate), and poly(3-hydroxybutyrate-co-3-hydroxyvalerate) are known scl-PHAs, and they having 3-4 carbon atoms, and they usually can be used in food packaging and disposable products (Hazer and Steinbüchel, 2007; Johnston et al., 2018; Grigore et al., 2019; Muneer et al., 2020).

In the preparation of these PHAs containing 3-hydroxyvalerate (3HV) or 4-hydroxybutyrate (4HB) monomers, copolymers containing a mixture of four carbon chain length subunits and bacteria synthesizing these polymers with valeric acid are used. The incorporation of HV into the PHB polymers results in a less hard and brittle poly(3-hydroxybutyrate-co-3-hydroxyvalerate) (Pachekoski et al., 2009; Sharma et al., 2017). Medium chain length PHAs have 6–14 carbon atoms, and they can be homopolymers like poly(3-hydoxyhexanoate), poly(3-hydroxyoctanoate) or P(3HO) (Hazer et al., 2007; Muneer et al., 2020). They are synthesized by various bacteria with β-oxidation or novo biosynthesis pathway (Coi et al., 2020). mcl-PHAs are flexible and elastic, having low crystallinity with low tensile strength and high elongation-to-break ratios. (Sharma et al., 2017). On the contrary to short chain (scl-PHAs), they are rare, and less used in process of bioplastics (Muneer et al., 2020). The difference between the two classes is mainly due to the substrate specificity. While a eutrophus can polymerize 3HAs consisting of 3-5 carbon atoms, *Pseudomonas oleovorans* can only use 3HAs consisting of 6-14 carbon atoms in PHA synthesis (Khanna and Srivastava, 2005; Anjum et al., 2016).

3.2. Properties of PHA

Hydrophobicity, melting point, glass transition temperature and degree of crystallinity, and some mechanical properties differ depending on the composition of the monomer (Crank et al., 2004; Anjum et al., 2016). Structural differences in the monomers that make up PHAs cause them to differ chemically as well. For example, poly3-hydroxybutyrate (PHB) has good moisture resistance compared to polypropylene, and barrier properties against gases. They also have a high degree of crystallinity, about 55-80%, and form fine crystals with melting points of about 175°C (Winnacker et al., 2019). Considering its tensile and impact strength, UV resistance, oxygen permeability, PHB is similar to isotactic polypropylene. This shows that it has the potential to be a packaging material (Winnacker et al., 2019). Having good resistance to hydrolytic attack, PHAs are insoluble in water and resistant to UV (Raza et al., 2018). In addition to these properties, they are biodegradable in nature (Bugnicourt et al., 2014). The main reason why the degradation of PHAs depends on their species and composition is that they have chiral molecules (Boyandin et al., 2013). The type of polymer, its composition, environmental conditions and microorganism species are effective in the biodegradation of PHAs. It is known that microorganisms produce different PHA-depolymerase enzymes to decompose PHAs and thus they are effective on its (Masood et al., 2014).

Some thermal and mechanical properties such as melting temperature, glass transition temperature, crystallinity, tensile strength, percent elongation determine the quality of PHA (Muneer et al., 2020). Some thermal properties of PHAs are crystallization and heat resistance, and they affect the polymer quality (Nduko and Taguchi, 2021). The glass transition temperature (Tg) for the amorphous phase and the melting temperature (Tm) for the crystalline phase are express (Muneer et al., 2020). Increasing the number of carbons in the side chain from 1 to 7 causes Tg to decrease and Tm to increase. As a result, the melting temperature rises from 45 to 69°C (Gopi et al., 2018; Grigore et al., 2019; Muneer et al., 2020). Medium chain length PHAs are crystalline, have more tensile strength, and high elongation at break. Due to this feature they show that diferent mechanical properties as compare to short chain length PHAs which have a high crystallinity usually 60–80%. Short chain length PHAs are more brittle and stif compare to medium chain length PHAs (Anjum et al., 2016; Muneer et al., 2020). Addition of different monomers or mixing with PHA are methods used to improve the fracture flexibility and elongation of the polymer (Sanhueza et al., 2019; Muneer et al.,

2020). One of the purposes of these processes is to reduce the difficulties created by the lack of flexibility of PHAs in terms of food packaging uses (Plackett and Siro, 2011). Blending with other polymers can decrease brittleness but its not enough to be competitive with fossil fuel-based polymers used for food packaging (Plackett and Siro, 2011). PHBs have similar water vapor permeability to thermoplastics such as PVC or PET, and their properties are seen as potential for food packaging applications. The fact that they do not swell and have lower hydrophilicity is seen as an advantage compared to various biopolymers such as starch and cellulose (Miguel et al., 1997; Plackett and Siro, 2011). In addition to these properties, PHAs also have good barrier properties to some organic solvents. They show relatively high permeability to the moderately polar solvents chloroform, acetone and toluene, while they have lower permeability to methanol, n-hexane, isopropyl ether (Miguel et al., 1997; Padermshoke et al., 2005).

3.3. The Microbial Production of PHA

The first discovered PHA, poly-3-hydroxybutyrate P(3HB), is produced by *Bacillus magaterium*, and French researcher Maurice Lemoigne isolated it between 1923 and 1927 (Laycock et al., 2013; Koller et al., 2019; Muneer et al., 2020), also more than 300 species have been reported to produce these polymers. These species include various Gram-positive and Gram-negative bacteria, fungi, and microalgae (Olivera et al., 2001; Chanprateep, 2010; Anjum et al., 2016). Biodegradable PHAs can be produced by many bacterial species and their different strains (Chodak, 2008; Muneer et al., 2020). Although many bacterial species are capable of producing a variety of biopolymers, only a few have high productivity and high production rate (Koller et al., 2017; Nahar et al., 2019; Muneer et al., 2020). Among these bacteria; *Wauteria eutropha*, *Azotobacter* spp. *Bacillus* sp., *Pseudomonas putida*, *Pseudomonas fuorescens*, *Pseudomonas oleovorans*, *Ralstonia eutropha*, *Cupriavidus necator*, *Burkholderia* sp., *Halomonas* sp., *Haloferax* sp., *Aeromonas* sp., *Thermus thermophilus*, *Hydrogenophagobacter*, *Saxogradanobacteria* de *Saxogradia Erwinia* sp., and recombinant *E. coli* (Koller et al., 2017; Nahar et al., 2019; Anjum et al., 2016). Buhwal et al. (2013) isolated bacteria that accumulate polyhydroxyalkonates (PHA) from pulp, paper and wastewater. The isolates Enterococcus sp. NAP11 and Brevundimonas sp. NAC1 showed maximum PHA production 79.27% and 77.63%, and they considered as good candidates for industrial production of

PHB. Preusting et al. (1993) investigated to high concentration of PHA and high productivity with *Pseudomonas oleovorans* by fed-batch and continuous culture, and they were reached in a continuous mode, and the culture productivity was 11.6 g/l and 0.58 g/(l h), respectively.

Another study by Guo-Qiang et al. (2001) was found to *Pseudomonas stutzeri* 1317 synthesised a variety of PHAs when grown in glucose and/or fatty acids. The use of recombinant *E. coli* is common in PHA production, as in many areas. Recombinant *E. coli* has been used to PHA biosynthesis, to synthesize the biopolymer to extremely high intracellular levels, and to produce the P(3HB-co-3HV) copolymer. As to PHB synthesiswith Recombinant *E. coli*, it depends on the amount of acetyl-CoA available. Some advantages of using recombinant E. coli for PHA production are rapid growth, high cell density, ability to use few inexpensive carbon sources (Khanna and Srivastava 2005). Masood et al., (2017) investigated various parameters on the yield of polyhydroxyalkanoates (PHAs) produced by Bacillus cereus. They determined that B. cereus was able to produce both PHAs copolymer and tercopolymer, and its depending on the type of substrates. Some fungi such as *Aspergillus fumigatus*, *Saccharomyces cerevisiae* and *Yarrowia lipolytica* can be producers (Muneer et al., 2020). Microalgae have the ability to produce pigments, carotenoids, proteins, enzymes, sugars, fatty acids, polysaccharides, vitamins as well as many bioactive compounds. They are ideal for PHA production, but the number of known is limited. Some of those are *Nostoc muscorum, Chlorella minutissima* and *Botryococcus braunii* (Afreen et al., 2021). Cyanobacteria are known to produce PHA by oxygenic photosynthesis, also various studies show that some cyanobacteria have natural capabilities to store PHAs. Although species specific, some Cyanobacteria produce predominantly PHB (de Philippis et al., 1992; Saharan et al., 2014). It is known to produce PHA in Archaea, which need salts to maintain their growth and can optimally tolerate 5% NaCl (w/v). First reported in 1970 from the Dead Sea, designated *Halobacterium marismortui* (Kirk and Ginzburg, 1972; Saharan et al., 2014), these halophilic archaea produce PHB under nutrient abundant carbon sources (Fernandez-Castillo et al., 1986; Saharan et al., 2014). Different bacteria produce different types of PHAs (Rehm, 2009), Fluorescent Pseudomonas species are preferred, because they have ability of mcl-PHA synthases, and can synthesis of PHAs with 6-14 carbon atoms (Kim et al., 2000). Figure 2 shows PHA biosynthesis process scheme.

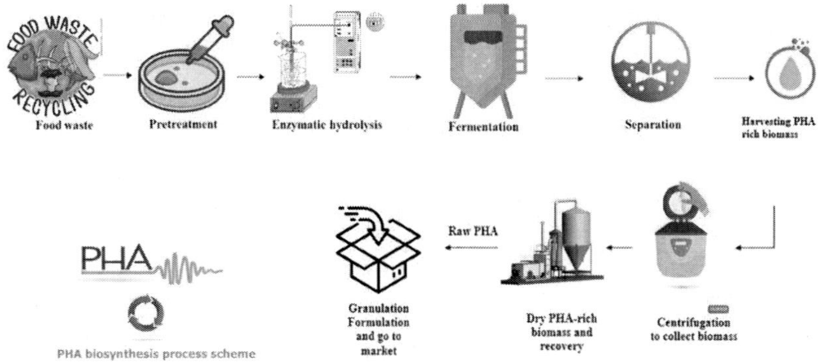

Figure 2. PHA biosynthesis process scheme.

3.4. Carbon Sources for the Production of PHAs

The substrates used in the biosynthesis of PHAs, which are synthesized by bacteria through a metabolic process, are generally small molecules. This is due to the rigid cell walls of bacteria. Large molecules cannot be transported into the cell, they must undergo an extracellular transformation by the microorganism or by a chemical process in order to be used. The substrates that can be used are simple sugars (monosaccharides), triacylglycerol and hydrocarbons. Most microorganisms use simple sugars. Triacylglycerol and hydrocarbon metabolism are less common. Different bacteria for the same substrate can produce PHAs with a different composition (Jiang et al., 2016). Monosaccharides and disaccharides do not need any hydrolysis to be used in the production of PHA, while polysaccharides cannot be fermented unless hydrolyzed first (Jiang et al., 2016). It has been reported by various researchers that *Azotobacter vinelandii, Alcaligenes latus* and *Hydrogenophaga pseudoflava* can hydrolyze sucrose, consisting of a glucose unit linked to a fructose, extracellularly to glucose and fructose, both of which can then be used for cell growth, and thus have the ability to produce PHA (Page et al., 1989; Grothe et al., 1999; Reid et al., 2005; Tanamool et al., 2010; Jiang et al., 2016). Lactose, a disaccharide composed of galactose and glucose, can be used by microorganisms. Whey is preferred as a source (Jiang et al., 2016).

Whey, rich media that is suited for microbial growth, is obtained by precipitation and removal of milk casein during cheese-making processes (Amaro et al., 2019). Being a good PHA producer does not mean that the microorganism can produce PHA directly from whey. Vandamme and

Coenye, (2004) stated that *Cupriavidus necator, Wautersia eutropha* or *Alcaligenes eutrophus* can accumulate up to 80% of its dry weight PHA. *Escherichia coli* cells, which can consume lactose as a solution for microorganisms that can produce high levels of PHA but cannot obtain it from whey, have been modified to express PHA biosynthesis genes from microorganisms that produce high PHA (Ahn et al., 2000; Amaro et al., 2019). Starch, the main component of maize, rice, potatoes, is a polymer of D-glucose and must be enzymatically or acid hydrolyzed to fine maltose and then to glucose for the industrial production of PHAs (Jiang et al., 2016). Triacylglycerols are the main components of animal fats and plant oils (Jiang et al., 2016), they were considered as a cheap and viable source for the biosynthesis of PHAs (Masood et al., 2017). In order for bacteria to use triacylglycerols in the production of PHA, they must be able to secrete lipase. The use of triacylglycerol as a carbon source was determined by Shiotani and Kobayashi in 1993 with Aeromonas caviae (Shiotani et al., 1993; Jiang et al., 2016). Several studies have reported that some Gram-negative bacteria can produce PHA using waste glycerol (Cavalheiro et al., 2012), or oleic acid (Cheema et al., 2012), palm kernel oil (Bhubalan et al., 2010).

3.5. Applications of PHA in the Food Industry

The packaging industry is responsible for the consumption of more than 40% of the total plastic produced worldwide. Food packaging accounts for about half of this percentage. The excessive use and unconscious disposal of disposable plastics cause large amounts of plastic waste. Therefore, packaging materials that are not easily recyclable, in particular, need to be converted from non-biodegradable plastics to biodegradable plastics such as PHAs (Bucci et al., 2005; Geyer et al., 2017). Figure 3 shows the different aplications of PHAs. Having good barrier properties to gases and water, PHAs are hydrophobic, thermoplastic and non-toxic to humans and nature. Because of these material properties, PHAs become suitable for packaging materials (Koller, 2014; Grumezescu and Oprea, 2017). The three most commonly used bio-based plastics for packaging are PLA, Starch-based plastics and Cellophane (Rahman et al., 2019).

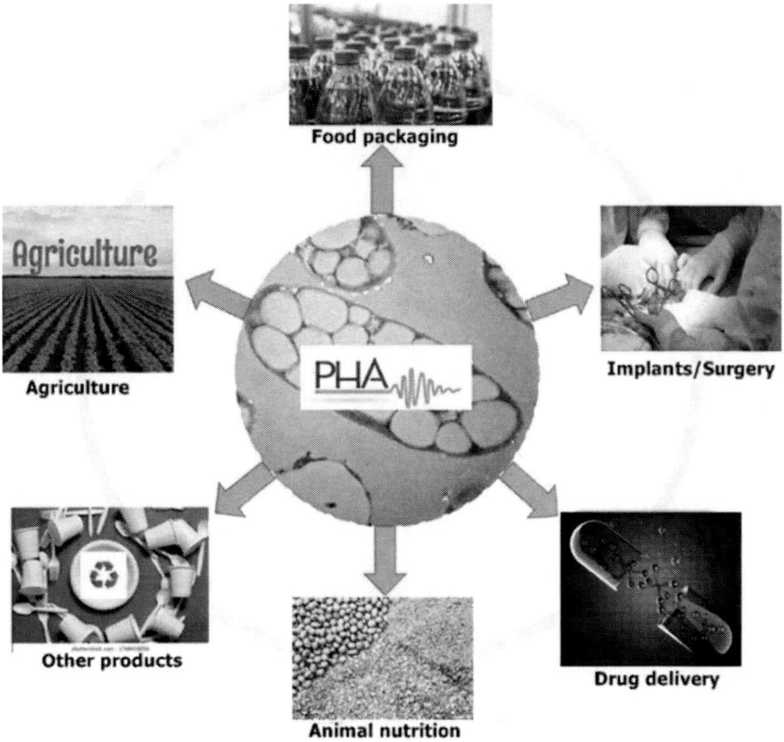

Figure 3. Varied applications of PHAs.

PHAs are an important biodegradable packaging material used as films, foils, etc, and such barrier-coated packaging keeps oxygen and moisture inside the packaging. (Montazer and Harifi, 2017; Basumatary et al., 2020). The film forming feature of PHA provides a protective barrier against oxygen and UV rays. PHA can be used as food packaging material and has similar properties to Polypropylene. PHA are insoluble in water and resistant to water and are hydrophobic. (Guillard et al., 2018; Rastogi and Samyn, 2015). Today, boxes, paper, boards, cardboard are available in the market as coated with PHA (Sharma et al., 2021). PHA is used in the food industry for interior surface coatings in beverage and milk packaging. Studies have been conducted to show that they can be used instead of PP and HDPE in the packaging of foods with high fat content such as ready-made sauces, mayonnaise, margarine cream.(Peelman et al., 2013; Singh and Kaur, 2015; Masood, 2017). Many companies are focusing on food packaging materials that combine conventional packaging materials with nanoparticles, thereby improving the mechanical performance of the packaging material and providing antimicrobial properties. (Montazer and Harifi, 2017) Castro-Mayorga et al. (2016) synthesized an antimicrobial polyhydroxyalkanoate containing silver nanoparticles and an active food that inhibits the growth of Salmonella enterica. They developed active food packaging.

In recent years, pHAs have become available as packaging materials for food-related products such as straws and bottles. Danimer Scientific who was developed first fully biodegradable plastic straw made of Nodax™ PHA in 2018. In addition, by the Cove were developed drink bottles which was made completely of PHAs in 2019, California. Containers mix the soil in about five years; however traditional PET plastic bottles can take around 500 years to decompose. Food and beverage manufacturing company Nestle has recently started a work with Danimer Scientific to develop biodegradable plastic bottles. (Cov, 2019; Boucher, 2019). Yield10 Agricultural Bioscience Company is the largest holding in the PHA industry with an annual production capacity of 50,000 tons (Obulisamy and Mehariya, 2021). Global biodegradable plastic materials are used in a variety of applications, and the packaging industry is the largest contributor, contributing 60 percent of global bioplastic production (Muthusamy and Pramasivam, 2019).

Conclusion and Future Outlook

The use of biodegradable packaging materials is becoming widespread day by day with the increase in environmental awareness, the desire to move away from the use of petroleum-derived packaging materials and the developing technology. The fact that natural resources such as starch, cellulose and protein are the raw materials of a significant portion of biodegradable plastics increases the usability of these packaging materials. The most important reason why bioplastics cannot compete with plastics yet is high research and development costs and low production capacity. In order to reduce the production cost of PHA, it is important to use some cheap carbon sources and to make the best use of the waste. It has been revealed that biodegradable films have weaker properties in terms of barrier and mechanical properties, which are important in food packaging, when compared to commercial plastics. In order to strengthen the barrier and mechanical properties of the films, montmorillonite, cellulose nanocrystals, nanoclay and similar nanofillers can be used as well as biocomposite application. It is clear that biodegradable packaging materials have potential for the food industry. It is thought that with the increase in crude oil prices in the future, renewable raw material sources will gain more importance and the production of environmentally friendly plastics will replace today's plastics. For this reason, it is thought that the use of environmentally friendly plastics, which do not have raw material shortages compared to petrochemical plastics, will be produced in much larger quantities with the help of new processes to be developed and more detailed studies should be done on these materials. By evaluating the food industrial wastes together, it can be achieved to develop high performance packaging materials with stronger mechanical and barrier properties. As biodegradable packaging materials develop, it is expected that the production and use of disposable materials such as biodegradable cups, cutlery and plates will become widespread.

References

Ahmadzadeh, S., Khaneghah, A. M. (2020). Role of green polymers in food packaging. *Encyclopedia of Renewable and Sustainable Materials*; Choudhury, I., Hashmi, S., Eds, 305-319.

Ahmed, J., Hiremath, N., Jacob, H. (2017). Antimicrobial efficacies of essential oils/nanoparticles incorporated polylactide films against L. monocytogenes and S. typhimurium on contaminated cheese. *Int J Food Properties, 20,*53-67.

Ahn, W. S., Park, S. J., Lee, S. Y. (2000). Production of poly (3- hydroxybutyrate) by fed-batch culture of recombinant *Escherichia coli* with a highly concentrated whey solution. *Appl. Environ. Microbiol. 66,*3624–3627. https://doi.org/10.1128/aem. 66.8.3624-3627.2000.

Al-Naamani, L., Dobretsov, S., Dutta, J. (2016). Chitosan-zinc oxide nanoparticle composite coating for active food packaging applications. *Innovat Food Sci Emerging Tech, 38,* 231-237.

Amache, R., Sukan, A., Safari, M., Roy, I., Keshavarz, T. (2009). Advances in PHAs production. *Chem. Eng. Technol, 32*, 931–936.

Aulin, C., Karabulut, E., Tran, A., Wågberg, L., Lindström, T. (2013). Transparent nanocellulosic multilayer thin films on polylactic acid with tunable gas barrier properties. *ACS applied materials & interfaces, 5,* 7352-7359.

Bajpai, P. (2019). *Biobased polymers: properties and applications in packaging*. (pp. 129-138). Elsevier.

Basumatary, I. B., Mukherjee, A., Katiyar, V., & Kumar, S. (2020). Biopolymer-based nanocomposite films and coatings: recent advances in shelf-life improvement of fruits and vegetables. *Critical Reviews in Food Science and Nutrition*, 1-24.

Bátori, V., Jabbari, M., Åkesson, D., Lennartsson, P. R., Taherzadeh, M. J., Zamani, A. (2017). Production of pectin-cellulose biofilms: a new approach for citrus waste recycling. *International Journal of Polymer Science.* 1-9.

Berthet, M. A., Angellier-Coussy, H., Guillard, V., Gontard, N. (2016). Vegetal fiber-based biocomposites: Which stakes for food packaging applications?. *Journal of Applied Polymer Science,* 133, (2).

Bhubalan, K., Rathi, D. N., Abe, H. (2010). Improved synthesis of P(3HB-co-3HV-co-3HHx) terpolymers by mutant Cupriavidus necator using the PHA synthase gene of Chromobacterium sp. USM2 with high affinity towards 3HV. *Polym Degrad Stab 95,*1436–1442. https://doi.org/10.1016/j.polymdegradstab.2009.12.018,

Boonniteewanich, J., Pitivut, S., Tongjoy, S., Lapnonkawow, S., Suttiruengwong, S. (2014). Evaluation of carbon footprint of bioplastic straw compared to petroleum based straw products. *Energy Procedia, 56,* 518-524.

Bostan, K., Aldemir, T., Aydın, A. (2007). Chitosan and its antimicrobial activity. *J Turk Mikrobiyol Cem, 37,*118-127.

Boucher, J. (2021). *Development of biodegradable water bottle*, https://www.foodpackagingforum.org/news/development-of-biodegradable-waterbottle (accessed: November 2021).

Boyandin, A. N., Prudnikova, S. V., Karpov, V. A., Ivonin, V. N., Do, N. L., Nguyen, T. H., Le, T. M. H., Filichev, N. L., Levin, A. L., Filipenko, M. L., Volova, T. G., Gitelson, I.I. (2013). Microbial degradation of polyhydroxyalkanoates in tropical soils. *Int. Biodeterior. Biodegr. 83,* 77–84. http://dx.doi.org/10.1016/j.ibiod.2013.04.014.

Bucci, D. Z., Tavares, L. B. B., Sell, I. (2005). PHB packaging for the storage of food products. *Polymer testing, 24*(5), 564-571.

Bugnicourt, E., Cinelli, P., Lazzeri, A., Alvarez, V., (2014). Polyhydroxyalkanoate (PHA): review of synthesis, characteristics, processing and potential applications in packaging. *Express Polym. Lett. 8,* 791–808. http://dx.doi.org/10.3144/expresspolymlett. 2014.82.

Canbolat, Ö., Kamalak, A., Kara, H. (2014). The effects of urea supplementation on pomegranate pulp (Punica granatum L.) silage fermentation, aerobic stability and in vitro gas production. *J Ankara Üniv Vet, 61,* 217-223.

Castro-Mayorga, J. L., Fabra, M. J., Cabedo, L., & Lagaron, J. M. (2016). On the use of the electrospinning coating technique to produce antimicrobial polyhydroxyalkanoate materials containing in situ-stabilized silver nanoparticles. *Nanomaterials*, *7*(1), 4.

Cavalheiro, J. M. B. T., Raposo, R. S., de Almeida, M. C. M. D. (2012). Effect of cultivation parameters on the production of poly (3- hydroxybutyrate-co-4-hydroxybutyrate) and poly (3-hydroxybutyrate-4-hydroxybutyrate-3-hydroxyvalerate) by Cupriavidus necator using waste glycerol. *Bioresour Technol 111,*391–397. https://doi.org/10.1016/j.biortech.2012.01.176.

Cha, D. S., Chinnan, M. S. (2004). Biopolymer-based antimicrobial packaging: a review. *Critical reviews in food science and nutrition, 44(4),* 223-237.

Chanprateep, S. J. (2010). *Biosci. Bioeng. 110,*621–632.

Cheema, S., Bassas-Galia, M., Sarma, P. M. (2012). Exploiting metagenomic diversity for novel polyhydroxyalkanoate synthases: production of a terpolymer poly (3-hydroxybutyrate-co3-hydroxyhexanoate-co-3-hydroxyoctanoate) with a recombinant Pseudomonas putida strain. *Bioresour Technol 103,*322–328. https://doi.org/10.1016/j.biortech.2011.09.098.

Chodak, I. (2008). Polyhydroxyalkanoates: origin, properties and applications. Monomers Polym Compos from Renew Resour. https://doi.org/10.1016/B978-0-08-045316-3.00022-3.

Chutia, M., Bhuyan, P. D., Pathak, M. G., Sarma, T. C., Boruah, P. (2009). Antifungal activity and chemical composition of Citrus reticulata Blanco essential oil against phytopathogens from North East India. *LWT-Food Science and Technology, 42*(3), 777-780.

Coltelli, M. B., Wild, F., Bugnicourt, E., Cinelli, P., Lindner, M., Schmid, M., Lazzeri, A. (2016). State of the art in the development and properties of protein-based films and coatings and their applicability to cellulose based products: An extensive review. *Coatings, 6*(1), 1.

Cov, The drink bottle made entirely of PHA, https://drinkcove.com (accessed: November 2021)

Crank, M., Patel, M., Marscheider-Weidemann, F., Schleich, J., Husing, B., Angerer, G., (2004). Techno-economic feasibility of large-scale production of bio-based polymers in Europe (PRO-BIP), Final Report Prepared for the European Commission's Institute for Prospective Technological Studies (IPTS).

Da Silva, I. S. V., Neto, W. P. F., Silvério, H. A., Pasquini, D., Zeni Andrade, M., Otaguro, H. (2017). Mechanical, thermal and barrier properties of pectin/cellulose nanocrystal nanocomposite films and their effect on the storability of strawberries (Fragaria ananassa). *Polymers for Advanced Technologies, 28*(8), 1005-1012.

de Philippis, R., Ena, A., Guastini, M., Sili, C., Vincenzini, M. (1992). Factors affecting poly-β-hydroxybutyrate accumulation in cyanobacteria and in purple non-sulfur bacteria. *FEMS Microbiology Reviews, 103,*187–194.

Deniz, E., Yeşilören, G., İşçi, N. Ö. (2015). Biomass And Biofuel Potential Of Food Industry In Turkey, *Food, 40*(1), 47-54.

Ding, D., Zhao, Y., Yang, S., Shi, W., Zhang, Z., Lei, Z., Yang, Y. (2013). Adsorption of cesium from aqueous solution using agricultural residue–walnut shell: equilibrium, kinetic and thermodynamic modeling studies. *Water research, 47*(7), 2563-2571.

Espitia, P. J. P., Du, W. X., de Jesús Avena-Bustillos, R., Soares, N. D. F. F., McHugh, T. H. (2014). Edible films from pectin: Physical-mechanical and antimicrobial properties-A review. *Food hydrocolloids, 35*, 287-296.

Fernandez-Castillo, R., Rodriguez-Valera, F., Gonzalez Ramos, J., Ruiz-Berraquero, F. (1986). Accumulation of poly (β-hydroxybutyrate) by Halobacteria, *Applied and Environmental Microbiology, 51,*214–216.

Geyer, R., Jambeck, J. R., Law, K. L. (2017). Production, use, and fate of all plastics ever made. *Science advances, 3*(7), e1700782.

González-Montelongo, R., Lobo, M. G., González, M. (2010). Antioxidant activity in banana peel extracts: Testing extraction conditions and related bioactive compounds. *Food Chemistry, 119*(3), 1030-1039.

Gopi, S., Kontopoulou, M., Ramsay, B. A., Ramsay, J. A. (2018). Manipulating the structure of medium-chain-length polyhydroxyalkanoate (mcl-PHA) to enhance thermal properties and crystallization kinetics. *Int J Biol Macromol 119,* 1248–1255. https://doi.org/10.1016/j.ijbiomac.2018.08.016

Grigore, M. E., Grigorescu, R. M., Iancu, L., Ion, R. M., Zaharia, C., Andrei, E. R. (2019) Methods of synthesis, properties and biomedical applications of polyhydroxyalkanoates: a review. *J Biomater Sci Polym Ed 30,*695–712. https://doi.org/10.1080/09205 063.2019.1605866.

Grigoriadi, K., Giannakas, A., Ladavos, A. K., Barkoula, N. M. (2015). Interplay between processing and performance in chitosan-based clay nanocomposite films. *Polymer Bulletin*, *72*(5), 1145-1161.

Grothe, E., Moo-Young, M., Chisti, Y. (1999). Fermentation optimization for the production of poly(β-hydroxybutyric acid) microbial thermoplastic. *Enzym. Microb. Technol. 25,* 132–141.

Grumezescu, A. M., Oprea, A. E. (Eds.) (2017). *Nanotechnology applications in Food: Flavor, stability, nutrition and safety*. Academic Press.

Guillard, V., Gaucel, S., Fornaciari, C., Angellier-Coussy, H., Buche, P., & Gontard, N. (2018). The next generation of sustainable food packaging to preserve our environment in a circular economy context. *Frontiers in nutrition*, 121.

Guo-Qiang, C., Jun, X., Qiong, W., Zengming, Z., Kwok-Ping, H. (2001). Synthesis of copolyesters consisting of medium-chain-length - hydroxyalkanoates by *Pseudomonas stutzeri* 1317. *React Funct Polym* 48:107–12.

Hanani, Z. N., Roos, Y. H., Kerry, J. P. (2014). Use and application of gelatin as potential biodegradable packaging materials for food products. *International journal of biological macromolecules*, *71*, 94-102.

Harini, K., Mohan, C. C., Ramya, K., Karthikeyan, S., Sukumar, M. (2018). Effect of Punica granatum peel extracts on antimicrobial properties in Walnut shell cellulose reinforced Bio-thermoplastic starch films from cashew nut shells. *Carbohydrate polymers, 184*, 231-242.

Hazer, B., Steinbüchel, A. (2007). Increased diversifcation of polyhydroxyalkanoates by modifcation reactions for industrial and medical applications. *Appl Microbiol Biotechnol 74,*1–12. https://doi.org/10.1007/s00253-006-0732-842.

Janjarasskul, T., Krochta, J. M. (2010). Edible packaging materials. *Annual Rev Food Sci Tech*, *1,* 415-448.

Johnston, B., Radecka, I., Hill, D., Chiellini, E., Ilieva, V .I., Sikorska, W., Musioł, M., Zięba, M., Marek, A. A., Keddie, D., Mendrek, B., Darbar, S., Adamus, G, Kowalczuk, M. (2018). The microbial production of polyhydroxyalkanoates from waste polystyrene fragments attained using oxidative degradation. *Polymers (Basel).* https://doi.org/10.3390/polym10090957.

Kanmaz, E. Ö., Saral, Ö. (2017). The relationship between antioxidant activities and phenolic compounds in subcritical water extracts from orange peel. *J Food, 42*(5), 485-93.

Karakuş, E., Ayhan, Z. (2019). Production Of Envıronmentally Friendly Biodegradable Packagıng Materials From Food Waste, *J Food, 44*(6), 1008-1019.

Karana, E. (2012). Characterization of 'natural'and 'high-quality'materials to improve perception of bio-plastics. *Journal of Cleaner Production*, *37*, 316-325.

Khanna, S., Srivastava, A. K. (2005). Recent advances in microbial polyhydroxyalkanoates. *Process Biochemistry* 40:607–619.

Khwaldia, K., Arab-Tehrany, E., Desobry, S. (2010). Biopolymer coatings on paper packaging materials. *Comprehensive Rev Food Sci Food Safety, 9*(1), 82-91.

Kim, D. Y., Kim, Y. B., Rhee, Y. H. (2000). Evaluation of various carbon substrates for the biosynthesis of polyhydroxyalkanoates bearing functional groups by *Pseudomonas putida. Int. J. Biol. Macromol. 28,* 23–29. http://dx.doi.org/10.1016/S0141-8130(00)00150-1

Kirk, R. G., Ginzburg, M. (1972). Ultrastructure of two species of halobacterium. *J Ultrasructure Res, 41*,80–94.

Koller, M. (2014). Poly (hydroxyalkanoates) for food packaging: Application and attempts towards implementation. *App. Food Biotech 1*,3-15.

Koller, M. (2017). Production of Polyhydroxyalkanoate (PHA) Biopolyesters by extremophiles?, *MOJ Polym. Sci.* 2,69-85. https://doi.org/10.15406/mojps.2017.01.00011

Koller, M. (2019). Switching from petro-plastics to microbial polyhydroxyalkanoates (PHA): the biotechnological escape route of choice out of the plastic predicament? *EuroBiotech J. 3*,32–44. https://doi.org/10.2478/ebtj-2019-0004.

Koller, M., Bona, R., Chiellini, E., Fernandes, E. G., Horvat, P., Kutschera, C., Braunegg, G. (2008). Polyhydroxyalkanoate production from whey by *Pseudomonas hydrogenovora. Bioresource technology, 99*(11), 4854-4863.

Koller, M., Salerno, A., Dias, M., Reiterer, A., Braunegg, G. (2010). Modern biotechnological polymer synthesis: A review. *Food Technol. Biotechnol, 48*, 255–269.

Kowalczyk, M., Piorkowska, E. (2012). Mechanisms of plastic deformation in biodegradable polylactide/poly (1, 4-cis-isoprene) blends. *Journal of applied polymer science, 124*(6), 4579-4589.

Laycock, B., Halley, P., Pratt, S., Werker, A., Lant, P. (2013). The chemomechanical properties of microbial polyhydroxyalkanoates. *Prog Polym Sci 38*,536–583. https://doi.org/10.1016/j.progpolymsci.2012.06.003.

Li, L., Zhao, C., Zhang, Y., Yao, J., Yang, W., Hu, Q., Cao, C. (2017). Effect of stable antimicrobial nano-silver packaging on inhibiting mildew and in storage of rice. *Food chemistry, 215*, 477-482.

Liu, L. (2006). Bioplastics in food packaging: Innovative technologies for biodegradable packaging. *San Jose State University Packaging Engineering, 13,* 1348-1368.

Lu, D. R., Xiao, C. M., Xu, S. J. (2009). Starch-based completely biodegradable polymer materials. *Express polymer letters, 3*(6), 366-375.

Masood, F. (2017). Polyhydroxyalkanoates in the food packaging industry. In *Nanotechnology Applications in Food* (pp. 153-177). Academic Press.

Masood, F., Abdul-Salam, M., Yasin, T., Hameed, A. (2017). Effect of glucose and olive oil as potential carbon sources on production of PHAs copolymer and tercopolymer by *Bacillus cereus* FA11. *3 Biotech* 7:87 https://doi.org/10.1007/s13205-017-0712-y.

Masood, F., Yasin, T., Hameed, A. (2014). Comparative oxo-biodegradation study of poly3-hydroxybutyrate-co-3-hydroxyvalerate/polypropylene blend in controlled environments. *Int. Biodeterior. Biodegr. 87,* 1–8. http://dx.doi.org/10.1016/j.ibiod.2013.09.023.

Miguel, O., Fernandez-Berridi, M. J., Iruin, J. J. (1997). Survey on transport properties of liquids, vapors, and gases in biodegradable poly(3-hydroxybutyrate) (PHB). *J. Appl. Polym. Sci., 64,*1849-1859.

Mohanty, A. K., Misra, M., Drzal, L. T. (2002). Sustainable bio-composites from renewable resources: opportunities and challenges in the green materials world. *Journal of Polymers and the Environment, 10*(1), 19-26.

Montazer, M., & Harifi, T. (2017). New approaches and future aspects of antibacterial food packaging: from nanoparticles coating to nanofibers and nanocomposites, with foresight to address the regulatory uncertainty. In *Food packaging* (pp. 533-565). Academic Press.

Morillon, V., Debeaufort, F., Blond, G., Capelle, M., Voilley, A. (2002). Factors affecting the moisture permeability of lipid-based edible films: a review. *Critical reviews in food science and nutrition, 42*(1), 67-89.

Muthusamy, M. S., Pramasivam, S. (2019). Bioplastics–an eco-friendly alternative to petrochemical plastics. *Current World Environment, 14*(1), 49.

Müller-Maatsch, J., Bencivenni, M., Caligiani, A., Tedeschi, T., Bruggeman, G., Bosch, M., Sforza, S. (2016). Pectin content and composition from different food waste streams. *Food Chemistry, 201*, 37-45.

Nahar, S., Jeong, M. H., Hur, J. S. (2019). Lichen-associated bacterium, a novel bioresource of polyhydroxyalkanoate (PHA) production and simultaneous degradation of naphthalene and anthracene. *J Microbiol Biotechnol 29,*79–90. https://doi.org/10.4014/jmb.1808.08037 64.

Obulisamy, P. K., & Mehariya, S. (2021). Polyhydroxyalkanoates from extremophiles: a review. *Bioresource Technology*, 325, 124653.

Oliveira, T. Í. S., Rosa, M. F., Cavalcante, F. L., Pereira, P. H. F., Moates, G. K., Wellner, N., Azeredo, H. M. (2016). Optimization of pectin extraction from banana peels with citric acid by using response surface methodology. *Food Chemistry, 198*, 113-118.

Oliveira, T. Í. S., Rosa, M. F., Ridout, M. J., Cross, K., Brito, E. S., Silva, L. M., Azeredo, H. M. (2017). Bionanocomposite films based on polysaccharides from banana peels. *Int J Biol Macromol*, 101, 1-8.

Oliveira, T. Í. S., Zea-Redondo, L., Moates, G. K., Wellner, N., Cross, K., Waldron, K. W., Azeredo, H. M. (2016). Pomegranate peel pectin films as affected by montmorillonite. *Food chemistry*, 198, 107-112.

Olivera, E., Carnicero, D., Jodra, R., Minambres, B., Garcia, B., Abraham, G. (2001) *Environ. Microbiol. 3,*612–618.

Pachekoski, W. M., Agnelli, J. A. M., Belem, L. P. (2009). Thermal, mechanical and morphological properties of poly(hydroxybutyrate) and polypropylene blends after processing. *Mater. Res, 12,* 159–164.

Padermshoke, A., Katsumoto, Y., Sato, H., Ekgasit, S., Noda, I., Ozaki, Y. (2005). Melting behavior of poly(3-hydroxybutyrate) investigated by two-dimensional infrared correlation spectroscopy. Spectrochim Acta-PartA *Mol. Biomol. Spectrosc. 61*,541–550. http://dx.doi.org/10.1016/j.saa.2004.05.004.

Page, W. J., (1989). Production of poly-β-hydroxybutyrate by Azotobacter vinelandii strains UWD during growth in molasses and other complex carbon sources. *Appl. Microbiol. Biotechnol. 31,* 329–333.

Peelman, N., Ragaert, P., De Meulenaer, B., Adons, D., Peeters, R., Cardon, L., Devlieghere, F. (2013). Application of bioplastics for food packaging. *Trends in Food Science & Technology, 32(*2), 128-141.

Peelman, N., Ragaert, P., De Meulenaer, B., Adons, D., Peeters, R., Cardon, L., & Devlieghere, F. (2013). Application of bioplastics for food packaging. *Trends in Food Science & Technology*, 32(2), 128-141.

Pereira, P. H. F., Oliveira, T. Í. S., Rosa, M. F., Cavalcante, F. L., Moates, G. K., Wellner, N., Azeredo, H. M. (2016). Pectin extraction from pomegranate peels with citric acid. *Int J Biol Macromol*, 88, 373-379.

Phillips, G. O., Williams, P. A. (Eds.). (2009). *Handbook of hydrocolloids*. Elsevier.

Podshivalov, A., Zakharova, M., Glazacheva, E., Uspenskaya, M. (2017). Gelatin/potato starch edible biocomposite films: Correlation between morphology and physical properties. *Carbohydrate Polymers*, *157*, 1162-1172.

Popović, S. Z., Lazić, V. L., Hromiš, N. M., Šuput, D. Z., Bulut, S. N. (2018). Biopolymer packaging materials for food shelf-life prolongation. *In Biopolymers for food design* (pp. 223-277). Academic Press.

Pötter, M., Steinbüchel, A. (2006). Biogenesis and structure of polyhydroxyalkanoate granules. *Microbiol Monogr* https://doi.org/10.1007/7171_005

Preusting, H., Hazenberg, W., Witholt B. (1993). Continuous production of poly(3-hydroxyalkanoates) by Pseudomonas oleovorans in a high cell density, two-liquid-phase chemostat. *Enzyme Microb Technol* 15:311–6.

Priya, B., Gupta, V. K., Pathania, D., Singha, A. S. (2014). Synthesis, characterization and antibacterial activity of biodegradable starch/PVA composite films reinforced with cellulosic fibre. *Carbohydrate polymers, 109*, 171-179.

Rahman, R., Sood, M., Gupta, N., Bandral, J. D., Hameed, F., Ashraf, S. (2019). Bioplastics for food packaging: a review. *Int J Curr Microbiol Appl Sci, 8*(3), 2311-2321.

Rastogi, V. K., & Samyn, P. (2015). Bio-based coatings for paper applications. *Coatings*, 5(4), 887-930.

Raza, Z. A., Abid, S., Banat, I. M. (2018) Polyhydroxyalkanoates: characteristics, production, recent developments and applications. *Int. Biodeterior. Biodegrad. 126*,45–56. https://doi.org/10.1016/j. ibiod.2017.10.001.

Reesha, K. V., Panda, S. K., Bindu, J., Varghese, T. O. (2015). Development and characterization of an LDPE/chitosan composite antimicrobial film for chilled fish storage. *Int J Biol Macromol, 79*, 934-942.

Rehm, B. (2009). *Microbial Production of Biopolymers and Polymer Precursors: Applications and Perspectives*. Caister Academic Press, Norfolk, UK.

Reid, S. J., Abratt, V. R. (2005). Sucrose utilisation in bacteria: Genetic organisation and regulation. *Appl. Environ. Microbiol, 67*,312–321.

Rogols, S., Sirovatka, D. M., Widmaier, R. G. (2003). U.S. Patent No. 6,524,639. Washington, DC: U.S. Patent and Trademark Office.

Romero-Bastida, C. A., Tapia-Blácido, D. R., Méndez-Montealvo, G., Bello-Pérez, L. A., Velázquez, G., Alvarez-Ramirez, J. (2016). Effect of amylose content and nanoclay incorporation order in physicochemical properties of starch/montmorillonite composites. *Carbohydrate Polymers*, *152*, 351-360.

Sanhueza, F., Acevedo, S., Rocha, P., Villegas, M., Seeger, R. (2019). Navia, Polyhydroxyalkanoates as biomaterial for electrospun scaffolds. *Int J Biol Macromol 124*:,102–110. https://doi.org/10.1016/j.ijbiomac.2018.11.068.

Sarasa, J., Gracia, J. M., Javierre, C. (2009). Study of the biodisintegration of a bioplastic material waste. *Bioresource technology*, *100*(15), 3764-3768.

Shah, M., Rajhans, S., Pandya, H. A., & Mankad, A. U. (2021). Bioplastic for future: A review then and now. *World J Adv ResRev*, *9*(2), 056-067.

Shankar, S., & Rhim, J. W. (2018). Bionanocomposite films for food packaging applications. *Reference module in food science*, 1, 1-10.

Sharma, V., Sehgal, R., & Gupta, R. (2021). Polyhydroxyalkanoate (PHA): properties and modifications. *Polymer*, *212*, 123161.

Shiotani, T., Kobayashi, G. (1993). Alternating Copolymer and Its Production Method. *Japanese Patent Application* 93049,16 April.

Singh, R. S., & Kaur, N. (2015). Microbial biopolymers for edible film and coating applications. *Advances in biotechnology*, 12, 187-216.

Siracusa, V., Lotti, N. (2018). Biobased plastics for food packaging. *Reference Module in Food Science*. Elsevier.

Siripatrawan, U., Vitchayakitti, W. (2016). Improving functional properties of chitosan films as active food packaging by incorporating with propolis. *Food Hydrocolloids, 61*, 695-702.

Staroszczyk, H., Malinowska-Pańczyk, E., Gottfried, K., & Kołodziejska, I. (2017). Fish gelatin-nanoclay films. Part I: Effect of a kind of nanoclays and glycerol concentration on mechanical and water barrier properties of nanocomposites. *J Food Process Preserv, 41*(5), e13211.

Sudesh, K., Abe, H., Doi, Y. (2000). Synthesis, structure and properties of polyhydroxyalkanoates: biological polyesters. *Prog Polym Sci 25,*1503–1555. https://doi.org/10.1016/S0079-6700(00)00035-6.

Sun, L., Sun, J., Chen, L., Niu, P., Yang, X., & Guo, Y. (2017). Preparation and characterization of chitosan film incorporated with thinned young apple polyphenols as an active packaging material. *Carbohydrate Polymers, 163*: 81-91.

Tanamool, V., Imai, T., Danvirutai, P., Kaewkannetra, P. (2010). Screening, identification and production of polyhydroxyalkanoates (PHAs) by sucrose utilizing microbes isolated from soil environments. *J. Biomech, 150,*361.

Trifol, J., Plackett, D., Sillard, C., Szabo, P., Bras, J., Daugaard, A. E. (2016). Hybrid poly (lactic acid)/nanocellulose/nanoclay composites with synergistically enhanced barrier properties and improved thermomechanical resistance. *Polymer International, 65*(8), 988-995.

Väisänen, T., Haapala, A., Lappalainen, R., Tomppo, L. (2016). Utilization of agricultural and forest industry waste and residues in natural fiber-polymer composites: A review. *Waste Management, 54,* 62-73.

Vandamme, P., Coenye, T. (2004). Taxonomy of the genus Cupriavidus: a tale of lost and found. *Int. J. Syst. Evol. Microbiol. 54,* 2285–2289. https://doi.org/10.1099/ijs.0.63247-0.

Van-Thuoc, D., Quillaguaman, J., Mamo, G., Mattiasson, B. (2008). Utilization of agricultural residues for poly (3-hydroxybutyrate) production by Halomonas boliviensis LC1. *J Appl Microbiol, 104*(2), 420-428.

Venkatachalam, H., Palaniswamy, R. (2020). Bıoplastıc World: A Revıew. *J Adv Sci Res, 11*(3).

Wikiera, A., Mika, M., Starzyńska-Janiszewska, A., Stodolak, B. (2016). Endo-xylanase and endo-cellulase-assisted extraction of pectin from apple pomace. *Carbohydrate polymers, 142*, 199-205.

Xie, F., Yu, L., Liu, H., & Chen, L. (2006). Starch modification using reactive extrusion. *Starch-Stärke, 58*(3-4), 131-139.

Yuliana, M., Huynh, L. H., Ho, Q. P., Truong, C. T., & Ju, Y. H. (2012). Defatted cashew nut shell starch as renewable polymeric material: Isolation and characterization. *Carbohydrate Polymers, 87*(4), 2576-2581.

Zepnik S, Kesselring A, Kopitzky R, Michels C. (2010). *Bioplastics Magazine. 1*:44-47.

Zhang, B., Carlson, R., Srienc, F. (2006). Engineering the monomer composition of polyhydroxyalkanoates synthesized in Saccharomyces cerevisiae. *Appl. Environ. Microbiol. 72*, 536–543. http://dx.doi.org/10.1128/AEM.72.1.536-543.2006.

Zink, J., Wyrobnik, T., Prinz, T., & Schmid, M. (2016). Physical, chemical and biochemical modifications of protein-based films and coatings: An extensive review. *Int J Mol Sci, 17*(9), 1376.

Section II: Optimization and Applications of Food Packaging Films

Chapter 6

Incorporation of Postbiotics into Biodegradable Films for Active Food Packaging

Saber Amiri*, Meysam Rajabi, Parasto Mohammadi and Mahmoud Rezazadeh Bari

Department of Food Science and Technology, Faculty of Agriculture,
Urmia University, Urmia, Iran

Abstract

In Section I, our colleagues have reviewed different types of materials applicable for the design and development of food packaging films. Starting from this chapter, we will focus our discussions more on optimization and applications of food packaging films. As the first chapter of Section II, we will start by discussing how postbiotics can be incorporated into biodegradable films for active food packaging. In fact, consumers are now looking to eat safer foods with lower levels of additives and chemicals and have linked food health to clean-labeled products. It has been important to study natural ingredients to ensure food safety and consumer satisfaction. The number of researches done on functional compounds that try to meet the needs of healthy and safe food is increasing due to the attention and importance of modern consumers. Probiotics, prebiotics, postbiotics, and parabiotics are some of the functional compounds that have received the most attention and have been used as biologically active substances in developing functional products. Postbiotics can be defined as soluble metabolites biosynthesized by probiotics released during growth and fermentation in microbiological culture media, food, or intestines, rich in metabolites. These are biologically active with high and low molecular weight and are

* Corresponding Author's Email: sa.amiri@urmia.ac.ir.

In: Food Packaging
Editor: Wing-Fu Lai
ISBN: 979-8-88697-249-8

very important for food safety. Postbiotics have many food safety functions, including packaging, biodegradation of harmful chemical contaminants (such as mycotoxins), and antibiofilm formed by pathogens. It is expected that this chapter can highlight the potential applications of postbiotics in natural food storage, food packaging, and biofilm control.

Keywords: postbiotics, biodegradable films, bioactivity, food packaging

1. Introduction

The demand has recently shifted to consuming safe, high-quality food with minimal processing. This has doubled the importance of studying natural compounds with different functional properties in the food industry. Antimicrobial agents are natural or chemical compounds that can inhibit the growth of pathogenic microorganisms or spoilage agents in food products [1]. Many microorganisms can be used as a source for the production of antimicrobial compounds, such as probiotics. In addition, probiotics are able to modulate human microbiota. So, the health benefits of these microorganisms such as the lower risk of infection, attract more attention to these microorganisms [2]. Probiotic microorganisms include various species such as lactic acid bacteria (LAB), *Bifidobacteria*, and yeasts [3].

The term "postbiotics" is a relatively new area in functional foods and there is no known definition for it yet. Postbiotics currently refer to bioactive compounds that do not conform to previous definitions of probiotics, prebiotics, and parabiotics. Antimicrobials secreted by probiotic bacteria can be identified as postbiotic, important for food safety. Many LABs are considered probiotics, and their postbiotic compounds have similar or additional health benefits to the consumer [4].

2. Current Understanding of Postbiotics

The live probiotics include many LAB species which have many benefits and if used consistently can balance the gut microbiota. However, using these microorganisms may alter the physicochemical, microbiological, and sensory properties of inoculated food. One of the benefits of adding probiotics to food is that they interact with pathogenic bacteria and compete with them for food and secrete antimicrobials that reduce the growth of pathogens and inhibit them [5]. A probiotic must be alive not to be used for dead bacterial cells or cellular components. Recently, terms such as postbiotics and parabiotics have emerged that have been rapidly used in the field of food science and human health and nutrition [6].

Current knowledge allows us to state that the survival of bacteria or their survival is not necessary to achieve all of their probiotic effects because not all mechanisms are related to their viability. Studies have shown that the survival of probiotic bacteria is not necessary to provide their health-oriented benefits, so new terms such as postbiotics and parabiotics, to indicate their health-oriented benefits beyond probiotics are alive, have been created [7]. Postbiotics can be called metabolic, biogenic, metabolite, supernatant, or cell-free extract. Unlike parabiotics, healthy microbial cells (probiotic or non-probiotic) are inactivated and inappropriate doses bring health benefits to consumers [8]. Although there is currently no known definition for postbiotics, it can be said that when probiotics are fed certain types of fiber molecules, they leave a residue that is generally, these compounds are called postbiotic compounds. There are several types of postbiotics, including lipopolysaccharides, muramyl dipeptide, indole, teichoic acid, and short-chain fatty acids such as butyrate, acetate, and propionates, which are the most widely studied [9].

In general, postbiotics can be based on a structural combination of lipids (such as butyrate, propionate, etc.), carbohydrates (such as galactose), vitamins (such as B vitamins), organic acids (such as propionic acids), and differentiated based on their physiological function, which includes immunomodulatory, anti-inflammatory, cholesterol-lowering, anti-hypertensive, anti-proliferative, and antioxidant effects [10]. Tsilingiri et al. (2012) showed that the use of some probiotics could induce a local inflammatory response that is similar to the response induced by Salmonella. In addition, theoretical concerns related to the administration of live probiotics (such as bloating, the possibility of antibiotic resistance gene transfer) in case reports, clinical trials, and experimental models in patients with major risk factors (such as immunosuppression, preterm infants), and minor risk factors (such as intestinal epithelial dysfunction) have been reported. Therefore, postbiotics may be a more valid and safer alternative to avoid the risks associated with live probiotic bacteria [11]. Recent research shows that postbiotics can have immune and clinical effects. The evidence of the use of postbiotics in healthy individuals to improve overall health and relieve symptoms in a wide range of diseases such as neonatal colic in adults and various causes of diarrhoea has been reported [12]. *In vitro* and *in vivo* studies have shown that postbiotics have anti-inflammatory, antioxidant, anti-proliferative, and immunomodulatory biological activities. These biological activities may promote the health observed in clinical and human trials, but the exact mechanisms of action are unclear. However, postbiotics and parabiotics have a very high potential for developing biotechnological products or developing functional compounds for the food industry [13].

3. Biological Activity of Postbiotics

Peroxidants result from the oxidative metabolism of cells, which in turn produce reactive oxygen species and free radicals, which in turn lead to oxidative damage [14]. Under normal circumstances, cells are protected by antioxidants, intracellular enzymes such as superoxide dismutase (SOD), glutathione peroxidase (GPX), and catalase, which remove peroxides and superoxides to form more reactive and active compounds [15]. Due to the high antioxidant activity of postbiotics, attention has been focused on the use of postbiotics as probiotic-based additives to combat oxidative damage. According to previous studies, postbiotics derived from LAB show higher antioxidant properties. This high antioxidant activity of postbiotics can modulate the antioxidant activity in the host [16]. In a study conducted by Izuddin et al. (2020), the results showed that postbiotics from *Bacillus licheniformis* and *Bacillus velezensis* increased serum antioxidants, SOD, GPX, and decreased serum malondialdehyde (MDA) was treated in rat models [17]. In another study conducted by Jia et al. (2020), the results showed that using *B. licheniformis* postbiotics as a supplement in treated areas could increase antioxidant capacity by increasing SOD enzymes. GPX, as well as increases immunity and effectively improve hepatic fermentation and microbial diversity in the host [18].

Apart from showing antioxidant activity, postbiotics show anti-cancer effects. The incidence of colorectal cancer (CRC) is increasing rapidly. Due to some side effects of conventional CRC therapies, microbial bioactive components, biomolecules derived from postbiotics have become important for adjuvant therapy inpatient with CRC. The "Postbiotics" include a wide range of complex micromolecules and macromolecules, such as cell fragments or metabolites that, when administered, bring a variety of physiological health benefits in sufficient quantities to the host. Postbiotics modulate the composition of intestinal microbiota and the function of the immune system, as well as improve the effectiveness of CRC treatment and reduce its side effects in CRC patients due to their antioxidant, anti-proliferative, anti-inflammatory, and anti-cancer activities. Scientific articles now confirm that postbiotics, with their unique characteristics in terms of clinical (safe origin), technological (sustainability), and economic (low production cost), can be a promising tool for both strategies (Table 1). Prevention and adjuvant therapy should be used in CRC patients (19-27).

Most *in vitro* and *in vivo* studies focus more on postbiotics and parabiotics derived from different strains of *Bifidobacterium*, *Lactobacillus,* and *Saccharomyces.* Although the mechanisms involved in the biological activities of postbiotics and parabiotics are not currently known, scientific evidence and research show that postbiotics and parabiotics have functional and bioactive properties such as antimicrobial and antioxidants activities and modulate the immune system through direct and indirect methods. These systems have inhibitory properties (such as antioxidants, antimicrobials, immunomodulatory) are indirect mechanisms due to postbiotics and parabiotics with different receptors [28-30]. However, it is involved in indirect mechanisms of microbial homeostasis or metabolic and signal pathways of the host, so this mechanism affects specific physiological reactions [31].

Table 1. Anti-cancer properties of some postbiotics

Postbiotic strain	**Derived postbiotic**	**Cancer cell lines**	**Effects**	**References**
Lactobacillus rhamnosus MD	Cell-free supernatant	Caco2,HT29	Decrease cell proliferation	**[20]**
Bacillus polyfermenticus	Indeterminate	HT29,DLD1,caco2	Decrease cell proliferation	**[21]**
Lactobacillus rhamnosus GG	Inactivated cell	Xaco2	Immune response interleukin8 production	**[22]**
Lactobacillus casei ATCC 393	Sonicated cell suspension	HT29	Antiproliferative	**[23]**
Lactobacillus salivarius FP25	Indeterminate	CACO2	Decrease cell proliferation activation of apoptosis	**[24]**
Lactobacillus paracasei SP	Cell wall protein	CACO2	Decrease cell proliferation induction of apoptosis	**[25]**
Bifidobacterium SPP	Cell free supernatant	SW 742	Decrease cell proliferation	**[26]**
Faecalibacterium prausnitzii A2	Cytosolic fraction	Caco2 cell	Immunomodulation	**[27]**

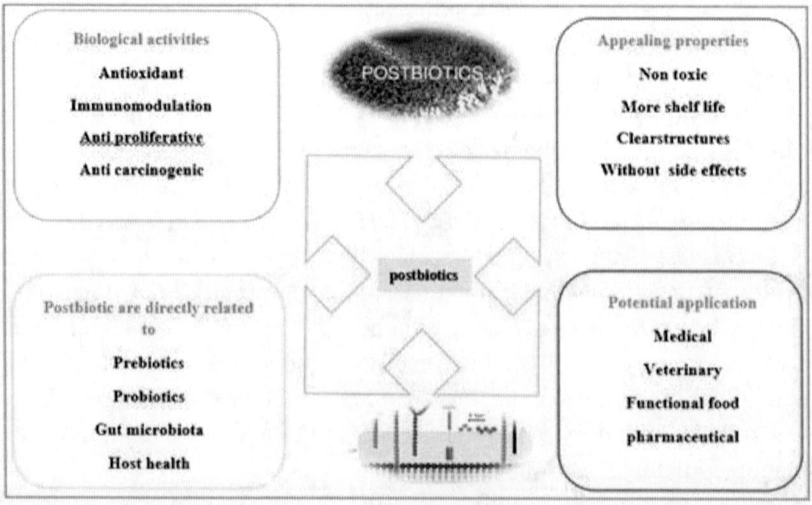

Figure 1. Some postbiotics and their positive systemic effects in the host.

Recently, the results of Chuah et al. (2019), showed that postbiotics derived from 6 strains of *Lactobacillus plantarum* (strains RG11, RL11, RS5, TL1, L-UL4, RG14) selectively select cytotoxic activity through Showed their anti-proliferative activity against cancer cells in a specific strain [32]. Karimi Ardestani et al. (2019) showed that postbiotics derived from *Lactobacillus acidophilus* LA5, *Lactobacillus casei* 431, and *Lactobacillus salivarius* strains had antimicrobial activity against *Listeria monocytogenes* in various food models such as meat and whole milk. They showed that it could be related to its organic acids and possibly similar components such as bacteriocin. Laboratorial studies have shown that postbiotics and parabiotics have antioxidant, anti-proliferative, anti-inflammatory, and immunomodulatory properties that these bioactive compounds may play a role in improving human health. However, there is no precise mechanism for them, and it requires further studies and research in this field [33]. *In vivo* study on the protective effects of parabiotics and postbiotics using an animal model conducted by Aguilar Toala et al. (2019) showed that intracellular administration of *L. casei* CRL431 was able to reduce oxidative stress Aflatoxin B1 was induced in mice by modulating the antioxidant activity of immune system [34]. Also, in another related study, these intracellular postbiotics and other intracellular antioxidant contents from different strains of *Lactobacillus* were able to protect red blood cells against oxidative damage caused by acrylamide. These studies have shown that postbiotics and parabiotics can interact with *in vivo* models as inanimate cells or intracellular components and provide health effects associated with a wide range of cases [35]. Clinical studies on the efficacy of postbiotics and parabiotics under *in vitro* and *in vivo* models have also confirmed this and allowed the use of these compounds in functional food and determine the health effect at the physiological level. It also shows that postbiotics and parabiotics may be effective in improving and preventing chronic diseases due to their biological activities and potential protective effects [36].

4. Food and Medicinal Uses of Postbiotics

Postbiotics and parabiotics are more stable and safe for food and medicine applications than the living microorganisms from which they are derived. The technology used to obtain postbiotics and parabiotics is based on the inactivation or inactivation of microorganisms. In general, in the production of parabiotics, living microorganisms must be exposed to factors that alter the microbial cell structure or physiological function of the cell without disturbing the bacterial membrane and cell wall so that the cell structure remains intact [8]. However, in order to produce postbiotics, we have to disrupt the bacterial membrane and cell wall of the microorganism by using compounds so that we can obtain intracellular metabolites or intracellular components. It is also possible in some cases to obtain a mixture of postbiotics and parabiotics [28]. The fermentation process is one of the most important sources in the natural production of postbiotics. In the process of microbial cell fermentation using prebiotic compounds (as raw materials for the production of some postbiotics) and naturally or in response to environmental conditions, a wide range of postbiotic compounds with different biological properties (antimicrobial, antioxidants, anti-cancer, etc.) and in turn, enrich the food matrix with these beneficial compounds [29].

In recent decades, consumed foods consisting of probiotics, prebiotics, and postbiotics have attracted the attention of many researchers. Probiotics and postbiotics are commonly used in formulations of pharmaceutical compounds or food products. These bioactive compounds can play a key role in maintaining and restoring host health by interacting with host eukaryotic cells. Despite the effectiveness of probiotic products, researchers have introduced the concept of postbiotics to optimize their beneficial effects and provide products with complete safety. Current scientific findings confirm that postbiotics as potential alternatives may be superior in safety to living maternal cells [37]. Many LABs are considered probiotics, and their postbiotic compounds have similar or additional health benefits (to probiotics). The current use of postbiotics is considered in the reduction and biodegradation of some food-related chemical contaminants (such as biogenic amines) [38].

Postbiotics produced from LAB contain various bioactive metabolites such as organic acids, short-chain fatty acids, carbohydrates, antimicrobial peptides, enzymes, vitamins, and immune signaling compounds [38]. Although LAB with antimicrobial properties have an advantage over antibiotics and chemical preservatives, their food applications also face problems [39]. One of the problems with using live cultures of LAB in food is their incompatibility with different matrices and environments that may impede their growth and survival in food. Therefore, postbiotics and parabiotics can be an excellent alternative to live microbial cultures. If used, they allow full utilization of antimicrobial activities and their thermal stability and other metabolites [40]. For industrial applications, postbiotics are superior to probiotics due to their defined chemical composition, safety, ease of use, temperature stability, broad-spectrum antimicrobial activity, and pH-wide stability. The effectiveness of postbiotics in food systems depends on (1) the LAB strains from which the postbiotic is prepared, (2) the type of target microorganism or contaminant, (3) the concentration and shape of application, and (4) the characteristics of the food matrix (8).

A systematic review of the benefits of using postbiotic drugs to prevent and treat common infectious diseases in children under 5 years of age has shown that while maintaining its benefits, it avoids the need to maintain bacterial viability. Furthermore, in terms of reducing the number of common infectious diseases in young children, it can be a good option when probiotics are inappropriate. Postbiotics play a vital role in the maturation of the immune system and, indirectly, the structure of microbiota. The boundaries between probiotics and postbiotics are blurred in some experiments, and their impact on results is often not evaluated [41].

5. Emerging Concepts of Food Biopreservation

Biopreservation is a new way to maintain safety and extend the shelf life of food products. In this method, the storage of certain microorganisms with the help of their metabolites ensures the safety and quality of the target food products [42]. Studies show the high antimicrobial activity of postbiotics, which are of great importance not only against pathogenic microorganisms but also against spoilage agents in the food industry. The resulting antimicrobial compounds, as postbiotics based on molecular weight, belong to low molecular weight compounds such as organic acids and low molecular weight compounds such as bacteriocins. Bacteriocins are heat-resistant antibacterial peptides with good immunity and biosynthesized by different LAB strains in the ribosome. This compound is known as an individual postbiotic with potential applications in the food industry [43]. In addition to the above, some new extracellular biological agents called exosaccharides (EPS) with antibacterial activity have received more attention [44].

LAB is indeed a highly heterogeneous group of gram-negative and catalase-positive microorganisms. LAB mainly includes *Bifidobacteria, Streptococcus, Lactococcus, and Lactobacilli. Lactobacillus* and *Bifidobacterium* species are among the most prominent probiotic bacteria. These agents are generally recognized as safe compounds (GRAS) [45]. LAB are somehow related to probiotics, prebiotics, postbiotics, and parabiotics, and this doubles the importance of studying LAB. LAB can produce beneficial compounds with bioactive properties to improve the safety and quality of food products [46]. Antimicrobial agents produced by LAB are of great importance for food safety and quality, and some of them are currently used in industry, such as nisin, a bacteriocin. In general, antimicrobial compounds derived from LAB can be divided into two groups based on their molecular weight: high molecular weight compounds (such as acetaldehyde, acetone, and H_2O_2) and low molecular weight compounds (such as alcohol, organic acids, bacteriocin, and enzymes) [47]. These compounds and major metabolites derived from LAB are responsible for the antimicrobial action of LAB, of which bacteriocins have received the most attention [48].

Bacteriocins were first found in the culture medium of Escherichia coli. Bacteriocins are peptides or proteins synthesized by ribosomes that have a wide range of antimicrobial activity and are produced by a wide range of bacteria. The use of these antimicrobial molecules in the food industry will meet the needs of the new century consumers who are looking for safe food with minimal chemical interference. Bacteriocins are low molecular weight (less than 10 kDa) synthesized complex proteins that can be produced by ribosomes in gram-positive and gram-negative bacteria and exerted their bactericidal and inhibitory effects against close or strain strains. Shows far from the generator of bacteriocins. The antimicrobial effect of bacteriocins on gram-positive bacteria is much greater than that of gram-negative bacteria, mainly due to an outer lipopolysaccharide layer in the cell wall of gram-negative bacteria. Bacteriocins can be classified into four categories according to various characteristics, including: genetic, biochemical, heat and enzymatic stability, post-translational modifications, and antimicrobial function: 1) Lantibiotics (class I) 2) Small heat-resistant peptides with amphiphilic helical structure (Class II) 3) Larger bacteriocins with complex activity and flexible protein structure (Class III) 4) Bacteriocins containing fat and carbohydrates [49]. Bacteriocins produced by LAB have several desirable properties that make them more important, including the following:

1. Most bacteriocins are generally known as safe compounds (GRAS).
2. They do not have a toxic effect on eukaryotic cells.
3. Easily inactivated by digestive proteases, thus not affecting intestinal microbiota.
4. Stable and active in a wide range of pH.
5. They have potent antimicrobial effects against many pathogens in food and spoilage microorganisms, even in low concentrations.
6. Most antimicrobial peptides have high thermal stability, and their integration with the polymer matrix is possible even through heat-dependent processing methods.

Due to these reasons, in the past decades, researchers have shown great interest in the production and evaluation of active packaging systems using bacteriocins [50].

Biopolymers are polymers obtained from renewable sources. These polymers are divided into three categories according to their origin: natural, synthetic, and microbial [51]. Polymers produced by living microorganisms have attracted much attention due to their potential applications in the food industry. Biopolymers can be produced by bacteria, molds, and yeasts [53]. EPS are polymers that are mainly produced by LAB. At present, these compounds are used to improve food texture, inhibit the growth of pathogens and spoilage agents, and increase the health benefits of food [53]. In addition to producing organic acids, LAB produce many other functional compounds such as short-chain fatty acids, enzymes, antioxidants, vitamins, cofactors, bioactive peptides, and EPS. In the food industry, these valuable biological compounds can be used as gelling, stabilizing, thickening, emulsifying compounds and in food packaging [54].

6. Safety and Health Effects of Postbiotics

Due to related safety issues, many clinical studies have been performed that have confirmed the proper absorption, metabolism, and distribution of postbiotics. These factors can also induce signaling to various host organs [55]. Regarding the safety of postbiotics, some of the potential side effects have been identified in a systematic review. In a study, seven RCT (Randomized controlled trials) with 1740 children examined the role of postbiotics in the prevention and treatment of common infectious diseases in children evaluated for 5 years. According to the results, only three RCTs evaluated the adverse effects of postbiotics; some of the reported side effects were higher rates of vomiting, abdominal distention, and severe dehydration in a group of people who received inactive *L. acidophilus* LB and micronutrients. The remaining RCTs reported no side effects. The authors state that several studies have reported potential adverse effects of prescribing postbiotics [56]. Postbiotics with excellent properties such as specific chemical structure, safety profiles, lack of lasting toxic effects, and stability in the gastrointestinal tract and market conditions can be a reliable alternative to live probiotic microbes, used in food and pharmaceutical industries for health benefits, disease prevention and achieve therapeutic goals [57]. However, maintaining the biological activity and stability of these postbiotic compounds during the preparation process and in delivery systems (medicines and functional foods), as well as related safety parameters, deserves further research. Evidence suggests an integrated link between beneficial intestinal microbiota (probiotics), optimal immune function, and homeostasis in the host. Most of the immunomodulatory effects of probiotics are related to the postbiotic compounds derived from them [29]. Lactic acid-derived postbiotics are associated with epithelial cells and the immune system and cause Immediate defensive response launches in the host [58]. The modulatory effect of the immune system was investigated by the probiotic *Lactobacillus gasseri* and its postbiotics in an exogenous model. Postbiotics derived from it showed a high ability to stimulate the immune system and produce interconline. The type of method of inactivation and maintenance of an effective dose of postbiotics are important factors that can affect the immunomodulatory property of postbiotics [59].

Beneficial microbiota, especially probiotics, are predominant in the intestinal tract and modulate food allergy symptoms in infected individuals by interacting with epithelial cells and the immune system. According to clinical studies, postbiotics, especially short-chain fatty acids such as butyrate, have a significant role in reducing food allergy symptoms in the host [60]. In recent years, much attention has been paid to healthy foods with the content of biological compounds such as probiotics, prebiotics, and postbiotics. These compounds can improve gastrointestinal function and immunity by modifying the composition of intestinal microbiota and developing metabolites derived from it, which has improved conventional therapies and reduced adverse therapeutic effects for cancer patients. The results showed that most beneficial effects were related to postbiotic compounds [6]. Furthermore, lactose intolerance is a common gastrointestinal disorder worldwide. The disease is diagnosed with an HBT test. In one study, prescribing postbiotic compounds in children with lactose intolerance was investigated, and the results showed a significant reduction in the HBT scale of patients. Postbiotic compounds have shown an acceptable potential in reducing the symptoms of the disease and improving the condition of children with lactose intolerance. These compounds can be entered into the industrial arena for this purpose and can be used to improve the health status of individuals [61].

7. Application of Postbiotics in Food Packaging

Active packaging is an innovative concept of food preservation. Recently, due to the demand of consumers to consume safe foods with high shelf life, they have been considered more than before. Active packaging changes the internal conditions of the packaging to a suitable level and increases the shelf life of food. These packages can be made in different shapes by combining functional and bioactive materials. Due to the very high antimicrobial potential of postbiotics, their direct use in the food matrix to maintain its quality and increase its shelf life can have disadvantages. Creating complexes and interactions between postbiotics and dietary compounds can reduce the function and effectiveness of postbiotics. Low mixing of some postbiotics in food is another drawback that can limit the use of postbiotics in food formulations. It is not economically feasible to add postbiotics to the inner part of the food matrix because it causes the growth of microorganisms, especially molds, on the surface of the food. A very promising solution to solve this problem is to use bioactive food packaging with the aim of controlling and maintaining safety and quality, as well as increasing the shelf life of food products. Active packaging is a system in which food products, packaging materials and the environment interact positively to increase the shelf life of food products [62].

Antimicrobial active packaging is a packaging system in which antimicrobial compounds of animal, plant and microbial origin are included in the package and protect food from destruction by microorganisms during transportation and storage [63]. Integration of bacterial cells with packaging materials can enhance the antimicrobial potential of packaging through the production and release of antimicrobials or competition between probiotics and spoilage microorganisms and pathogens at the food surface [36]. However, the direct use of live microorganisms in packaging materials has disadvantages. The short and low viability of microbial cells reduces the antimicrobial potential of active packaging during storage. The viability of probiotics in bioactive packaging and the secretion of antimicrobial compounds depends on environmental conditions such as relative humidity, temperature, light, etc. [64]. Therefore, according to the above, postbiotics can be a good alternative for living microorganisms to remove barriers. However, many studies have been done on EPS (such as cellulose, pololan, gelan, and xanthan) derived from bacteria other than LAB and their application in food packaging. However, studies indicate that these compounds may provide more advantages in food packaging. EPS from LAB include dextran, levan, kefiran, and hyaluronic acid [65].

Each food film or coating comprises three different components, including structural polymers, solvents, and additives. EPS can be used both in structure and as an additive with various functions in developing edible films. The use of EPS alone or in combination with other functional substances in food films can increase not only the safety and quality of food products but also increase their shelf life [66]. A study by Razavi et al. (2020) entitled "The production of antimicrobial packaging nano paper to protect minced meat using lyophilized postbiotics impregnated with *ex-situ* method in bacterial nanocellulose (BNC)" have reported follow results. Results showed that the antimicrobial activity of bacterial nanocellulose (P-BNC)-based postbiotic films against *L. monocytogenes* showed that P-BNC at a concentration of 21:21% and a saturation time of 28 minutes showed a significant reduction. The count showed *L. monocytogenes* in minced meat at the end of the storage period (9 days at 4°C). During this study, P-BNC film in minced meat showed a significant reduction of mesophiles, psychrophiles, and total bacterial count compared to the control sample.

The antimicrobial potential of LAB has been linked to some major compounds such as bacteriocins, lactic acid, and hydrogen peroxide, and some minor compounds such as bacteriocins, cyclopentane, and benzoic acid that are secreted during bacterial growth [67]. In a study by Moradi et al. (2019)

entitled "Study of functional and antibacterial properties (CFS) of cell-free supernatants or postbiotics derived from *L. acidophilus*, *L. casei* 431, and *L. salivarius* against *L. monocytogenes* under *in vitro* condition," it showed that most of the antimicrobial activity of CFS studied was associated with the compounds Pyrrolo [1,2a] pyrazine-1,4 dione and, the presence of organic acids. Postbiotics studied in this study showed antibacterial activity against *L. monocytogenes* that this activity was highly dependent on the type of postbiotic and contact time. The minimum effective concentrations of postbiotics (MEC) have been reported in different food models. Studies have shown that this index was significantly lower for postbiotics derived from *L. salivarius*. The results of various studies indicate that most of the antimicrobial activity of lactic acid bacteria is due to the secretion of substances such as bacteriocins, organic acids, and hydrogen peroxide, which inhibits the growth of pathogens, especially gram-positive bacteria. The results of this study indicate that the antimicrobial activity of postbiotics derived from *L. acidophilus* LA5 is not associated with the production of bacteriocins, while *L. salivarius* bacteriocin produces organic acids and bacteriocin-like substances responsible for its potent antimicrobial activity. In another study, Moradi et al. (2019) investigated "the antimicrobial properties of cell-free supernatant (CFS) derived from *L. salivarius* (LS-BU2) against *E. coli* and increased the shelf life of ground beef", and the following results were obtained. The antimicrobial properties of these compounds were evaluated using the agar disk diffusion method against *E. coli* [68]. Recently, Mirnejad et al. (2013) also showed that *L. casei431* showed the lowest anti-listeria activity among the various postbiotics studied, indicating that weak antibacterial activity has been attributed to the presence of organic acids [69].

As far as the use of postbiotics in food packaging is concerned, the antibiofilm effect of postbiotics, indeed, should not be ignored. Biofilms are a group-bound population created by a population of microorganisms on a specific environment or surface that can greatly increase the level and strength of microbial contamination. In other words, biofilms are organized structures of microbial cells trapped in an extracellular matrix with variable density and composition [70]. In nature, most biofilms are composed of primary microbial binding and then expand by forming cell clusters, causing the growth and stabilization of microcolonies. Biofilms are significantly resistant to antibiotics and host defense mechanisms and pose a major challenge in treating and eradicating infections. For example, the severity of Pseudomonas aeruginosa infections in humans relates to their ability to form biofilms [71].

Biofilms are one of the most important issues in the quality control and safety of food products [72]. Various studies have shown that postbiotics derived from lactic acid bacteria can be used as an antibiofilm [73]. Two mechanisms can be expressed to counteract postbiotics with biofilms: 1) prevent biofilm formation, 2) remove extended biofilm. Postbiotics affect biofilms because of their compounds, including bacteriocins, organic acids, biosurfactants, exopolysaccharides, and biofilms. Postbiotics can affect both strategies by type, even simultaneously [73]. The following study was performed by Kouhestani et al. (2018) on the stability of antimicrobial properties of cell-free suspension (CFS) from Lactobacillus acidophilus LA5 and Lactobacillus casei 431 against Staphylococcus. In this study, the effectiveness of CFS obtained from the mentioned lactic acid bacteria against the biofilm formed by the target bacteria was investigated using a fluorescence microscope. The results showed that the inhibitory region by CFS of L. acidophilus (50.26 mm) was more than L. casei (37.06 mm). The addition of both CFSs in a concentration-dependent process removed the biofilm formed on both surfaces tested. It was also reported that the removal property of CFS biofilm from L. acidophilus was generally higher than that of L. casei, which was confirmed by fluorescence microscopy [74]. Therefore, using probiotic species (including LAB) as antibacterial compounds to control the growth of foodborne pathogens can be very effective.

Conclusion

Some postbiotics have been introduced in medicine, veterinary medicine, and food and show the unique ability to prevent and treat some diseases, improve animal health and develop functional foods. In this regard, advanced functional food science leads to the formation of a product. New healthy, including probiotic products with beneficial effects on host health. On the other hand, postbiotic components must be stable to live probiotic cells and have safe properties, so they can be a safe alternative to live probiotic cells in functional foods. Postbiotics and parabiotics refer to non-living microorganisms and bacterial-free extracts that can benefit to the host by providing additional biological activity to the probiotics. These biological activities can play a role in promoting the health observed in clinical trials. However, despite the investigations, the mechanisms of action and signaling pathways involved have not been fully defined. Therefore, in the future, more studies and studies will be needed to determine the issue more accurately. In addition, development of active edible films as probiotics carriers will also be a direction that is worth paying special attention to because it can significantly enhance the efficiency and versatility in food preservation. This will be discussed in the next chapter.

References

[1] Amiri S., Kohneshahri S. R., Nabizadeh F. The effect of unit operation and adjunct probiotic culture on physicochemical, biochemical, and textural properties of Dutch Edam cheese. *LWT.* 2021 Nov 24:112859.

[2] Vickers, N. J. (2017). Animal communication: when i'm calling you, will you answer too? *Current biology*, 27(14), R713-R715.

[3] Moradi, M., Molaei, R., & Guimarães, J. T. (2021). A review on preparation and chemical analysis of postbiotics from lactic acid bacteria. *Enzyme and Microbial Technology,* 143, 109722.

[4] Amiri, S., Mokarram, R. R., Khiabani, M. S., Bari, M. R., & Khaledabad, M. A. (2022). Characterization of antimicrobial peptides produced by Lactobacillus acidophilus LA-5 and Bifidobacterium lactis BB-12 and their inhibitory effect against foodborne pathogens. *LWT*, *153*, 112449.

[5] Amiri S, Rezazadeh Bari M, Alizadeh Khaledabad M, Rezaei Mokarram R, Sowti Khiabani M. Co-production of parabiotic metabolites by Lactobacillus acidophilus LA5 and Bifidobacterium animalis subsp. lactis BB12 in dairy effluents. *Chemical Review and Letters.* 2021 Apr 1;4(2):66-76.

[6] Aguilar-Toalá, J. E., Garcia-Varela, R., Garcia, H. S., Mata-Haro, V., González-Córdova, A. F., Vallejo-Cordoba, B., & Hernández-Mendoza, A. (2018). Postbiotics: An evolving term within the functional foods field. *Trends in Food Science & Technology*, 75, 105-114.

[7] Amiri S, Teymorlouei MJ, Bari MR, Khaledabad MA. Development of Lactobacillus acidophilus LA5-loaded whey protein isolate/lactose bionanocomposite powder by electrospraying: A strategy for entrapment. *Food Bioscience.* 2021 Oct 1; 43:101222.

[8] Barros, C. P., Guimarães, J. T., Esmerino, E. A., Duarte, M. C. K., Silva, M. C., Silva, R., & Cruz, A. G. (2020). Paraprobiotics and postbiotics: concepts and potential applications in dairy products. *Current Opinion in Food Science*, 32, 1-8.

[9] Moludi, J., Alizadeh, M., Yagin, N. L., Pasdar, Y., Nachvak, S. M., Abdollahzad, H., & Tabaei, A. S. (2018). New insights on atherosclerosis: A cross-talk between endocannabinoid systems with gut microbiota. *Journal of cardiovascular and thoracic research,* 10(3), 129.

[10] Izuddin, W. I., Loh, T. C., Foo, H. L., Samsudin, A. A., & Humam, A. M. (2019). Postbiotic L. plantarum RG14 improves ruminal epithelium growth, immune status and upregulates the intestinal barrier function in post-weaning lambs. *Scientific reports*, *9*(1), 1-10.

[11] Tsilingiri, K., & Rescigno, M. (2013). Postbiotics: what else?. *Beneficial microbes, 4*(1), 101-107.

[12] Fleming, P. F., Berrington, J. E., & Jacobs, S. E. (2019). Addressing safety concerns of probiotic use in preterm babies. *Early Human Development, 135*, 72-74.

[13] Schmidt, E. K., Torres-Espin, A., Raposo, P. J., Madsen, K. L., Kigerl, K. A., Popovich, P. G., & Fouad, K. (2020). Fecal transplant prevents gut dysbiosis and anxiety-like behaviour after spinal cord injury in rats. *PloS one,* 15(1), e0226128.

[14] Schogor, A. L. B., Palin, M. F., dos Santos, G. T., Benchaar, C., Lacasse, P., & Petit, H. V. (2013). Mammary gene expression and activity of antioxidant enzymes and oxidative indicators in the blood, milk, mammary tissue and ruminal fluid of dairy cows fed flax meal. *British Journal of Nutrition*, 110(10), 1743-1750.

[15] Sohrabpour S, Rezazadeh Bari M, Alizadeh M, Amiri S. Investigation of the rheological, microbial, and physicochemical properties of developed synbiotic yogurt containing Lactobacillus acidophilus LA-5, honey, and cinnamon extract. *Journal of Food Processing and Preservation.* 2021 Apr;45(4): e15323.

[16] Shimamura, S., Abe, F., Ishibashi, N., Miyakawa, H., Yaeshima, T., Araya, T., & Tomita, M. (1992). Relationship between oxygen sensitivity and oxygen metabolism of Bifidobacterium species. *Journal of Dairy Science*, 75(12), 3296-3306.

[17] Izuddin, W. I., Humam, A. M., Loh, T. C., Foo, H. L., & Samsudin, A. A. (2020). Dietary postbiotic Lactobacillus plantarum improves serum and ruminal antioxidant activity and upregulates hepatic antioxidant enzymes and ruminal barrier function in post-weaning lambs. *Antioxidants,* 9(3), 250.

[18] Jia, P., Cui, K., Ma, T., Wan, F., Wang, W., Yang, D. & Diao, Q. (2018). Influence of dietary supplementation with Bacillus licheniformis and Saccharomyces cerevisiae as alternatives to monensin on growth performance, antioxidant, immunity, ruminal fermentation and microbial diversity of fattening lambs. *Scientific reports*, 8(1), 1-10.

[19] Sadreddini, S., Baradaran, B. Aghebati-Maleki, A. Sadreddini, S. Shanehbandi, D. Fotouhi, A. and Aghebati-Maleki L. 2019. Immune checkpoint blockade opens a new way to cancer immunotherapy. *Journal of Cellular Physiology* 234 (6):8541–8549. doi: 10.1002/jcp.27816.

[20] Sharma, M., Chandel, D., & Shukla, G. (2020). Antigenotoxicity and cytotoxic potentials of metabiotics extracted from isolated probiotic, Lactobacillus rhamnosus MD 14 on Caco-2 and HT-29 human colon cancer cells. *Nutrition and cancer*, 72(1), 110-119.

[21] Ma, E. L., Choi, Y. J., Choi, J., Pothoulakis, C., Rhee, S. H., & Im, E. (2010). The anticancer effect of probiotic Bacillus polyfermenticus on human colon cancer cells is mediated through ErbB2 and ErbB3 inhibition. *International journal of cancer,* 127(4), 780-790.

[22] Van Hoffen, E., Korthagen, N. M., De Kivit, S., Schouten, B., Bardoel, B., Duivelshof, A., & Willemsen, L. E. (2010). Exposure of intestinal epithelial cells to UV-killed Lactobacillus GG but not Bifidobacterium breve enhances the effector immune response *in vitro. International archives of allergy and immunology*, 152(2), 159-168.

[23] Tiptiri-Kourpeti, A., Spyridopoulou, K., Santarmaki, V., Aindelis, G., Tompoulidou, E., Lamprianidou, E. E., & Chlichlia, K. (2016). Lactobacillus casei exerts anti-proliferative effects accompanied by apoptotic cell death and up-regulation of TRAIL in colon carcinoma cells. *PloS one*, 11(2), e0147960.

[24] Motalebi Moghanjougi Z, Rezazadeh Bari M, Alizadeh Khaledabad M, Amiri S, Almasi H. Microencapsulation of Lactobacillus acidophilus LA-5 and Bifidobacterium animalis BB-12 in pectin and sodium alginate: A comparative study on viability, stability, and structure. *Food Science & Nutrition*. 2021 Sep;9(9):5103-11.

[25] Nozari, S., Y. Faridvand, A. Etesami, M. Ahmad Khan Beiki, S. A. Miresmaeili Mazrakhondi, and J. Abdolalizadeh. 2019. Potential anticancer effects of cell wall protein fractions from Lactobacillus paracasei on human intestinal Caco-2 cell line. *Letters in Applied Microbiology* 69 (3):148–54. doi: 10.1111/lam.13198.

[26] Bahmani, S., Azarpira, N., & Moazamian, E. (2019). Anti-colon cancer activity of Bifidobacterium metabolites on colon cancer cell line SW742. *The Turkish Journal of Gastroenterology*, 30(9), 835.

[27] Jafari, B., R. A. K. Nejad, F. Vaziri, and S. D. Siadat. 2019. Evaluation of the effects of extracellular vesicles derived from Faecalibacterium prausnitzii on lung cancer cell line. *Biologia* 74 (7):889–98. doi: 10. 2478/s11756-019-00229-8.

[28] Aguilar-Toalá, J. E., Estrada-Montoya, M. C., Liceaga, A. M., Garcia, H. S., González-Aguilar, G. A., Vallejo-Cordoba, B., & Hernández-Mendoza, A. (2019). An insight on antioxidant properties of the intracellular content of Lactobacillus casei CRL-431. LWT, 102, 58-63.

[29] Xu, Y., Zang, J., Regenstein, J. M., & Xia, W. (2021). Technological roles of microorganisms in fish fermentation: a review. *Critical reviews in food science and nutrition*, 61(6), 1000-1012.

[30] de Almada, C. N., Almada, C. N., Martinez, R. C., & Sant'Ana, A. S. (2016). Paraprobiotics: evidences on their ability to modify biological responses, inactivation methods and perspectives on their application in foods. *Trends in food science & technology*, 58, 96-114.

[31] Sharma, M., & Shukla, G. (2016). Metabiotics: one step ahead of probiotics; an insight into mechanisms involved in anticancerous effect in colorectal cancer. *Frontiers in microbiology*, 7, 1940.

[32] Chuah, L. O., Foo, H. L., Loh, T. C., Alitheen, N. B. M., Yeap, S. K., Mutalib, N. E. A., & Yusoff, K. (2019). Postbiotic metabolites produced by Lactobacillus plantarum strains exert selective cytotoxicity effects on cancer cells. *BMC complementary and alternative medicine*, 19(1), 1-12.

[33] Karimi Ardestani, S., Tafvizi, F., & Tajabadi Ebrahimi, M. (2019). Heat-killed probiotic bacteria induce apoptosis of HT-29 human colon adenocarcinoma cell line via the regulation of Bax/Bcl2 and caspases pathway. *Human & experimental toxicology*, 38(9), 1069-1081.

[34] Aguilar-Toalá, J. E., Astiazarán-García, H., Estrada-Montoya, M. C., Garcia, H. S., Vallejo-Cordoba, B., González-Córdova, A. F., & Hernández-Mendoza, A. (2019). Modulatory effect of the intracellular content of Lactobacillus casei CRL 431

against the aflatoxin B 1-induced oxidative stress in rats. *Probiotics and antimicrobial proteins*, 11(2), 470-477.

[35] Cuevas-González, P. F., Aguilar-Toalá, J. E., García, H. S., González-Córdova, A. F., Vallejo-Cordoba, B., & Hernández-Mendoza, A. (2020). Protective effect of the intracellular content from potential probiotic bacteria against oxidative damage induced by acrylamide in human erythrocytes. *Probiotics and antimicrobial proteins,* 1-12.

[36] Maruyama, M., Abe, R., Shimono, T., Iwabuchi, N., Abe, F., & Xiao, J. Z. (2016). The effects of non-viable Lactobacillus on immune function in the elderly: a randomised, double-blind, placebo-controlled study. *International journal of food sciences and nutrition,* 67(1), 67-73.

[37] Salminen, S., & Szajewska, H. (2022). Postbiotics. In *Textbook of Pediatric Gastroenterology, Hepatology and Nutrition* (pp. 733-736). Springer, Cham.

[38] Rodríguez-González, F., de Oca, M. P. M., Ávila-Reyes, S. V., Camacho-Díaz, B. H., Alamilla-Beltrán, L., Jiménez-Aparicio, A. R., & Arenas-Ocampo, M. L. (2019). A rheological study of chicory and agave tequilana fructans for use in foods. *LWT,* 115, 108137.

[39] Shakiba, N., & Zandstra, P. W. (2017). Engineering cell fitness: lessons for regenerative medicine. *Current Opinion in Biotechnology*, 47, 7-15.

[40] O'bryan, C. A., Crandall, P. G., Ricke, S. C., Ndahetuye, J. B., & TAYLOR, T. (2015). 7-Lactic acid bacteria (LAB) as antimicrobials in food products: Analytical methods and applications. *Handbook of natural antimicrobials for food safety and quality*, 137-151.

[41] Chua, K. J., Kwok, W. C., Aggarwal, N., Sun, T., & Chang, M. W. (2017). Designer probiotics for the prevention and treatment of human diseases. *Current Opinion in Chemical Biology*, *40*, 8-16.

[42] Remuzzi, A., & Bozzetto, M. (2017). Biological and physical factors involved in the maturation of arteriovenous fistula for hemodialysis. *Cardiovascular engineering and technology*, 8(3), 273-279.

[43] Šušković, J., Kos, B., Beganović, J., Leboš Pavunc, A., Habjanič, K., & Matošić, S. (2010). Antimicrobial activity–the most important property of probiotic and starter lactic acid bacteria. *Food Technology and Biotechnology*, 48(3), 296-307.

[44] Li, C., Li, W., Chen, X., Feng, M., Rui, X., Jiang, M., & Dong, M. (2014). Microbiological, physicochemical and rheological properties of fermented soymilk produced with exopolysaccharide (EPS) producing lactic acid bacteria strains. *LWT-Food Science and Technology*, 57(2), 477-485.

[45] Amiri, S., Rezazadeh-Bari, M., Alizadeh-Khaledabad, M., Rezaei-Mokarram, R., & Sowti-Khiabani, M. (2021). Fermentation optimization for co-production of postbiotics by Bifidobacterium lactis BB12 in cheese whey. *Waste and Biomass Valorization*, *12*(11), 5869-5884.

[46] Amiri, S., Moghanjougi, Z. M., Bari, M. R., & Khaneghah, A. M. (2021). Natural protective agents and their applications as bio-preservatives in the food industry: An overview of current and future applications. *Italian Journal of Food Science*, *33*(SP1), 55-68.

[47] Amiri, S., Mokarram, R. R., Khiabani, M. S., Bari, M. R., & Alizadeh, M. (2020). Optimization of food-grade medium for co-production of bioactive substances by Lactobacillus acidophilus LA-5 for explaining pharmabiotic mechanisms of probiotic. *Journal of Food Science and Technology*, 1-12.

[48] Rohani, S. M. R., Moradi, M., Mehdizadeh, T., Saei-Dehkordi, S. S., & Griffiths, M. W. (2011). The effect of nisin and garlic (Allium sativum L.) essential oil separately and in combination on the growth of Listeria monocytogenes. *LWT-Food Science and Technology*, 44(10), 2260-2265.

[49] Amiri, S., & Rajabi, M. (2022). An overview of the application of natural antimicrobial compounds from plant, animal and microbial origin in foods. *Journal of food science and technology* (Iran), 18(119), 143-156.

[50] Santos, J. C., Sousa, R. C., Otoni, C. G., Moraes, A. R., Souza, V. G., Medeiros, E. A., & Soares, N. F. (2018). Nisin and other antimicrobial peptides: Production, mechanisms of action, and application in active food packaging. *Innovative Food Science & Emerging Technologies*, 48, 179-194.

[51] Niaounakis, M. (2014). Biopolymers: processing and products. William Andrew.

[52] Nehal, F., Sahnoun, M., Smaoui, S., Jaouadi, B., Bejar, S., & Mohammed, S. (2019). Characterization, high production and antimicrobial activity of exopolysaccharides from Lactococcus lactis F-mou. *Microbial pathogenesis*, 132, 10-19.

[53] Guimarães, J. T., Balthazar, C. F., Silva, R., Rocha, R. S., Graça, J. S., Esmerino, E. A., & Cruz, A. G. (2020). Impact of probiotics and prebiotics on food texture. *Current Opinion in Food Science,* 33, 38-44.

[54] Saadat, Y. R., Khosroushahi, A. Y., & Gargari, B. P. (2019). A comprehensive review of anticancer, immunomodulatory and health beneficial effects of the lactic acid bacteria exopolysaccharides. *Carbohydrate polymers*, 217, 79-89.

[55] Shenderov, B. A. (2013). Metabiotics: novel idea or natural development of probiotic conception. *Microbial ecology in Health and Disease*, 24(1), 20399.

[56] Malagón-Rojas, J. N., Mantziari, A., Salminen, S., & Szajewska, H. (2020). Postbiotics for preventing and treating common infectious diseases in children: a systematic review. *Nutrients*, 12(2), 389.

[57] Amiri, S., Rezazadeh-Bari, M., Alizadeh-Khaledabad, M., Rezaei-Mokarram, R., & Sowti-Khiabani, M. (2021). Fermentation optimization for co-production of postbiotics by Bifidobacterium lactis BB12 in cheese whey. *Waste and Biomass Valorization,* 1-16.

[58] Piqué, N., Berlanga, M., & Miñana-Galbis, D. (2019). Health benefits of heat-killed (Tyndallized) probiotics: An overview. *International journal of molecular sciences,* 20(10), 2534.

[59] Chuang, L., Wu, K. G., Pai, C., Hsieh, P. S., Tsai, J. J., Yen, J. H., & Lin, M. Y. (2007). Heat-killed cells of lactobacilli skew the immune response toward T helper 1 polarization in mouse splenocytes and dendritic cell-treated T cells. *Journal of agricultural and food chemistry*, *55*(26), 11080-11086.

[60] Nowak-Węgrzyn, A., & Chatchatee, P. (2017). Mechanisms of tolerance induction. *Annals of Nutrition and Metabolism*, *70*(Suppl. 2), 7-24.

[61] Rampengan, N. H., Manoppo, J., & Warouw, S. M. (2010). Comparison of efficacies between live and killed probiotics in children with lactose malabsorption. *Southeast Asian J Trop Med Public Health, 41*(2), 474-481.

[62] Vilela, C., Kurek, M., Hayouka, Z., Röcker, B., Yildirim, S., Antunes, M. D. C., & Freire, C. S. (2018). A concise guide to active agents for active food packaging. *Trends in Food Science & Technology*, 80, 212-222.

[63] Amiri S, Mokarram RR, Khiabani MS, Bari MR, Khaledabad MA. Characterization of antimicrobial peptides produced by Lactobacillus acidophilus LA-5 and Bifidobacterium lactis BB-12 and their inhibitory effect against foodborne pathogens. *LWT*. 2022 Jan 1; 153:112449.

[64] Moradi, M., Kousheh, S. A., Almasi, H., Alizadeh, A., Guimarães, J. T., Yılmaz, N., & Lotfi, A. (2020). Postbiotics produced by lactic acid bacteria: The next frontier in food safety. *Comprehensive Reviews in Food Science and Food Safety*, 19(6), 3390-3415.

[65] Mbye, M., Baig, M. A., AbuQamar, S. F., El-Tarabily, K. A., Obaid, R. S., Osaili, T. M., & Ayyash, M. M. (2020). Updates on understanding of probiotic lactic acid bacteria responses to environmental stresses and highlights on proteomic analyses. *Comprehensive Reviews in Food Science and Food Safety*, 19(3), 1110-1124.

[66] Razavi, R., Tajik, H., Moradi, M., Molaei, R., & Ezati, P. (2020). Antimicrobial, microscopic and spectroscopic properties of cellulose paper coated with chitosan sol-gel solution formulated by Epsilon-Poly-L-lysine and its application in active food packaging. *Carbohydrate research*, 489, 107912.

[67] Koohestani, M., Moradi, M., Tajik, H., & Badali, A. (2018). Effects of cell-free supernatant of Lactobacillus acidophilus LA5 and Lactobacillus casei 431 against planktonic form and biofilm of Staphylococcus aureus. *In Veterinary Research Forum* (Vol. 9, No. 4, p. 301). Faculty of Veterinary Medicine, Urmia University, Urmia, Iran.

[68] Moradi, M., Mardani, K., & Tajik, H. (2019). Characterization and application of postbiotics of Lactobacillus spp. on Listeria monocytogenes *in vitro* and in food models. *LWT,* 111, 457-464.

[69] Mirnejad, R., Vahdati, A. R., Rashidiani, J., Erfani, M., & Piranfar, V. (2013). The antimicrobial effect of lactobacillus casei culture supernatant against multiple drug resistant clinical isolates of Shigella sonnei and Shigella flexneri *in vitro*. *Iranian Red Crescent Medical Journal,* 15(2), 122.

[70] Branda, S. S. (2005). Ã. s. Vik, L. Friedman, and R. Kolter. *Biofilms: The matrix revisited*, 20-26.

[71] Paredes, J., Alonso-Arce, M., Schmidt, C., Valderas, D., Sedano, B., Legarda, J., & Arana, S. (2014). Smart central venous port for early detection of bacterial biofilm related infections. *Biomedical microdevices*, *16*(3), 365-374.

[72] Moradi, M., & Tajik, H. (2017). Biofilm removal potential of neutral electrolysed water on pathogen and spoilage bacteria in dairy model systems. *Journal of applied microbiology,* 123(6), 1429-1437.

[73] Moradi, M., Tajik, H., Mardani, K., & Ezati, P. (2019). Efficacy of lyophilized cell-free supernatant of Lactobacillus salivarius (Ls-BU2) on Escherichia coli and shelf

life of ground beef. In *Veterinary Research Forum* (Vol. 10, No. 3, p. 193). Faculty of Veterinary Medicine, Urmia University, Urmia, Iran.

[74] Kim, Y., & Kim, S. H. (2009). Released exopolysaccharide (r-EPS) produced from probiotic bacteria reduce biofilm formation of enterohemorrhagic Escherichia coli O157: H7. *Biochemical and biophysical research communications*, 379(2), 324-329.

Chapter 7

Opportunities and Advantages of Active Edible Films for Delivery of Probiotics

Saber Amiri*, Parviz Ahmadi Gheshlagh, Ayla Elmi Kashtiban and Mahmoud Rezazadeh Bari

Department of Food Science and Technology, Faculty of Agriculture, Urmia University, Urmia, Iran

Abstract

In the previous chapter, we talked about how postbiotics can be incorporated into biodegradable films for active food packaging. Based on what we have discussed; in this chapter we will review more deeply how active edible films can be used potentially for delivery of probiotics. Practically, edible film or layer seems to be an effective carrier in microorganism survival, improving the viability of probiotics during the food and storage process. Several parameters, such as the existence of a microorganism, plasticizer type, and basic materials affect the physical and chemical properties of the layers and films. Different methods are used to combine probiotics in these edible films and coatings. These methods directly affect the survival and effective distribution of probiotics in the polymer matrix. In addition, various studies have indicated that the combination of microorganisms with antimicrobial properties applied in food packaging limits the growth of pathogens. However, further studies are needed to explain the effect of different process variables on the development of these edible films and coatings. New methods have been developed to increase the use of microorganisms in edible films and coatings. For instance, a combination of biocontrol microorganisms with specific compounds that can stimulate the secretion

---

* Corresponding Author's Email: sa.amiri@urmia.ac.ir.

In: Food Packaging
Editor: Wing-Fu Lai
ISBN: 979-8-88697-249-8

of microbial metabolites or a combination of probiotics that can maintain sufficient probiotic populations is beneficial to the host. New formulations are being developed, combining emulsion films and coatings with nano-silver to distribute and enhance microbial stability homogeneously. Nevertheless, more investigations about new film formation methods and microorganism composition are required.

Keywords: probiotics, active edible film, prebiotics, active packaging

1. Introduction

Consumers around the world are concerned about a healthy diet. Also, the nutritional benefits of probiotic foods have attracted their attention. Between 2010 and 2011, global product sales increased to 6.21 and 23.24 milliard dollars, respectively. So far, it has reached approximately 96 milliard dollars [1]. This has attracted the attention of food researchers regarding the production of new food products containing probiotics and their effects on human health. The term "probiotics" refers to microorganisms that, if consumed in sufficient quantities, have health benefits for the host [1-4]. A real probiotic should preferably have a human origin, and be safe and free of carriers capable of transmitting resistance to antibiotics and pathogens or toxicity. Furthermore, a probiotic must have a large capacity to survive in intestinal conditions (acidic pH, enzymes, and bile salts, etc.). Also, a probiotic must have a contradiction against pathogens and stimulate the immune system. Ultimately, it should have provable beneficial effects on the host, and the maintenance of activity, viability, and effectiveness of probiotic growth in technological treatment must be demonstrated [1, 5.[

Ilya Ilyich Mechnikov conducted the first research on lactic acid-producing bacteria (LAB) and their health effects on human beings, and the results of this first research indicated that consuming LAB improves the host's health [1, 6]. LABs are a heterogeneous group of microorganisms that, are often found in the human gut and are imported by eating fermented foods like yogurt, other fermented milk products, various cheeses, and fermented meat by-products. *Bifidobacterium*, *Enterococcus*, *Lactobacillus*, and *Saccharomyces* strains are the most widely used probiotic bacteria [1]. However, other strains like *Lactococcus*, *Leoconostoc*, *Pediococcus*, and *Streptococcus* are also used as probiotics [7-9]. In 2014, the International Scientific of Probiotics and Prebiotics announced that developing metabolic by-products, dead microorganisms, or other non-viable microbial-based products has potential. However, they are not in the probiotic category [2]. Nevertheless, several studies have shown that dead bacteria and bacterial molecular components have probiotic properties [5, 10, 11]. In the beginning, probiotics were defined as microorganisms that had beneficial effects through single or mixed cultures on humans and animals by improving the properties of the native microflora of the host digestive system. In other words, these compounds are considered as one or more species of live microorganisms which are beneficial to the host's health when they are sufficiently available to the body. Probiotics are mainly bacteria, and the most commercially available bacteria in the world belong to the *Lactobacillus* and *Bifidobacterium* species. However, *Streptococcus thermophilus* and *Saccharomyces boulardii* are also commercially available in some dairy products [2].

2. The Effect of Probiotics on Human Health

The effects of probiotics on host's health have been reported in many articles, surveys, and systematic reviews [12-14]. These studies have documented the role of probiotics in preventing health problems, including gastrointestinal disorders such as diarrhea caused by infection [1, 5], diarrhea caused by antibiotics [1, 15], Irritable Bowel Syndrome (IBS) [16], Diarrhea in adults and children [17], inflammatory bowel disease (IBD), only in ulcerative colitis [18], allergic disorders such as atopic dermatitis (eczema) [19], allergic rhinitis [20].

Although many probiotic strains are well-documented as "safe" or "generally recognized as safe," the European Food Safety Authority (EFSA) and the FDA do not attribute the ability to prevent or treat disease to probiotics. Probiotics are purchased as dietary supplements in many countries and are followed by current market policies [1]. EFSA has not endorsed any products with health claims related to probiotic prescribing. More than 300 approval applications have been submitted for 200 probiotic strains or combinations of strains claiming to have more than 60 beneficial effects [21]. The main reasons for rejecting these approval requests were as follows: inadequate description, undefined claims, useless claims, lack of relevant human studies, lack of measurable results that reflect direct benefits to the human beings, and finally, the quality of the presented studies [21].

In addition, the FDA may regulate probiotic strains as a dietary supplement, food, or drug [1]. Like the EFSA, the FDA hasn't approved any probiotics to prevent or treat health problems [22]. Both food agencies emphasize the following: each health claim is unique to each probiotic strain. Scientific requirements must be considered in the context of each program. Past guidelines and evaluations are valuable sources of information. Understanding the logic behind the principles applied is important. There is no instruction for success. Moreover, researchers and companies must endeavor, fail, learn, and try again [21]. Furthermore, many studies on probiotics have lacked insight into the potential mechanism of action [1]. It is now accepted that intestinal dysbiosis refers to changes in the quantitative and qualitative microbiota composition. These changes may lead to changes in the microbial interaction of the host, which can contribute to a disease condition often linked to inflammation, which is associated with disease [1]. Many non-communicable human diseases, the mechanism by which homeostasis is maintained, are not yet fully known [5, 23]. Recent researchers have demonstrated that during homeostasis, epithelial hypoxia limits the availability of oxygen to the colon and leads to maintaining a balanced microbiota that acts as a microbial cell, and produces metabolites that contribute to host nutrition and safety training, and site security [23].

3. Enzymatic Activity, Cell Adhesion and Mucin Production by Probiotics

The enzymatic activity of probiotics in the intestinal tract can play a role in the biological effects of these probiotics. *Lactobacilli* and *bifidobacteria* show more than 20 different enzymatic actovoties, and the most common one is beta-galactosidase activity. On the other hand, intestinal bacterial beta-galactosidase hydrolyzes glucuronidated metabolites in the intestine in a toxic way, resulting from damage to the intestine [24]. In addition, low activity of beta-galactosidase in feces is associated with increased levels of carcinogens in the colon's lumen [24, 25]. *B. longum*, when added to the diet, contributes to changes in the intestinal microbial complex, reduces beta-glucuronidas activity, which is aligned with inhibition of crypt abnormal formation, and is a primary indicator of pre-neoplastic malignancy in the carcinogenic process [24, 26]. Furthermore, a systematic review of randomized clinical trials (RCTs) that tested probiotics, prebiotics, or both (synbiotics) for the treatment of non-alcoholic fatty liver disease (NAFLD) in adult patients led to a reduction in aminotransferase activity [24, 27].

Probiotics interact with bile acids in the intestinal tract, alter bile acid metabolism, and affect cholesterol absorption [24]. Bile salt hydrolase (BSH) is an enzyme produced by bacterial species of several gastrointestinal tracts and the most known probiotics. This enzyme may participate in the first deconjugation reaction of bile salts [28]. Due to these beneficial effects of BSH-containing bacteria, BSH activity is included in FAO/WHO guidelines to evaluate probiotics for food application [24, 28]. Enzymatic decongestion of bile acids by BSH from probiotics has been considered as one of the main mechanisms of hypocholesterolemia attributed to probiotics [28, 29].

Apart from those mentioned above, when a microbial strain is shown to be a probiotic, certain prerequisites must be considered. One is adhesion to the intestinal mucosa for colonization and further interaction between the prescribed probiotic strains and host [7]. This particular interaction is required to adjust the contrast against pathogens and act on the immune system [30]. Intestinal epithelial cells secrete mucin to prevent pathogenic bacteria from adhesion [31]. It has been indicated that several lactobacillus proteins enhance this adhesion [32], and surface adhesions facilitate attachment to the mucosal layer [33]. Over the past few years, the Caco-2 cell line has been widely used to determine the adhesion capacity of probiotics *in vitro* conditions [34]. These cells form a homogeneous monolayer that resembles adult enterocytes in the human small intestine and form crypts that are typical single-layer epithelial structures [7, 35, 36].

4. Importance of the Probiotics Microencapsulation Process

Prebiotics are food components (such as oligosaccharides) that preferentially stimulate the growth of beneficial bacteria in the gut rather than harmful bacteria. Probiotics are live beneficial bacteria that cross the upper gastrointestinal tract (GIT) and colonize the colon. Some investigations have shown that oral administration of probiotics can promote a healthier gut microbiome, leading to an increase in probiotic fortified foods and supplements [37-40]. Some researchers have evaluated the viability of some commercial probiotics during GIT transportation. Surprisingly, the results showed that all tested commercial products endured more than 106 times reduction in colony-forming units (CFU) during 5 min incubation in gastric fluids [39, 40]. Therefore, there are concerns that many commercial probiotics products are ineffective because beneficial bacteria do not survive during digestion, storage, and passage through the upper gastrointestinal tract. In addition, even if they reach the large intestine, they may not be able to stabilize as part of the intestine microbiome and may simply pass out in the feces [40]. Many of the probiotics currently used in commercial products belong to the *Lactobacillus* or *Bifidobacterium* generation, and they are susceptible to the harsh conditions in many foods [40, 41]. *Bacillus*, *pediococcus*, and some yeasts have been reported as a suitable probiotic candidates [42, 43]. Some next-generation probiotic candidates, such as *Akkermansia muciniphila* and *Faecalibacterium prausnitzii* are very sensitive to oxygen, stomach acids, and bile salts which also limits their effectiveness in improving human health through dietary interventions [44]. As a result, effective strategies are needed to improve the stability of probiotics in foods and during their passage through the human intestine [40].

Microencapsulation has been proposed as an effective means of protecting probiotics against degradation [39, 41, 45, 46]. An effective microencapsulation system must keep the stability of probiotics during storage, protect them from harsh conditions in the upper gastrointestinal tract, release them in the large intestine, and then enhance their ability to colonize mucosal surfaces [47-49]. Some recent studies have focused on different types of oral delivery systems developed to enclose late probiotics [38, 40, 50-53]. However, many of these systems cannot adequately protect probiotics from degradation in the human gut due to inherent limitations such as their permeability to acids, enzymes, or bile salts. Therefore, the main purpose of the present study is to focus on recent advances in the development of stronger oral transmission systems specially designed to increase the stress tolerance of probiotics in foods and the human intestine. Moreover, emerging *in vitro* and *in vivo* methods have recently been developed to evaluate the efficiency of probiotic delivery systems [40]. In other words, microencapsulation may be defined as the packaging process of functional compounds (such as probiotic cultures) in microcapsules made of encapsulating material. When the edible film is used for probiotics, micro-coating methods are used to transport and protect them from factors like oxygen, pH, and light, which can be harmful to probiotics and minimize damage to them. These methods include spray drying, extrusion, emulsion, and other methods [40, 54].

5. Edible Films Containing Probiotics

Edible films are among those food products that as well as protecting food as packaging, are consumed with packaged food [1]. On the other hand, due to the many disadvantages of plastic films and packaging, edible films have become popular in the scientific world and have attracted the attention of eco-friendly authorities and consumers. The main components of edible films are biopolymers and lipids or a mixture. The performance of the components on physical and chemical properties are different and effective [55]. These biopolymers contain proteins of plant or animal origins, such as wheat gluten proteins, corn zein, soybeans, sunflower proteins, gelatin, whey, casein, and creatine, and polysaccharides like cellulose derivatives, starches, alginates, pectins, chitosans, carrageenans, gums, and fiber [56]. In other words, it can be said that these compounds make it possible to prepare edible films in fresh or processed food packaging to increase the shelf life of food. In the formulation of edible films, in addition to the above compounds, plasticizers are added to improve the flexibility of the protein network. Sometimes, hydrophobic compounds like beeswax or oil are added to overcome water permeability and improve crystallinity, hydrophobicity, surface charge, and molecular size [1].

Edible films are produced from food suspensions and are usually obtained and dried in the form of solid sheets on inert molding surfaces, and they come into contact with food through further processing as packaging in the form of bags, capsules, bags, or containers. Active edible films can act by releasing active substances into the surrounding atmosphere or by absorbing components such as free radicals, moisture, and oxygen which can deteriorate the food quality [1, 55]. Consumer awareness around the world about health, nutrition, food safety, and environmental issues leading research into the film-forming properties of biopolymers for the production of edible films for food packaging [1]. However, edible films do not replace external packaging (irrefrangible) and they can act as a good oxygen barrier against probiotics, but packaging, in terms of its function in food protection helps to reduce the rate of moisture and gas transfer between food and the environment to increase food stability. Apart from the inactive protective function, edible films may play some active or bioactive roles [57].

The role of probiotics in human health depends on survival and metabolic activity throughout the process and food supply chain, as well as in the human digestive system [58]. Several methods have been tested to increase the quality of probiotic cultures like osmotic stress, exposure to deadly heat or cold, low pH, redox potential reduction (through the use of oxygen absorbers, oxygen consuming microorganisms), and modified atmosphere packaging [59]. In this regard, microbial encapsulation in various studies has shown successful results [60]. Recently, microencapsulation with active packaging of edible films has been integrated as an alternative to probiotic products or probiotics. This method (probiotic microencapsulation) has become a promising alternative to antibiotics because this process allows the desired probiotics to be released to preserve food during storage [61].

In active edible films containing probiotic microorganisms, besides their positive effects on human health, they can protect food against pathogenic bacteria, which leads to an increase in food safety. They also increase the shelf life of food by controlling microorganisms involved in food spoilage. On the other hand, some studies on the storage of bioactive products in edible film packaging have shown that some probiotics like *Listeria monocytogenes* hurt beef slices by producing compounds such as bacteriocin [1, 62].

The concepts of active and bioactive packaging are sometimes misused, but there are some differences. Active food packaging systems are systems beyond the traditionally passive role of food protection and include desirable interactions with foods, as it is associated with increasing and maintaining food quality and safety [1, 63]. A typical active packaging is an antimicrobial package that interacts with the product to reduce, prevent, or delay microbial growth on food surfaces [64]. In previous studies, biopolymers have been used to carry natural antimicrobial compounds such as essential oils [1, 63, 65-67], organic acids [68], enzymes (like lysozyme) [67], and bacteriocin [64, 69, 70]. Active food packaging is created in response to consumer demand for safer and less processed foods with a longer shelf life [1, 71].

6. Selection of Probiotic Strains for Edible Films

Traping cells using edible films can lead to the use of probiotics in food. The use of edible films seems to be an alternative to encapsulation techniques [72]. One of the most important groups of bacteria that are used as probiotics in the food industry is Lactic Acid Bacteria (LAB) [73]. LAB are good candidates as probiotics because several strains of LAB are native to the human oral cavity and intestines and have a positive effect on the health of the human body [74]. Yeasts like *Saccharomyces cerevisiae*, *Debaryomyces hansenii*, *Torulaspora delbrueckii*, *Kluyveromyces lactis*, *Saccharomyces boulardii*, *Yarrowia lipolytica*, *Kluyveromyces marxianus*, *Kluyveromyces lodderae*, are also beneficial to human health due to their probiotic properties. Only *Saccharomyces boulardii* is considered very effective in terms of probiotic ability [75]. Yeasts, compared to bacteria, have special benefits as probiotics and as an effective treatment for diarrhea and intestinal infections caused by antibiotics, and can also produce vitamins. Determination of species and probiotic strains is necessary due to the specificity of the probiotic effects by microorganisms with probiotic properties in terms of specific aspects of health [1].

Traditionally, systems designed to provide probiotic cultures have both a conventional approach (pharmaceutical products) and a non-conventional approach (e.g., food-based products) [76]. Active edible films and coatings may show both aspects because they may be expected to be consumed alongside a regular food product [1]. Regardless of the mechanism, the main role (i.e., providing health benefits) of probiotics containing films and coatings is more or less efficient depending on their ability to deliver alive bacteria to the gastrointestinal tract. The stability and survival of several probiotic strains under different conditions have been extensively studied and reviewed [1, 77-80]. Fewer studies, although significant, have focused on evaluating the viability of probiotic bio-strains, especially in films, to evaluate their stability in terms of active/bioactive properties. Pullulan and various starches (from potatoes, tapioca, and corn) have been used to produce new edible films with a combination of probiotic cultures (*Lactobacillus reuteri* ATCC 55730, *L. rhamnosus* GG ATCC 53103, and *L. acidophilus* DSM) at an initial concentration of 12.9 log CFU.mL^{-1} [1, 72]. The treatments were tested with films prepared from pure pullulan and a mixture of different starches and pullulan kept at 25°C and 4°C. The viability of probiotics in pure pullulan film after about 10 days of storage at 25°C was about 80%, which was reduced to 35% after 20 days. However, films combined with starch reduce cell viability, the higher the starch content, the lower the survival. On the other hand, the viability of pullulan and pullulan/potato starch films (weight ratio 75:25) up to 30 days of storage was almost 90% higher [1, 72].

Prebiotics may be defined as "selected fermented ingredients that cause specific changes in the composition or activity of the gastrointestinal microbiota that help the health of the host" [4]. They are also included in edible films to improve the stability of probiotic strains added to film-forming dispersions. Romano et al. (2014) prepared methylcellulose (MC) films containing two probiotic strains (*Lactobacillus delbrueckii* subsp. *Bulgaricus* CIDCA 333 and *Lactobacillus plantarum* CIDCA 83114) and fructooligosaccharide (FOS) as a prebiotic. Bacterial strains are completely embedded in the film matrix, but the presence of these probiotic bacteria does not affect its integrity. Since FOS was found to show not only positive effects (protection of *L. delbrueckii*), but also negative effects (reduction of the glass transition temperature of films), FOS concentration was selected according to the balance between these effects. Films containing *L. delbrueckii* were added with 3% (w/v) of FOS, while films containing *L. plantarum* with 1% of FOS were added. *L. plantarum* is stable for longer periods at higher RH values than *L. delbrueckii*, which may be useful for practical applications [81]. Similarly, inulin, chitosan oligosaccharide (COS), galactooligosaccharide (GtOS), and FOS have been added to corn starch-based edible films for prebiotic purposes, and have been shown that they increase the growth of probiotic *Bifidobacterium infantis* ATCC 15697. In addition to the prebiotic effect, these compounds also affect the physical properties of the edible layers, which show weak tensile strength compared to non-prebiotic films (control). According to the authors, this plasticizing effect may be attributed to the much smaller molecular weight of prebiotics than starch [1, 82].

7. Antimicrobial Compounds in Active Edible Films

In non-food packaging systems, any kind of preservatives such as organic acids and their salts, fungicides, antibiotics, enzymes, alcohol, antioxidants, metals, and disinfectants can be used. However, in the case of edible films, the choice of antimicrobial agent is limited to edible compounds because these compounds are consumed with food, and their safety is essential [1]. Characteristics of antimicrobial compounds used in active packaging must have:

1. Approved by regulatory agencies and allowed to apply in food
2. Has low price and affordable
3. To be effective on a wide range of microorganisms
4. To be effective on microorganisms in low concentration
5. Do not have a negative effect on food sensory properties [83].

The antimicrobial compounds that enter the polymer directly are listed in Table 1 [83-86]. Besides antimicrobial compounds, important components are used in antimicrobial packaging, especially in edible films. They are listed in Table 2 [87].

Table 1. Classification of antimicrobial compounds in active packaging

Factor	Effect
Mass transfer coefficient and controlled antimicrobial release	By determining the mass transfer coefficients and creating a model for the mass transfer of antimicrobial substances from the polymer film, it is possible to determine how long antimicrobial substance remains above the critical point and thus figure out the shelf life of food [83]
Interaction between food antimicrobial agent	Each antimicrobial agent has a unique growth inhibition mechanism against the target microorganism. Most environmental factors are the nature of food and storage conditions. Physicochemical properties of food compounds affect antimicrobial activities. Therefore, the composition of the food affects the effectiveness of the antimicrobial agent and the susceptibility of microorganisms [83]
Storage temperature	Storage temperature can affect antimicrobial activity and the migration rate of microbial and antimicrobial substances. As the temperature increases, the rate of diffusion and migration from the package increases, while the release rate in the refrigerator decreases [84].
Process conditions	High temperature and pressure processes in film production reduce antimicrobial activity. Antimicrobial agents are very unstable and decompose quickly. Antimicrobial agents with low heat resistance must be used in polymers produced at low temperatures [85].
Physical and chemical characteristics of films	When an antimicrobial agent is added to the packaging polymer, it is likely to change the physical properties of the polymer. For instance, these compounds may decrease transparency and tensile properties and increase the permeability of the polymer film [86].
Organoleptic properties of antimicrobial agents and edible films	Antimicrobial edible films are part of the edible part of packaged food. Thus, the composition of the edible film should not affect organoleptic properties [86].

Table 2. Antimicrobial compounds with examples [87]

Category	Effect
Benomyl	Fungicides
Heptyl paraben	Parabens
Allyl isothiocyanate, SO2	Gases
EDTA	Chelating agents
Thymol	Herbal extracts
Nicin	Bacteriocins
Imazalil	Antibiotics
Lysozyme, Peroxidase	Enzymes
Allicin	Thiosulfates
Silver	Metals
Sorbic acid, Propionic acid, Benzoic acid	Organic acids
Conalbumin	proteins

Table 3. The studies related to active edible films as probiotic carrier packages

Film composition	Result	Reference
Pectin-based edible film containing berry leaf extract and Pseudomonas aeruginosa and Bacillus cereus bioactive compounds	Expanded pectin based films combined with bioactive compounds showed better antioxidant, antimicrobial, and mechanical properties.	[89]
Bioactive edible film based on cassava starch, inulin, and L. casei.	Addition of inulin had a lubricating effect on cassava starch layers and increased the protective effect of L. casei and survival in simulated gastric conditions.	[90]
Physicochemical properties of edible film containing Lactococcus lactis	Addition of Lactococcus lactis reduced film transparency, a slight decrease in tensile strength, increased elongation at break (EB), and increased survival of Lactococcus lactis in sodium alginate, carboxymethyl cellulose.	[91]
Survival of probiotic bacteria in carboxymethyl cellulose edible films and evaluation of qualitative parameters	Survival of L. acidophilus and L. rhamnosus better preserved at 4°C than 25°C for 40 days. Probiotic films showed less water vapor permeability and turbidity than the control sample. The addition of probiotics to CMC edible films causes a reduction in tensile strength (TS) and increased elongation at break (EB).	[92]
Bioactive edible film based on Lactobacillus plantarum: Physicochemical properties and kiwi durability	At the end of the maintained period, Lactobacillus survived in all groups treated with high levels of probiotics. In probiotic films compared to the control sample, L. plantarum reduces the rate of rot and discoloration, and chlorophyll and ascorbic acid content was preserved. Probiotic films, compared to the control sample, had higher acceptability. Probiotic treatment compared to the control group significantly reduced mold and yeast.	[93]
Formation and evaluation of physicochemical properties of probiotic edible films based on alginate, casein	Probiotic edible films showed significant improvement in microbial, antioxidant, structural, optical, and thermal properties. Furthermore, it was found that this film could keep probiotics alive and increase the survival of probiotics.	[94]
Application of LAB in active edible films	LAB provides safety of consumption, structural reinforcement, application in a bioactive probiotic carrier in edible film	[95]
Preparation of L-glutamic acid active antimicrobial packaging materials based on L-lysine to protect probiotic bacteria	Addition of L-lysine causes an increase in density, thickness, moisture content, L, b, and whiteness index, elongation at break (EB) in probiotic film, and simultaneous reduction of overall color changes tensile strength of the edible film. It did not affect the survival of probiotics.	[96]

Film composition	Result	Reference
Evaluation of the stability of films containing Lactobacillus rhamnosus GG	Negative effect of the film-forming solution on cell survival. Edible films based on pectin, sodium alginate, chiti bean gum, and carrageenan showed more cell death than the film containing whey concentrate. Physical entrapment of probiotic cells in the inner space of hydrogels caused an increase in viability of Lactobacillus rhamnosus GG.	[97]
Evaluation of the survival rate in edible films based on sodium alginate in contact with pork pieces during storage (66 days at 4°C, 47 days at 8°C, and 40 days at 12°C)	No population dependence or survival rate of microbial strains on storage temperature. The decrease in a probiotic population at the sampling stage after phosphorus uptake for all temperatures was probably due to the dry process and stress.	[98]
Freezing carrot juice powder containing dried probiotics for better storage of probiotics.	The formulation based on Arabic gum and maltodextrin as coating agents could increase the shelf life of probiotics by one month (6-7 Log CFU/g)	[99]
Effect of chitosan-coated with low-fat milk nanoparticles-alginate on the shelf life of probiotics in simulated intestinal and gastric conditions.	Edible film obtained in a month, causes 96% survival of Lactobacillus plantarum and 98% Lactobacillus rhamnosus gg.	[100]
Probiotic edible films as a new strategy for the development of bakery products	Whey proteins cause a reduction of L. rhamnosus GG survival during drying and storage. Edible films based on sodium alginate showed a protective role in probiotic survival.	[101]

8. Survival of Probiotics in Edible Films

The most important parameters in developing probiotic carrier the edible films are survival of the probiotic strains and reaching the gastrointestinal tract. The chemical structure of the film and the film formation process are two important parameters affecting the survival of probiotic microorganisms [75]. Distribution temperature and storage conditions of the edible film are the most important factors affecting the survival of probiotics [75, 88]. Table 3 shows the research related to active edible films as probiotic carrier packages [89-101].

Conclusion

Edible films and coatings have attracted much attention as an alternative to synthetic films in reducing the environmental pollution. Edible films and coatings can also contain live microorganisms. One of the recent advances in this field is the production of probiotic films and edible coatings that act as potential carriers for these microorganisms by increasing their survival. The potential use of edible films and coatings as stabilizing matrices is to preserve probiotic microorganisms, neutralize the disadvantages of their direct application, and enhance the functional properties of edible coatings through the probiotic mixture. The combination of probiotics in edible films and coatings allows them to reach the consumer's intestine adequately. Microorganisms such as bacteria or yeasts can be used as bio-control agents to increase the shelf life of food products. Based on the evidence we have presented in this chapter, hopefully we can convince you that combining them with edible films and coatings is a promising approach. In the next chapter, our colleagues will continue to present to you the use of supercritical solvent impregnation in food packaging film production. This technique will enhance the versatility and flexibility of packaging films for food preservation.

References

[1] Amiri S, Kohneshahri SR, Nabizadeh F. The effect of unit operation and adjunct probiotic culture on physicochemical, biochemical, and textural properties of Dutch Edam cheese. *LWT*. 2022 Feb 1;155:112859.

[2] Amiri S, Teymorlouei MJ, Bari MR, Khaledabad MA. Development of Lactobacillus acidophilus LA5-loaded whey protein isolate/lactose bionanocomposite powder by electrospraying: A strategy for entrapment. *Food Bioscience.* 2021 Oct 1;43:101222.

[3] Amiri, S., Mokarram, R. R., Khiabani, M. S., Bari, M. R., & Khaledabad, M. A. (2022). Characterization of antimicrobial peptides produced by Lactobacillus acidophilus LA-5 and Bifidobacterium lactis BB-12 and their inhibitory effect against foodborne pathogens. *LWT*, 153, 112449.

[4] Sohrabpour S, Rezazadeh Bari M, Alizadeh M, Amiri S. Investigation of the rheological, microbial, and physicochemical properties of developed synbiotic yogurt containing Lactobacillus acidophilus LA-5, honey, and cinnamon extract. *Journal of Food Processing and Preservation*. 2021 Apr;45(4):e15323.

[5] Amiri S, Rezazadeh Bari M, Alizadeh Khaledabad M, Rezaei Mokarram R, Sowti Khiabani M. Co-production of parabiotic metabolites by Lactobacillus acidophilus LA5 and Bifidobacterium animalis subsp. lactis BB12 in dairy effluents. *Chemical Review and Letters.* 2021 Apr 1;4(2):66-76.

[6] Metchnikoff II. *The prolongation of life: optimistic studies*. Springer Publishing Company; 2004.

[7] Fontana L, Bermudez-Brito M, Plaza-Diaz J, Munoz-Quezada S, Gil A. Sources, isolation, characterisation and evaluation of probiotics. *Br J Nutr.* 2013;109(S2):S35–50.

[8] Gibson GR, Roberfroid MB. Dietary modulation of the human colonic microbiota: introducing the concept of prebiotics. *J Nutr*. 1995;125(6):1401–12.

[9] Ouwehand AC, Salminen S, Isolauri E. Probiotics: an overview of beneficial effects. *Lact acid Bact Genet Metab Appl*. 2002;279–89.

[10] Plaza-Díaz, J, Robles-Sánchez, C, Abadía-Molina, F, Morón-Calvente, V, Sáez-Lara, MJ, Ruiz-Bravo, A, Jiménez-Valera, M, Gil, Á, Gómez-Llorente, C, & Fontana, L. Adamdec1, Ednrb and Ptgs1/Cox1, inflammation genes upregulated in the intestinal mucosa of obese rats, are downregulated by three probiotic strains. *Sci Rep*. 2017;7(1):1–10.

[11] Plaza-Díaz, J, Robles-Sánchez, C, Abadía-Molina, F, Sáez-Lara, MJ, Vilchez-Padial, LM, Gil, Á, Gómez-Llorente, C, & Fontana, L. Gene expression profiling in the intestinal mucosa of obese rats administered probiotic bacteria. *Sci data*. 2017;4(1):1–10.

[12] Didari T, Solki S, Mozaffari S, Nikfar S, Abdollahi M. A systematic review of the safety of probiotics. *Expert Opin Drug Saf*. 2014;13(2):227–39.

[13] Liu J, Fan LT, Seib P, Friedler F, Bertok B. Downstream process synthesis for biochemical production of butanol, ethanol, and acetone from grains: generation of optimal and near-optimal flowsheets with conventional operating units. *Biotechnol Prog*. 2004;20(5):1518–27.

[14] Tsilingiri, K, Barbosa, T, Penna, G, Caprioli, F, Sonzogni, A, Viale, G, & Rescigno, M. Probiotic and postbiotic activity in health and disease: comparison on a novel polarised *ex-vivo* organ culture model. *Gut*. 2012;61(7):1007–15.

[15] Horváth R, Pedersen HC, Skivesen N, Selmeczi D, Larsen NB. Optical waveguide sensor for on-line monitoring of bacteria. *Opt Lett*. 2003;28(14): 1233–5.

[16] Alicia R Maher, Zhen Wang, Jeremy NV Miles, Roberta Shanman, Hempel S, Newberry SJ, Maher AR, Wang Z, Miles JNV, Shanman R, Breanne J, PG Shekelle. Probiotics for the prevention and treatment of antibiotic-associated diarrhea: a systematic review and meta-analysis. *Jama*. 2012;307(18):1959–69.

[17] Goldenberg JZ, Yap C, Lytvyn L, Lo CK, Beardsley J, Mertz D, Johnston BC. Probiotics for the prevention of Clostridium difficile-associated diarrhea in adults and children. *Cochrane Database Syst Rev*. 2017;(12).

[18] Saez-Lara MJ, Gomez-Llorente C, Plaza-Diaz J, Gil A. The role of probiotic lactic acid bacteria and bifidobacteria in the prevention and treatment of inflammatory bowel disease and other related diseases: a systematic review of randomized human clinical trials. *Biomed Res Int*. 2015;2015.

[19] Rather IA, Bajpai VK, Kumar S, Lim J, Paek WK, Park Y-H. Probiotics and atopic dermatitis: an overview. *Front Microbiol*. 2016;7:507.

[20] Berings, M, Karaaslan, C, Altunbulakli, C, Gevaert, P, Akdis, M, Bachert, C, & Akdis, CA. Advances and highlights in allergen immunotherapy: on the way to sustained clinical and immunologic tolerance. *J Allergy Clin Immunol*. 2017;140(5):1250–67.

[21] EFSA Panel on Dietetic Products N and A (NDA). Guidance on the scientific requirements for health claims related to the immune system, the gastrointestinal tract and defence against pathogenic microorganisms. *EFSA J*. 2016;14(1):4369.

[22] Degnan FH. The US Food and Drug Administration and probiotics: regulatory categorization. *Clin Infect Dis*. 2008;46(Supplement_2):S133–6.

[23] Byndloss MX, Bäumler AJ. The germ-organ theory of non-communicable diseases. *Nat Rev Microbiol*. 2018;16(2):103–10.

[24] Plaza-Diaz J, Ruiz-Ojeda FJ, Gil-Campos M, Gil A. Mechanisms of action of probiotics. *Adv Nutr*. 2019;10(suppl_1):S49–66.

[25] Kim DH, Jin YH. Intestinal bacterial β-glucuronidase activity of patients with colon cancer. *Arch Pharm Res*. 2001;24(6):564–7.

[26] Kulkarni N, Reddy BS. Inhibitory effect of Bifidobacterium iongum cultures on the azoxymethane-induced aberrant crypt foci formation and fecal bacterial β-glucuronidase. *Proc Soc Exp Biol Med*. 1994;207(3):278–83.

[27] Buss C, Valle-Tovo C, Miozzo S, de Mattos AA. Probiotics and synbiotics may improve liver aminotransferases levels in non-alcoholic fatty liver disease patients. *Ann Hepatol*. 2014;13(5):482–8.

[28] Pavlović N, Stankov K, Mikov M. Probiotics—interactions with bile acids and impact on cholesterol metabolism. *Appl Biochem Biotechnol*. 2012;168(7): 1880–95.

[29] Kumar S, Sharma HK, Sarkar BC. Effect of substrate and fermentation conditions on pectinase and cellulase production by *Aspergillus niger* NCIM 548 in submerged (SmF) and solid state fermentation (SSF). *Food Sci Biotechnol*. 2011;20(5):1289–98.

[30] Yadav, AK, Tyagi, A, Kumar, A, Panwar, S, Grover, S, Saklani, AC, Hemalatha, R, & Batish, VK. Adhesion of lactobacilli and their anti-infectivity potential. *Crit Rev Food Sci Nutr.* 2017;57(10):2042–56.

[31] Collado MC, Gueimonde M, Salminen S. Probiotics in adhesion of pathogens: mechanisms of action. In: *Bioactive foods in promoting health.* Elsevier; 2010. p. 353–70.

[32] Van Tassell ML, Miller MJ. Lactobacillus adhesion to mucus. *Nutrients.* 2011;3(5):613–36.

[33] Buck BL, Altermann E, Svingerud T, Klaenhammer TR. Functional analysis of putative adhesion factors in Lactobacillus acidophilus NCFM. *Appl Environ Microbiol.* 2005;71(12):8344–51.

[34] Dicks L, Botes M. Probiotic lactic acid bacteria in the gastro-intestinal tract: health benefits, safety and mode of action. Benef Microbes. 2010;1(1):11–29.

[35] Lenaerts K, Bouwman FG, Lamers WH, Renes J, Mariman EC. Comparative proteomic analysis of cell lines and scrapings of the human intestinal epithelium. *BMC Genomics.* 2007;8(1):1–14.

[36] Huang SH, He L, Zhou Y, Wu CH, Jong A. Lactobacillus rhamnosus GG suppresses meningitic E. coli K1 penetration across human intestinal epithelial cells *in vitro* and protects neonatal rats against experimental hematogenous meningitis. *Int J Microbiol.* 2009;2009.

[37] Domingo JJS. Review of the role of probiotics in gastrointestinal diseases in adults. *Gastroenterol y Hepatol* (English Ed. 2017;40(6):417–29.

[38] Sarao LK, Arora M. Probiotics, prebiotics, and microencapsulation: A review. *Crit Rev Food Sci Nutr.* 2017;57(2):344–71.

[39] Dodoo CC, Wang J, Basit AW, Stapleton P, Gaisford S. Targeted delivery of probiotics to enhance gastrointestinal stability and intestinal colonisation. *Int J Pharm.* 2017;530(1–2):224–9.

[40] Yao M, Xie J, Du H, McClements DJ, Xiao H, Li L. Progress in microencapsulation of probiotics: A review. *Compr Rev Food Sci Food Saf.* 2020;19(2):857–74.

[41] Cook MT, Tzortzis G, Charalampopoulos D, Khutoryanskiy V V. Microencapsulation of probiotics for gastrointestinal delivery. J Control release. 2012;162(1):56–67.

[42] Lv B, Zhou H, Liu C, Guo X, Zhang C, Liu Z, Luo, L. The relationship between mother–child discrepancies in educational aspirations and children's academic achievement: The mediating role of children's academic self-efficacy. Child Youth Serv Rev. 2018;86:296–301.

[43] Mingmongkolchai S, Panbangred W. Bacillus probiotics: an alternative to antibiotics for livestock production. *J Appl Microbiol.* 2018;124(6):1334–46.

[44] Kolida S, Gibson GR. Synbiotics in health and disease. *Annu Rev Food Sci Technol.* 2011;2:373–93.

[45] Del Piano, M, Carmagnola, S, Andorno, S, Pagliarulo, M, Tari, R, Mogna, L, Strozzi, GP, Sforza, F, & Capurso, L. Evaluation of the intestinal colonization by microencapsulated probiotic bacteria in comparison with the same uncoated strains. *J Clin Gastroenterol.* 2010;44:S42–6.

[46] Speranza B, Campaniello D, Bevilacqua A, Altieri C, Sinigaglia M, Corbo MR. Viability of Lactobacillus plantarum on fresh-cut chitosan and alginate-coated apple and melon pieces. *Front Microbiol*. 2018;2538.

[47] Anselmo AC, McHugh KJ, Webster J, Langer R, Jaklenec A. Biomaterials: Layer-by-Layer Encapsulation of Probiotics for Delivery to the Microbiome (Adv. Mater. 43/2016). *Adv Mater*. 2016;28(43):9442.

[48] Fakhrullin RF, Lvov YM. "Face-lifting" and "make-up" for microorganisms: layer-by-layer polyelectrolyte nanocoating. *ACS Nano*. 2012;6(6):4557–64.

[49] Tripathi, P, Beaussart, A, Alsteens, D, Dupres, V, Claes, I, von Ossowski, I, de Vos, WM, Palva, A, Lebeer, S, Vanderleyden, J, & Dufrêne, YF. Adhesion and nanomechanics of pili from the probiotic Lactobacillus rhamnosus GG. *ACS Nano*. 2013;7(4):3685–97.

[50] Liu C, Hao G, Su M, Chen Y, Zheng L. Potential of multispectral imaging combined with chemometric methods for rapid detection of sucrose adulteration in tomato paste. *J Food Eng*. 2017;215:78–83.

[51] Heidebach T, Först P, Kulozik U. Microencapsulation of probiotic cells for food applications. *Crit Rev Food Sci Nutr*. 2012;52(4):291–311.

[52] De Prisco A, Mauriello G. Probiotication of foods: A focus on microencapsulation tool. *Trends Food Sci Technol*. 2016;48:27–39.

[53] Ramos PE, Cerqueira MA, Teixeira JA, Vicente AA. Physiological protection of probiotic microcapsules by coatings. *Crit Rev Food Sci Nutr*. 2018;58(11):1864–77.

[54] Tian H, Guo G, Fu X, Yao Y, Yuan L, Xiang A. Fabrication, properties and applications of soy-protein-based materials: A review. *Int J Biol Macromol*. 2018;120:475–90.

[55] Pavli F, Tassou C, Nychas GJE, Chorianopoulos N. Probiotic incorporation in edible films and coatings: Bioactive solution for functional foods. *Int J Mol Sci*. 2018;19(1):150.

[56] Heinemann RJB, Carvalho RA, Favaro-Trindade CS. Orally disintegrating film (ODF) for delivery of probiotics in the oral cavity—Development of a novel product for oral health. *Innov Food Sci Emerg Technol*. 2013;19:227–32.

[57] Zoghi A, Khosravi-Darani K, Mohammadi R. Application of edible films containing probiotics in food products. *J Consum Prot Food Saf*. 2020;15(4):307–20.

[58] Śliżewska K, Markowiak-Kopeć P, Śliżewska W. The role of probiotics in cancer prevention. *Cancers* (Basel). 2020;13(1):20.

[59] Mohammadi R, Sohrabvandi S, Mohammad Mortazavian A. The starter culture characteristics of probiotic microorganisms in fermented milks. *Eng Life Sci*. 2012;12(4):399–409.

[60] Reque PM, Brandelli A. Encapsulation of probiotics and nutraceuticals: Applications in functional food industry. *Trends Food Sci Technol*. 2021;114:1–10.

[61] Atia A, Gomaa A, Fliss I, Beyssac E, Garrait G, Subirade M. A prebiotic matrix for encapsulation of probiotics: physicochemical and microbiological study. *J Microencapsul*. 2016;33(1):89–101.

[62] Islam MM, Yang CJ. Efficacy of mealworm and super mealworm larvae probiotics as an alternative to antibiotics challenged orally with Salmonella and E. coli infection in broiler chicks. *Poult Sci.* 2017;96(1):27–34.

[63] Botre DA, Soares N de FF, Espitia PJP, Sousa S de, Renhe IRT. Avaliação de filme incorporado com óleo essencial de orégano para conservação de pizza pronta. *Rev Ceres.* 2010;57(3):283–91.

[64] Soares CMD, Alves RC, Oliveira MBPP. Chapter 69 - Acrylamide in Coffee: Influence of Processing. In: *Preedy VBT-P and I on AC in F,* editor. San Diego: Academic Press; 2015. p. 575–82. Available from: https://www.sciencedirect.com/science/article/pii/B978012404699300069X.

[65] Agarwal A, Jaiswal N, Tripathi AD, Paul V. Downstream processing; applications and recent updates. In: *Bioprocessing for Biofuel Production.* Springer; 2021. p. 29–55.

[66] Tripathi P, Dubey NK. Exploitation of natural products as an alternative strategy to control postharvest fungal rotting of fruit and vegetables. *Postharvest Biol Technol.* 2004;32(3):235–45.

[67] Appendini P, Hotchkiss JH. Immobilization of lysozyme on food contact polymers as potential antimicrobial films. *Packag Technol Sci An Int J.* 1997;10(5):271–9.

[68] Schirmer, BC, Heiberg, R, Eie, T, Møretrø, T, Maugesten, T, Carlehøg, M, & Langsrud, S. A novel packaging method with a dissolving CO_2 headspace combined with organic acids prolongs the shelf life of fresh salmon. *Int J Food Microbiol.* 2009;133(1–2):154–60.

[69] Han JH. *Innovations in food packaging*. Elsevier; 2005.

[70] Lopez-Rubio A, Gavara R, Lagaron JM. Bioactive packaging: turning foods into healthier foods through biomaterials. *Trends Food Sci Technol.* 2006;17(10): 567–75.

[71] Ahvenainen R. *Novel food packaging techniques*. Elsevier; 2003.

[72] Kanmani P, Lim ST. Development and characterization of novel probiotic-residing pullulan/starch edible films. *Food Chem.* 2013;141(2):1041–9.

[73] Hirsch, E, Pantea, E, Vass, P, Domján, J, Molnár, M, Suhajda, Á, Andersen, SK, Vigh, T, Verreck, G, Marosi, GJ, & Nagy, ZK. Probiotic bacteria stabilized in orally dissolving nanofibers prepared by high-speed electrospinning. *Food Bioprod Process.* 2021;128:84–94.

[74] Joute JR, Chawhan P, Rungsung S, Kirthika P. Food additives and their associated health risks. *Int J Vet Sci Anim Husb.* 2016;1(1):1–5.

[75] Pandhi S, Kumar A, Alam T. Probiotic edible films and coatings: Concerns, applications and future prospects. *J Packag Technol Res.* 2019;3(3):261–8.

[76] da Silva BV, Barreira JCM, Oliveira MBPP. Natural phytochemicals and probiotics as bioactive ingredients for functional foods: Extraction, biochemistry and protected-delivery technologies. *Trends Food Sci Technol.* 2016;50:144–58.

[77] Corona-Hernandez RI, Álvarez-Parrilla E, Lizardi-Mendoza J, Islas-Rubio AR, de la Rosa LA, Wall-Medrano A. Structural stability and viability of microencapsulated probiotic bacteria: a review. *Compr Rev Food Sci Food Saf.* 2013;12(6):614–28.

[78] Dianawati D, Mishra V, Shah NP. Survival of microencapsulated probiotic bacteria after processing and during storage: a review. *Crit Rev Food Sci Nutr.* 2016;56(10):1685–716.

[79] Paseephol T, Sherkat F. Probiotic stability of yoghurts containing Jerusalem artichoke inulins during refrigerated storage. *J Funct Foods.* 2009;1(3):311–8.

[80] Ranadheera CS, Evans CA, Adams M, Baines SK. Co-culturing of probiotics influences the microbial and physico-chemical properties but not sensory quality of fermented dairy drink made from goats' milk. *Small Rumin Res.* 2016;136:104–8.

[81] Romano N, Tavera-Quiroz MJ, Bertola N, Mobili P, Pinotti A, Gómez-Zavaglia A. Edible methylcellulose-based films containing fructo-oligosaccharides as vehicles for lactic acid bacteria. *Food Res Int.* 2014;64:560–6.

[82] Tang, Y, Xie, F, Zhang, D, Zhu, M, Liu, L, Liu, P, & Gu, C. Physical properties and prebiotic activity of maize starch-based functional films. *Starch-Stärke.* 2015;67(1–2):124–31.

[83] Bahmid NA, Heising J, Dekker M. Multiresponse kinetic modelling of the formation, release, and degradation of allyl isothiocyanate from ground mustard seeds to improve active packaging. *J Food Eng.* 2021;292:110370.

[84] Lalnunthari C, Devi LM, Badwaik LS. Extraction of protein and pectin from pumpkin industry by-products and their utilization for developing edible film. *J Food Sci Technol.* 2020;57(5):1807–16.

[85] Dyshlyuk L, Babich O, Belova D, Prosekov A. Comparative analysis of physical and chemical properties of biodegradable edible films of various compositions. *J Food Process Eng.* 2017;40(1):e12331.

[86] Ansorena MR, Pereda M, Marcovich NE. Edible films. In: *Polymers for food applications.* Springer; 2018. p. 5–24.

[87] Lück E, Jager M. *Antimicrobial food additives: characteristics, uses, effects.* Vol. 2. Springer Science & Business Media; 1997.

[88] Juvi, D, Sthefanie, M, Sugata, M, Jap, L, Andrian, D, Rizkinata, D, Michelle, K, & Tan, TJ. Evaluation of Microencapsulated Local Isolates Lactobacillus casei 97/L3 and Lactobacillus delbrueckii 94/L4 for Improved Probiotic and Yogurt Starter Culture Application. *Microbiol Biotechnol Lett.* 2019;47(2):211–9.

[89] Shivangi S, Dorairaj D, Negi PS, Shetty NP. Development and characterisation of a pectin-based edible film that contains mulberry leaf extract and its bio-active components. *Food Hydrocoll.* 2021;121:107046.

[90] Orozco-Parra J, Mejía CM, Villa CC. Development of a bioactive synbiotic edible film based on cassava starch, inulin, and Lactobacillus casei. *Food Hydrocoll.* 2020;104:105754.

[91] Ma D, Jiang Y, Ahmed S, Qin W, Liu Y. Physical and antimicrobial properties of edible films containing Lactococcus lactis. *Int J Biol Macromol.* 2019;141:378–86.

[92] Ebrahimi B, Mohammadi R, Rouhi M, Mortazavian AM, Shojaee-Aliabadi S, Koushki MR. Survival of probiotic bacteria in carboxymethyl cellulose-based edible film and assessment of quality parameters. *LWT.* 2018;87:54–60.

[93] Hashemi SMB, Jafarpour D. Bioactive edible film based on Konjac glucomannan and probiotic Lactobacillus plantarum strains: Physicochemical properties and shelf life of fresh-cut kiwis. *J Food Sci.* 2021;86(2):513–22.

[94] Namratha S, Sreejit V, Preetha R. Fabrication and evaluation of physicochemical properties of probiotic edible film based on pectin–alginate–casein composite. *Int J Food Sci Technol.* 2020;55(4):1497–505.

[95] Moradi M, Guimarães JT, Sahin S. Current applications of exopolysaccharides from lactic acid bacteria in the development of food active edible packaging. *Curr Opin Food Sci.* 2021;40:33–9.

[96] Karimi M, Yazdi FT, Mortazavi SA, Shahabi-Ghahfarrokhi I, Chamani J. Development of active antimicrobial poly (l-glutamic) acid-poly (l-lysine) packaging material to protect probiotic bacterium. *Polym Test.* 2020;83:106338.

[97] Soukoulis C, Behboudi-Jobbehdar S, Macnaughtan W, Parmenter C, Fisk ID. Stability of Lactobacillus rhamnosus GG incorporated in edible films: Impact of anionic biopolymers and whey protein concentrate. *Food Hydrocoll.* 2017;70:345–55.

[98] Pavli, F, Kovaiou, I, Apostolakopoulou, G, Kapetanakou, A, Skandamis, P, Nychas, GJ, Tassou, C, & Chorianopoulos, N. Alginate-based edible films delivering probiotic bacteria to sliced ham pretreated with high pressure processing. *Int J Mol Sci.* 2017;18(9):1867.

[99] Rishabh D, Athira A, Preetha R, Nagamaniammai G. Freeze dried probiotic carrot juice powder for better storage stability of probiotic. *J Food Sci Technol.* 2021; 1–9.

[100] Padhmavathi V, Shruthy R, Preetha R. Chitosan coated skim milk-alginate microspheres for better survival of probiotics during gastrointestinal transit. *J Food Sci Technol.* 2021;1–7.

[101] Soukoulis C, Yonekura L, Gan HH, Behboudi-Jobbehdar S, Parmenter C, Fisk I. Probiotic edible films as a new strategy for developing functional bakery products: The case of pan bread. *Food Hydrocoll.* 2014;39:231–42.

Chapter 8

Production of Active Packaging Films by Using Supercritical Solvent Impregnation

Cristina Cejudo Bastante* **and Lourdes Casas Cardoso**
Chemical Engineering and Food Technology Department, Wine and Agrifood Research Institute (IVAGRO), University of Cadiz, Cadiz, Spain

Abstract

Consumers are increasingly interested in preservative-free and healthy food; however, lifestyle has drastically changed compared to previous generations. The current trend tends to store food and buy less frequently, but in the case of fresh food and fourth-range products, the shelf life is limited. To cover this demand, the food industry needs to provide food with a longer expiration date while maintaining quality and sensory properties. Active packaging has become one of the most interesting alternatives of packaging aimed to retard food spoilage. The research in this field is focused mainly on two perspectives; i) the study of different techniques to develop the active material and ii) the study of the bioactive compounds suitable to include into the package and/or in the polymer formulation. The different techniques for the development of active packaging are nowadays one of the most studied topics, being supercritical fluids a competitive alternative due to several advantages reported, such as lower use of solvents, better bioactivity of the material, or rapid action compared to the materials obtained by other techniques. On the other hand, extracts from agri-food by-products offer a great number of natural preservatives to be included again in the food chain, providing green materials with a preservation power competitive with the use of other synthetic preservers. In the last two chapters, our colleagues

* Corresponding Author's Email: cristina.cejudo@uca.es.

In: Food Packaging
Editor: Wing-Fu Lai
ISBN: 979-8-88697-249-8

have introduced to you how postbiotics and probiotics can be incorporated into packaging films for active food packaging. In this chapter, we will have a focused discussion on the use of supercritical solvent impregnation in food packaging research, and will evaluate the most studied natural extracts for food preservation.

Keywords: supercritical impregnation, agricultural by-products, natural extracts, food preservation, antioxidant activity, antimicrobial activity

1. Introduction and Background

According to the Food Loss and Waste database published by FAO [1], an estimated 1/3 of all food produced globally is lost or goes to waste, which represents 1.3 billion tons of food. As the United Nations Environment Programme (UNEP) published in the Food Waste Index Report of 2021 [2]), the most food waste occurs in households (61%), followed by food service (26%) and retails (13%). Just in Spain, and according to data published in 2020 by The Spanish Ministry of Agriculture, Fisheries and Food (MAPA), the households wasted 1.36 million kg, being approximately 1 million kg of unprocessed products, representing 44%, fruits and vegetables, and 325 million kg to cooked dishes [3]. This situation worsened during the COVID-19 pandemic, since the limitation in purchases and, therefore, the need for food storage, urged by the health alert situation. Food preservation plays a key role in reducing economic losses in different sectors of the population, from cultivation, production, and household economy, as well as reducing the environmental impact. Considering this alarming data, reducing food waste and food loss is included among the United Nations Sustainable Development Goals (SDGs), as Goal 12.3 aims to reduce food loss along the food chain.

Heat treatment or the use of preservatives gives the food industry room to work on food products. However, fresh or minimally processed foods, such as convenience products, have a very limited shelf life, and extending the expiration date is a challenge. The adequacy of packaging materials to each kind of food, as well as modification of oxygen on packages, are very useful tools for limiting the growth of microorganisms and retarding degradation reactions such as fatty acid oxidation. The great efforts of engineering of materials for food packaging led to the study and optimization of new generation biomaterials from renewable sources, the study of new polymers and fillers to obtain a material suitable regarding physical and mechanical properties, as well as the emergence of edible films [4]. However, the majority of food packaging publications lead to the inclusion of bioactive compounds that helped in the preservation. Thus, active and intelligent packaging seems to be the promising future of the food packaging industry.

Active packaging provides more than just a simple barrier to external detrimental conditions; they play an active role in food and quality preservation during the marketing process. They modify factors associated with food preservation, namely absorbing/scavenging, releasing/emitting, and moisture, temperature, microbial, and quality control [5]. In this sense, active packaging improves food preservation and prolongs its shelf-life, while consumers are more informed about its quality, which eventually contributes to reducing food waste. Although chemicals have long been used as preservers, including active substances from natural sources is one of the most ground-breaking alternatives. Polyphenols are substances with well-known antioxidant and antimicrobial capacities among others that can be obtained from agricultural by-products. Natural extracts from by-products have already been used in different applications [6]: as coloring agents (such as chlorophyll, carotenoid, anthocyanin, betalain, or curcumin), flavors and fragrances (especially essential oils), edible oils (applied as shortenings or salad dressing), functional food and nutraceuticals (mainly phytosterols, tocopherols, carotenoids or polyphenols) and preserving agents (being polyphenols and essential oils the main compounds used). In this sense, researchers have focused on the study of emerging techniques that could be used for food formulation and its incorporation into food-grade polymers, not only because they are natural resources, generally classified as GRAS (Generally Recognized As Safe) in the food industry [7], but also for being a mixture of compounds with a broader spectrum of antioxidant and antibacterial activity than that provided by a single compound [8].

By incorporating these compounds into the wrapping materials, the food industry is indirectly adding value to its by-products. However, the inclusion of these compounds in the formulation is not an easy task. One of the factors to bear in mind when using these compounds is their thermolability [9]. Sometimes, the methods used to incorporate them into polymers involve the use of high temperatures that may result in a significant mass loss and, therefore, the "active" character of the polymer might be compromised. In this sense, new strategies should be studied to incorporate active agents into porous matrices.

Several authors have classified active packaging into four groups depending on their mode of action [10] as i) absorbing systems (scavengers)

(Type I), those which absorb or retain undesirable substances from the atmosphere inside the packaging; ii) the releasing systems (emitters), (Type II) where beneficial substances act on the product surface or headspace; iii) the thermal systems (Type III), which transfer heat inside the packaging and iv) the regulatory systems (Type IV), which regulate the entry or exit of undesirable gases. The active films developed using natural extracts generally behave as emitters. The successful performance of this kind of active packaging depends on several factors:

1. The nature of the active substance: migration capacity depending on the volatility of the compounds. Volatile compounds exert their action in the headspace and generally lower concentrations are required to achieve their effect. On the other hand, these compounds can have a greater sensory effect.
2. Interactions between the raw material and the active compound: the composition of the food in terms of pH, water activity, lipids, influences the migration kinetics of the compounds.
3. The interactions between the film and the active substance, in which not only the nature of the film and the amount of active substance are involved in the modulation of migration, but also the manufacturing technique will have a strong influence on the migration kinetics of the compounds. For example, the addition of different polymer formulation, plasticizers or cross-linkers, respectively facilitate or hinder the crystallinity of the polymer and its flexibility and, therefore, its ability to transfer compounds [11]. The fundamental objective of active packaging is that it modulates the migration of substances. High content of volatile active compounds leads to high migration as the partition coefficient favors their migration. However, higher migration is not always desirable since it can lead to prooxidant effects [12]. On the other hand, non-volatile compounds that have a higher affinity for the polymer may require a higher concentration in the polymer to favor their migration.

Taking these three premises into account, the active packaging process can be adjusted to the needs of the food. Among the more named methodologies can be found casting/coating, thermoforming, spinning, spraying, or blow molding, among others [13]. Table 1 shows a review of some literature using active films with natural extracts developed by different procedures.

Table 1. Active packaging developed by different functionalization techniques using natural extracts (antioxidant activity (AO), antibacterial activity (AB), antifungal (AF))

Extract	Functionalization technique	Matrix	*In vitro* bioactivity	Food tested	Reference
Green Tea	Encapsulation/extrusion	Polyethylene	AO (DPPH, ORAC)	-	[14]
	Encapsulation/casting	Hydroxypropyl methylcellulose/Polylactic acid	AO (DPPH)	-	[15]
Zataria multiflora	Coating	Paper	AO (DPPH, ORAC) AB (*S. aureus, L. inocua, E. coli, P. aeruginosa, S. enteritidis*)	-	[16]
Curcumin	Coating	Paper	AO (DPPH, ORAC) AB (*S. aureus, L. inocua, E. coli, P. aeruginosa, S. enteritidis*)	-	[16]
Thyme	Electrospinning	Gelatin nanofibers	AB (*C. jejuni)*	-	[17]
Rosemary	Electrospinning	Polyvinyl Alcohol	AO (DPPH)	-	[18]
Anthocyanin from red cabbage	molding	Fish gelatin film	AO (DPPH)	-	[19]
Garlic	Coating	Polyethylene/Ethylene vinyl-alcohol	AF (*P. expansum)*	-	[20]
Grapefruit seed	Extrusion	Low-density polyethylene and Poly-lactic acid	AB (*L. monocytogenes, E. coli)*	-	[21]
Grape pomace	Supercritical Solvent Impregnation	Jute	AO (DPPH) AB (*S. aureus, E. coli, P. aeruginosa, S. enteritidis*)	-	[22]
Mango leaves	Supercritical Solvent Impregnation	Polyethylene Terephthalate/ Polypropylene film	AO (DPPH)	Tangerine and lettuce	[23]
	Casting and Supercritical Solvent Impregnation	Nanofibrillated Cellulose	AO (DPPH) AB (*S. aureus, E. coli*)	-	[24]

Table 1. (Continued)

Extract	Functionalization technique	Matrix	*In vitro* bioactivity	Food tested	Reference
Olive leaves	Supercritical Solvent Impregnation	Polyethylene Terephthalate /Polypropylene film	AO (DPPH) AB (*S. aureus, E. coli, P. aeruginosa, S. enteritidis*)	Cherry Tomato	[25]
	Supercritical Solvent Impregnation	Polyethylene Terephthalate / Polypropylene film	Peroxide Value	Peeled sunflower seeds	[26]
Apricot kernel oil	Casting	Chitosan	AO (H_2O_2, DPPH) AM (*B. subtilis, E. coli*)	Bread slices	[27]
Banana peels extract	Casting and coating	Chitosan	AO (DPPH, FRAP, ABTS)	Apple	[28]
Lavender essential oil	Casting	Gelatin	AO (ABTS) AB (*S. aureus, E. coli*)	Shrimps	[29]
Clove	Infusion	Linear Low-density polyethylene	AB (S. *Typhimurium*, L. monocytogenes)	-	[30]
	Casting	Carboxymethyl cellulose /Polyvinyl alcohol	*AB (B. cereus, S. aureus, E. coli, P. fluorescens)*	Chicken meat	[31]
Tomato seeds	Coating	Gelatin	AO (DPPH, FRAP, ABTS) TBARS	Pork meat	[32]
Lemongrass	Casting	Poly-lactic acid	AB (*L. monocytogenes*)	Pork sausages	[33]
Artemisia essential oil	Edible coating	Chitosan	AB, AF (Coliforms, molds, and yeast)	Chicken meat	[34]
Oregano and thyme essential oils	Edible coating	Sodium alginate	AO (DPPH, ABTS) TBARS	Lamb patties	[35]

DPPH: 2,2-diphenyl-1-picrilhidrazil; FRAP: Ferric Reducing Antioxidant Power; TBARS: Thiobarbituric Acid Reactive Substances; ABTS: 2,2′-Azino-bis-3-ethylbenzthiazoline-6-sulfonic acid.

Edible packaging has been highly explored in the last years since it supposes an alternative eco-friendly to traditional packaging, especially in terms of solid waste reduction. Its effect in preservation is very high, as it is the sensory impact over food, so this fact has to be highly considered when commercializing this product [4]. On the other hand, casting is a procedure in which a thin layer of polymer solution is spread on a mold and is formed after a drying process. As can be seen in Table 1, there is a deep investigation of this film approach at laboratory scale, due to the easy process and the mild temperature used in the drying step. However, there are important drawbacks to consider. One of the main important ones is the addition of the active substance is limited to the good solubility on the polymeric solution, to guarantee good homogeneity and mechanical properties after polymerization. If not, wrinkling and breaking might result from the peeling of the mold [36]. Besides, the retention rate of active compounds is in some cases relatively high due to higher binding extracts-film [37]. Moreover, extrusion is one of the traditional film processing methods. It is based on the melting of the polymer at high temperatures and extrusion for its formation. Therefore, it uses high temperatures which can result in a lack of functionality of the active films. If the mixture is further blow-formed, another inconvenience can arise, that is the modification of the rheological properties of the mixture can hinder the blowing, so the formulation of the film has to be carefully adjusted [38]. Electrospinning is a technique for producing polymeric fibers that can be prepared functionalized and added to other packaging plastics as fillers or coatings. Although functional fibers have been obtained with this technique, the density of the mixture for fiber production is a factor limited by the compound/polymer ratio, which generates certain limitations. Furthermore, in comparison with casting methodologies, Tampau et al. (2020) have reported a higher release of encapsulated carvacrol in poly-lactic acid films made by casting versus electrospinning fibers, which retains more strongly the active compounds, which may be a limitation for the packaging of highly perishable foods [39]. All in all, the functionalization techniques before the polymerization process have a common limitation, that is the limited amount of active substance to include in the formulation to preserve the mechanical properties, and the higher retention of the compound in the matrix, which could lead to a limited preservation action.

The list of active packaging processing methodologies has been joined in recent years by Supercritical Solvent Impregnation (SSI). This technology allows the functionalization of polymers once polymerized using supercritical CO_2, modulating the polymer loading through the operating conditions. This technique has been widely used in biomedical applications, such as dental floss [40], biodegradable patches [41], intraocular lenses [42], filaments for 3D printing [43] or wound dressings [44] as well as in food packaging and natural fibers, as will be further disclosed.

2. Supercritical Solvent Impregnation (SSI) of Food Polymers

Supercritical fluids are substances at pressure and temperature levels above their critical point. CO_2 is the most commonly used supercritical fluid due to its numerous advantages, such as its low cost, natural abundance, non-flammability, and non-toxic nature [45], all of which makes it be considered as a "green solvent". In fact, the use of CO_2 as a solvent, together with other solvents such as ethanol or water, is permitted for the development of food products or ingredients.

With moderate critical temperature (T_c = 31.06 °C) and critical pressure (P_c = 73.8 bar), CO_2 supercritical region is easily reached, so it is possible to develop processes at near ambient temperatures without any degradation of thermolabile substances. Furthermore, CO_2 provides an inert medium that is suitable to process substances that are susceptible to oxidation [46]. Although their solvent power is the highest for non-polar or slightly polar components and decreases as molecular weight increases, the density, viscosity, and thus solvent power of supercritical CO_2 can be modulated by changing some of the operational parameters. Therefore, operational parameters such as pressure, temperature, or the addition of low amounts of co-solvent would allow selective extraction of compounds [47]. However, the uses of supercritical CO_2 ($scCO_2$) go beyond extraction and matrices can be impregnated with active compounds under high pressure to serve different purposes such as the alimentary one.

SSI benefits from the high diffusivity and low viscosity of CO_2 when used as a carrier to deliver bioactive molecules into polymer matrices. Furthermore, SSI offers some additional advantages such as polymer surfaces do not need to be modified to improve their adhesion properties [30] or the use of moderate operating temperatures does not compromise the stability of heat-sensitive substances. This is a substantial advantage when compared to other methods to incorporate active substances into polymers, such as thermal extrusion, where plastic matrix processes require temperatures over 200 °C [48], which would compulsorily force to include the active substance after processing the film.

From a technical point of view, SSI comprises three main steps [49, 50] that are represented in Figure 1: i) Solubilisation of the active substance into the $scCO_2$ and its saturation into the vessel, ii) Swelling of the polymer matrix, in which the polymer opens its chains and let the impregnation of the active substance –mainly in the amorphous region of the polymer–, and iii) Depressurization of the system and recover of the functionalized polymer ready to use.

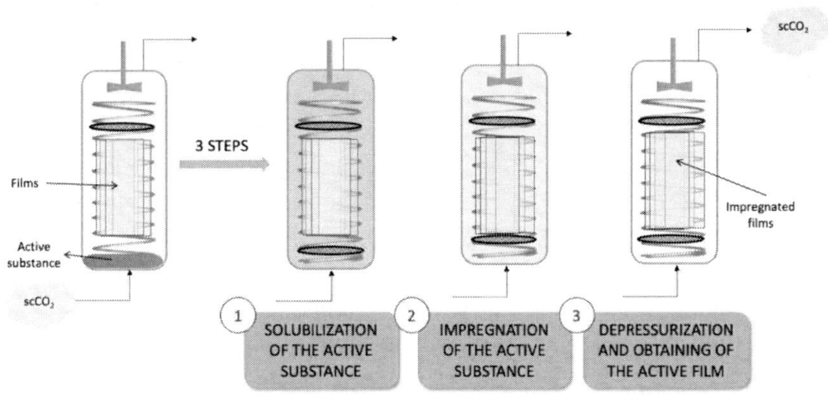

Figure 1. Schematic representation of the impregnation process of films.

Several aspects have an impact on the impregnation process results. Pressure and temperature are operational parameters that affect not only the solubility of the active substance into $scCO_2$ but also the polymer's sorption capacity. Furthermore, time and depressurization rate, the use of modifiers, each particular polymer's properties as well as the rest of the interactions or affinities of the ternary system CO_2-active substance-polymer also have an impact on the efficiency of the process [51]. In this sense, the loading rate can be controlled by modifying these operating parameters [52].

Regarding pressure and temperature ranges employed in supercritical impregnation, 90–200 bar and 35–55 °C are the most frequently found in the literature. However, higher pressure and temperature levels have also been studied due to the suitability of high pressure on impregnation yields. Champeau et al. [52] (2015) reported better yields on the impregnation of polyethylene terephthalate and polypropylene suture threats as pressure and temperature are increased. Probably, the employment of higher pressures can be adequate mainly in more crystalline polymers, due to the increase of swelling effect, which promotes impregnation [53].

As far as process time is concerned, different values have been reported in the literature depending on the matrix and active substance used [54, 55]. The impregnation procedure is both a thermodynamic and a kinetic process, so time is a dependent parameter in impregnation efficiency. The dissolution of the active substance and its diffusion into the matrix depend on the $scCO_2$ density because its lubricant effect [56]. Although generally high times are employed to not be a limiting factor of the process, in the case of the impregnation of film polymers, the processing time could be faster than in other polymers such as filaments, due to the width of the matrix is just some micrometers. Cejudo et al. (2017) [57] studied the variation of loading with time when impregnating olive leaf extract in polyethylene terephthalate/polypropylene films and demonstrated that impregnation started as soon as the extract is dissolved, since 5 minutes offered results comparable to 1 h, and even superior to those at 22 h. On the contrary, Rosales et al. (2021) [43] observed that the impregnation of mango leaf extracts in poly-lactic acid was superficial at 1 h of impregnation, and much superior at 24 h, observing the diffusion of the extract in the inner part of the filament.

Regarding the use of modifiers, minor amounts of polar solvents can be added when the active substances do not solve satisfactorily in CO_2. Therefore, polar and non-polar compounds can be impregnated [58]. Besides, modifiers act as enhancers of the swelling effect on polymers. However, depending on the other operational parameters named pressure, temperature, impregnation time, and density of supercritical CO_2 among others, the amount of co-solvent can be reduced or it may not even be required [59]. Hence, the relation CO_2/co-solvent is a parameter that has to be studied together with the rest of the operational parameters, particularly when employing substances that present low solubility in pure CO_2. It should also be taken into consideration that polymers' loading capacity is limited, and consequently, the amount of active substance that can be impregnated effectively is limited too. These conclusions have already been discussed by several authors [60-62].

Another important factor for guaranteeing the impregnation efficiency is the depressurization rate. Once the impregnation time is over, the depressurization starts and different velocities can be modulated to favor the impregnation rate while maintaining the mechanical parameters of polymers. Generally, the depressurization step is a crucial factor in polymers, since the CO_2 adsorbed can be expelled drastically from the inner of the polymer, which can cause nucleation and eventually foaming, especially if the temperature exceeds the glass transition temperature (*Tg*) of the polymer [63]. In films case, the small width limited the CO_2 sorption and hence the foaming probability, but microbubbles can be observed on films if this parameter is not correctly controlled [24]. In this sense, it is possible to see mechanical damages on films treated at higher depressurization rates [51]. On the other hand, the affinity solute-polymer has to be considered when studying the depressurization rate, to avoid causing dragging of the active substance at the end of the process if the partition coefficient favour the dissolution in the supercritical phase [52].

Concerning polymer properties such as composition, mobility chain, or crystallinity influence the impregnation load. Semi-crystalline polymers have both crystalline structure and amorphous regions. Several studies have reported that it is in the amorphous regions where impregnation mainly takes place due to a lower reorganization of the chains, which allows better penetration and sorption of $scCO_2$ [64]. This penetration causes a temporary increase of the polymer's free volume (swelling) since the CO_2 acts as a sort of lubricant by augmenting the segmental chain mobility inside the polymeric matrix and thus plasticizing the polymer. This phenomenon enhances the diffusion of the active compound into the matrix, which favors an easier and faster impregnation process [23]. On the other hand, the presence of functional groups in the polymer chain, such as carbonyl or amine groups, can also favor impregnation by reacting with the CO_2 despite their low mobility chain [65]. CO_2 sorption is related to pressure and temperature, as well as to the kind and amount of co-solvent or modifier used for the impregnation [66]. A partition equilibrium of the bioactive substance between polymer and CO_2 in the supercritical phase takes place [67] and the impregnation occurs if the partition coefficient is in favor of the polymer. Otherwise, it will be precipitated onto the polymer surface, the bottom of the vessel, or dragged to the cyclonic separator during the depressurization phase.

In polymer impregnation, different impregnation procedures have been reported in the bibliography:

- via static or batch mode (BM), the amount of matrix, active substance, and CO_2 used in the experiments remains constant.
- via the dynamic or semi-continuous mode (SM), where the amount of matrix is invariable but the active substance and/or CO_2 are constantly renewed.
- An integrated supercritical fluid extraction-impregnation procedure (SFE-SSI).

While BM and SM can be carried out in one or two vessels, the integrated SFE-SSI can only be performed in two vessels. BM and SM procedures employ an extract that is obtained from a previous step, while in the SFE-SSI procedure, extraction and impregnation occur simultaneously. BM is the most commonly used of the supercritical impregnation methods, although SM processes are being explored focused on the feasibility of the scaling of the process [68]. For BM the extract is loaded inside the vessel and the process takes place while the vessel's outlet is closed. The $scCO_2$ is gradually introduced until the pre-set pressure conditions are reached inside the vessel. In contrast, in SM different devices have been reported, since generally the CO_2 is pumped into the system together or not with the extract, to renew the active substance and the carrier into the system [69]. The integrated process SFE-SSI has been scarcely reported but holds a potential interest for the production of highly active polymers at an industrial scale. This process uses a supercritical device formed by two vessels. The first one contains the material to be extracted and the supercritical extraction is carried out. Then, the outlet of the extraction vessel is coupled to the impregnation vessel's inlet, and the extract is directly introduced to impregnate the polymer. There is a recirculation of the CO_2+extract supply from the impregnation vessel's outlet to the extraction inlet so that the extract's loss in the tubes, vessels, and exchangers of the equipment is minimized [69]. This system is limited by the fact that generally, supercritical conditions are the same for both vessels, i.e., extraction and impregnation, which forces a compromise between optimal extraction and impregnation conditions. However, results found in the literature for SM and the integrated process pointed to a great decrease in the impregnation yield. For instance, Maya et al. (2021) compared the impregnation in BM and SM in two vessels in PP for agronomic used impregnated with pyrethrin extract [70], pointing to BM as a more convenient methodology. On the other hand, Valor et al. (2021) achieved the same conclusion when impregnating alginate wound dressings with mango leaf extract [71]. Hence, the alternative process to BM impregnation has to be deeply explored to achieve comparable results.

3. Active Packaging Developed by Supercritical Impregnation with Natural Extracts

Among the type of active packaging described above, those developed by SSI has focused on the emitters system, especially those with an antimicrobial and antioxidant capacity [9], offering very promising results both *in vitro* and food tests. The basic of these formulations is to guarantee that the film possesses the bioactivity of the initial active substance and it can exert its action over the food, so migration studies are of great interest. Although migration studies are complex and very dependent on the active substance, the polymer affinity, and the simulant formulation, valuable information can be obtained when comparing the migration among different functionalization techniques. Although there is scarce bibliography comparing the kinetic migration of active films developed by SSI and other different methodologies, the majority of them pointing better results when using this technology. For instance, Cejudo et al. (2021) compared the antioxidant capacity of active nanofibrillated cellulose films manufactured by casting and SSI with mango leaf extract, observing that those developed by SSI exert their action slightly faster and at a higher rate according to their antioxidant capacity kinetic reaction [24]. Similar conduct was also observed by López di Castillo et al. (2018), who compared the kinetic migration of poly-lactic acid electrospun mats functionalized with cinnamaldehyde by direct addition and SSI. In this case, the cinnamaldehyde fibers migrated faster but at a minor rate than the SSI ones in a 50% ethanol simulant [72]. In agreement, Coutinho et al. (2021) observed a higher loading rate of aspirin in poly-lactic acid films submitted to supercritical impregnation than traditional soaking, as well as a more progressive release on time [73].

The research on supercritical impregnation of food polymers aims mainly on the impregnation of standards of essential oils [74-76] and inorganic compounds [58, 77] but there is a high interest in the impregnation of polyphenol rich-natural extracts [23, 57, 78]. Although the optimal conditions have to be studied for each system, some general conclusions can be obtained. In general, essential oils or the major compounds of these (thymol, carvacrol, etc.) are the best options to impregnate polymers, due to the good solubility with CO_2 and the good swelling properties. They are better impregnated at moderate temperatures and pressures (100-150 bar, 35-40 ºC), where the solubility of the compounds is not maximized in CO_2 and, therefore, the substance-polymer affinity is favored [69]. In fact, some polymers start to degrade at high pressures and temperatures, causing foaming or even melting or deformation [43]. However, when using crystalline polymers or compounds with a lower solubility in CO_2, higher CO_2 density offers better results [53, 75]. In the case of polymer-rich extracts, the optimization of the impregnation is more complex due to the heterogeneous composition, where the molecular weight and the solubility of the compounds are very variable, and the impregnation can be very selective. For example, Belizón et al. [23] despite observing a high impregnation efficiency and a good antioxidant capacity of the films, observed a great selectivity of the impregnated compounds of mango leaf extract, identifying only methyl gallate in the impregnated polyethylene terephthalate/polypropylene films. On the other hand, Cejudo et al. (2018) [26] observed the impregnation of a great number of compounds from olive leaf extract in the same polymer, although there was a great impregnation percentage of oleuropein. It has been stated previously that the effectiveness of the impregnation depends on a great extension of the polymer composition and characteristics, so if the desired polymer does not possess favorable properties and/or affinity for the extract, some activation techniques are also explored to increase the impregnation yield. These treatments are focused on creating free radicals and additional polar groups that promote impregnation, as well as decrease the crystallinity of polymers. Some of the options explored are the corona discharge treatment [79, 80], the plasma, the UV treatment, or the addition of acids or oxidizers [81] that can be employed before the supercritical treatment. The conditions employed in SSI could lead to think to a greater change in the mechanical properties of the film compared to other techniques. However, despite the great pressures employed, these are not substantial compared to the untreated films [53]. Thus, conclusions are comparable to those found with other techniques, where the inclusion of a different compound in the film formulation influences the change in the mechanical properties to a greater extent than the CO_2 presence [82].

Despite the good results, the SSI technique has also some aspects to consider. The direct addition of compounds has a great advantage over the SSI, i.e., knowing the amount of the compound loaded into the film. However, sometimes the determination of the loading and the bioactivity is not easy when they are developed by supercritical impregnation. Different alternatives have been explored for the determination of the drug loading of films. The first articles about impregnation quantify the loading gravimetrically before and after the treatment. However, this measurement sometimes has substantial deviations when the polymer has great CO_2 sorption. In this regard, a diffusion time after treatment has to be considered before calculating the loading gravimetrically to not over-dimension, as some authors have observed [75, 83]. Besides, lipophilic compounds can be extracted from polymer formulation during depressurization, providing uncertain results [84]. Another alternative is the indirect determination through a reaction as DPPH standard [82]. This could be applied to films, but not to other greater polymers where the impregnation takes place deeper. The most certain methodology is the dissolution of the treated polymers in organic solvents and its further determination by chromatography or UV-Vis [69]. In this sense, non-destructive techniques rapid and accurate should be explored.

Regarding the bioactive activity of natural extracts, it is well known the broad spectra in terms of antioxidant and antibacterial activity. A great number of studies reported both the inherent activity of the natural extract and the acquired activity of the films after impregnation. Both Gram+ and Gram – bacteria have been inhibited on *in vitro* assays in presence of impregnated polymers [22, 24, 85-87], being *Staphylococcus aureus* and *Escherichia coli* the most used as representative foodborne bacteria. However, although the assays in real food have been very reported in other active packaging developed by other technologies (*viz.* Table 1), it has not been as much in films developed by supercritical impregnation, where the main analysis is still focused on the controlling of the process and the film properties modification after treatment. Regarding the antioxidant capacity, Cejudo et al. [26] reported a significant delay in the lipid oxidation of peeled sunflower seeds packaged into polyethylene terephthalate/polypropylene impregnated with olive leaf extract. Peeled raw nuts are extremely sensitive to oxidation, and to find alternatives to chocolate covering or fried treatment that preserve nuts is really interesting in the offering of commercial healthier alternatives. The impregnation conditions used to produce the polyethylene terephthalate/polypropylene -OLE film was 400 bar and 35 ºC during 1 h by batch mode, which offers films with a high oleuropein content. The product was packaged with no use of modified atmosphere, to evaluate just the influence of the active film. The storage of 15 days revealed a significantly lower peroxide of the seeds package into the impregnated film value (~60 meq O_2/Kg) respecting to those into no impregnated (~80 meq O_2/Kg). Another application of supercritical impregnation is the direct impregnation of the antioxidants over peeled nuts [88, 89]. Nevertheless, the use of high pressures has some limitations in this field since it can only be applied on foods capable of withstanding the operating conditions, finding some breakings in nuts after depressurization. In this sense, the application on ground nuts, which are even more sensitive to oxidation, could be more appropriate to direct impregnation. In the case of fresh vegetables and fruits, Cejudo et al. [25] reported a study of the effectiveness of impregnated polyethylene terephthalate/polypropylene with olive leaf extract at 400 bar and 35 ºC during 1 h by batch mode. First, the antimicrobial capacity was proven *in vitro* against different pathogenic bacteria, i.e., *E. coli, S. aureus, and P. aeruginosa* observing growth inhibition in *ca.* 75, 81, and 89%

respectively. Then, the efficiency of the packaging was verified in the cherry tomato preservation. In this experiment, 5 cherry tomatoes were wrapped into a package with any atmosphere modification. The study monitored the Senescence of cherry tomato for 50 days, observing the visual alteration and the observation of mycelia growth. It was found an increase in the shelf life of cherry tomatoes in 20 days respecting the control wrapped into the untreated films. A similar study was also conducted by Belizón et al. [23] using polyethylene terephthalate/polypropylene impregnated with mango leaf extract at 100 bar and 35 ºC during 22 h. The experiment was carried out wrapping one lettuce leaf into the impregnated film, using as control one wrapped into non impregnated as well as one unwrapped. In parallel, the tangerine preservation was studied carrying out the same procedure. Results demonstrated the increase of shelf life in both foods, lettuce and tangerine, increasing the shelf life in 14 and 39 days respectively, respecting to those wrapped into non impregnated films. Both olive and mango leaf extracts are rich in phenolic compounds and not in volatile ones as essential oils extracts, which are more likely to migrate into the headspace of the package. However, both experiments confirm the action of this emitter system in food preservation.

Conclusion

The SSI technology supposes the use of a sustainable methodology to create a new packaging material that minimizes food waste. Although impregnation is a complex process whose variables can be studied, it can be a great green alternative for the creation of emitters due to high impregnation yields as well as improved controlled release. The first step in the study of active packaging is the control of the technology and the process, to ensure the efficacy of the packaging, both in *in vitro* analysis and real food. This is the direction in which the current studies reported in this chapter have focused. However, there is still a long way ahead regarding the scaling of the process, mainly attending to the design of the reactors and the inclusion of the film to guarantee homogeneity in the impregnations. Another issue that has to be resolved before commercialization is the sensory impact of these packages on foods. This impact can be especially important when impregnated with essential oils, whose highly volatile nature and particularly low perception threshold can cause a high sensory impact.

The creation of active packaging materials could enhance diverse aspects such as the economy of the agricultural companies and the food industry and also the health of consumers. Regarding the first one, the recycling of natural agro-food by-products contributes to the circular economy and to low the environmental impact, revalorizing its value and giving the agricultural companies other economical input. Moreover, the decrease in food waste optimizes the food industry's resources, improving its profits and reducing its ecological footprint. On the other hand, the extended shelf life of fresh products opens a door to alternative marketing channels, such as vending machines, which can offer a wider range of fresh products such as fruits and vegetables, instead of other more caloric products with longer shelf life, such as processed bakery products. Having said so, preserving fresh products is not an easy task. Our colleagues will specifically talk about this topic and present to you how food packaging films can be designed and applicable to package fresh products (particularly green vegetables and mushrooms) in the next two chapters.

References

[1] Food loss and waste database. Food and Agriculture Organization of the United Nations. Online. April 2022. Available from URL: https://www.fao.org/food-loss-and-food-waste/flw-data/en/.

[2] Food Waste Index Report 2021. Food and Agriculture Organization of the United Nations. Online. April 2022. Available from URL: https://www.fao.org/platform-food-loss-waste/resources/detail/en/c/1378978/.

[3] Household food waste 2021. Ministry of Agriculture, Fisheries and Food (MAPA) Spanish Gobernment. Online. Available April 2022 from URL: https://www.lamoncloa.gob.es/serviciosdeprensa/notasprensa/agricultura/Paginas/2021/050821-desperdicio_alimentos.aspx.

[4] Verma, M. K., S. Shakya, P. Kumar, J. Madhavi, J. Murugaiyan, M. V. R. Rao, Trends in packaging material for food products: historical background, current scenario, and future prospects, *Journal of Food Science and Technology* 58(11) (2021) 4069-4082.

[5] Amaral, R. A., C. A. Pinto, V. Lima, J. Tavares, A. P. Martins, L. G. Fidalgo, A. M. Silva, M. M. Gil, P. Teixeira, J. Barbosa, F. J. Barba, J. A. Saraiva, Chemical-Based Methodologies to Extend the Shelf Life of Fresh Fish—A Review, *Foods* (Basel, Switzerland) 10(10) (2021).

[6] Cavalcanti, R. N. T. Forster-Carneiro, M. T. M. S. Gomes, M. A. Rostagno, J. M. Prado, M. A. A. Meireles, CHAPTER 1 Uses and Applications of Extracts from Natural Sources, Natural Product Extraction: Principles and Applications, *The Royal Society of Chemistry* 2013, pp. 1-57.

[7] Özge Erdohan, Z., B. Çam, K. N. Turhan, Characterization of antimicrobial polylactic acid based films, *Journal of Food Engineering* 119(2) (2013) 308-315.

[8] Mousavi Khaneghah, A., S. M. B. Hashemi, S. Limbo, Antimicrobial agents and packaging systems in antimicrobial active food packaging: An overview of approaches and interactions, *Food and Bioproducts Processing* 111 (2018) 1-19.

[9] Rojas, A., A. Torres, M. José Galotto, A. Guarda, R. Julio, Supercritical impregnation for food applications: a review of the effect of the operational variables on the active compound loading, *Crit Rev Food Sci Nutr* 60(8) (2020) 1290-1301.

[10] Wyrwa, J., A. Barska, Innovations in the food packaging market: active packaging, *European Food Research and Technology* 243(10) (2017) 1681-1692.

[11] Göksen, G., M. J. Fabra, A. Pérez-Cataluña, H. I. Ekiz, G. Sanchez, A. López-Rubio, Biodegradable active food packaging structures based on hybrid cross-linked electrospun polyvinyl alcohol fibers containing essential oils and their application in the preservation of chicken breast fillets, *Food Packaging and Shelf Life* 27 (2021) 100613.

[12] Kuai, L., F. Liu, B. S. Chiou, R. J. Avena-Bustillos, T. H. McHugh, F. Zhong, Controlled release of antioxidants from active food packaging: A review, *Food Hydrocolloids* 120 (2021) 106992.

[13] Silva, A. O., R. S. Cunha, D. Hotza, R. A. F. Machado, Chitosan as a matrix of nanocomposites: A review on nanostructures, processes, properties, and applications, *Carbohydrate Polymers* 272 (2021) 118472.

[14] Wrona, M., C. Nerín, M. J. Alfonso, M. Á. Caballero, Antioxidant packaging with encapsulated green tea for fresh minced meat, *Innovative Food Science & Emerging Technologies* 41 (2017) 307-313.

[15] Wrona, M., M. J. Cran, C. Nerín, S. W. Bigger, Development and characterisation of HPMC films containing PLA nanoparticles loaded with green tea extract for food packaging applications, *Carbohydrate Polymers* 156 (2017) 108-117.

[16] Akrami, F., A. Rodríguez-Lafuente, K. Bentayeb, D. Pezo, S. R. Ghalebi, C. Nerín, Antioxidant and antimicrobial active paper based on Zataria (Zataria multiflora) and two cumin cultivars (Cuminum cyminum), LWT - *Food Science and Technology* 60(2, Part 1) (2015) 929-933.

[17] Lin, L., Y. Zhu, H. Cui, Electrospun thyme essential oil/gelatin nanofibers for active packaging against Campylobacter jejuni in chicken, *LWT* 97 (2018) 711-718.

[18] Estevez-Areco, S., L. Guz, R. Candal, S. Goyanes, Release kinetics of rosemary (Rosmarinus officinalis) polyphenols from polyvinyl alcohol (PVA) electrospun nanofibers in several food simulants, *Food Packaging and Shelf Life* 18 (2018) 42-50.

[19] Uranga, J., A. Etxabide, P. Guerrero, K. de la Caba, Development of active fish gelatin films with anthocyanins by compression molding, *Food Hydrocolloids* 84 (2018) 313-320.

[20] Heras-Mozos, R., V. Muriel-Galet, G. López-Carballo, R. Catalá, P. Hernández-Muñoz, R. Gavara, Development and optimization of antifungal packaging for sliced pan loaf based on garlic as active agent and bread aroma as aroma corrector, *International Journal of Food Microbiology* 290 (2019) 42-48.

[21] Wang, L. F., J. W. Rhim, Grapefruit seed extract incorporated antimicrobial LDPE and PLA films: Effect of type of polymer matrix, *LWT* 74 (2016) 338-345.

[22] Cejudo Bastante, C., P. Arjona Mudarra, M. T. Fernández-Ponce, L. Casas, C. Mantell, E. J. Martínez de la Ossa, C. Pereyra, Application of a natural antioxidant from grape pomace extract in the development of bioactive jute fibers for food packaging, *Antioxidants* 10(2) (2021) 1-15.

[23] Belizón, M., M. T. Fernández-Ponce, L. Casas, C. Mantell, E. J. Martínez de la Ossa Fernández, Supercritical impregnation of antioxidant mango polyphenols into a multilayer PET/PP food-grade film, *Journal of CO_2 Utilization* 25 (2018) 56-67.

[24] Cejudo Bastante, C., N. H. C. S. Silva, L. Casas Cardoso, C. Mantell Serrano, E. J. Martínez de la Ossa, C. S. R. Freire, C. Vilela, Biobased films of nanocellulose and mango leaf extract for active food packaging: Supercritical impregnation versus solvent casting, *Food Hydrocolloids* 117 (2021).

[25] Cejudo Bastante, C., L. Casas Cardoso, M. T. Fernández-Ponce, C. Mantell Serrano, E. J. Martínez de la Ossa, Supercritical impregnation of olive leaf extract to obtain bioactive films effective in cherry tomato preservation, *Food Packaging and Shelf Life* 21 (2019) 100338.

[26] Cejudo Bastante, C., L. Casas Cardoso, M. T. Fernández Ponce, C. Mantell Serrano, E. J. Martínez de la Ossa-Fernández, Characterization of olive leaf extract polyphenols loaded by supercritical solvent impregnation into PET/PP food packaging films, *The Journal of Supercritical Fluids* 140 (2018) 196-206.

[27] Priyadarshi, R., Sauraj, B. Kumar, F. Deeba, A. Kulshreshtha, Y. S. Negi, Chitosan films incorporated with Apricot (Prunus armeniaca) kernel essential oil as active food packaging material, *Food Hydrocolloids* 85 (2018) 158-166.

[28] Zhang, W., X. Li, W. Jiang, Development of antioxidant chitosan film with banana peels extract and its application as coating in maintaining the storage quality of apple, *International Journal of Biological Macromolecules* 154 (2020) 1205-1214.

[29] Wang, J., X. Sun, H. Zhang, M. Dong, L. Li, H. Zhangsun, L. Wang, Dual-functional intelligent gelatin based packaging film for maintaining and monitoring the shrimp freshness, *Food Hydrocolloids* 124 (2022) 107258.

[30] Mulla, M., J. Ahmed, H. Al-Attar, E. Castro-Aguirre, Y. A. Arfat, R. Auras, Antimicrobial efficacy of clove essential oil infused into chemically modified LLDPE film for chicken meat packaging, *Food Control* 73 (2017) 663-671.

[31] Muppalla, S. R., S. R. Kanatt, S. P. Chawla, A. Sharma, Carboxymethyl cellulose–polyvinyl alcohol films with clove oil for active packaging of ground chicken meat, *Food Packaging and Shelf Life* 2(2) (2014) 51-58.

[32] Gallego, M., M. Arnal, P. Talens, F. Toldrá, L. Mora, Effect of gelatin coating enriched with antioxidant tomato by-products on the quality of pork meat, *Polymers* 12(5) (2020).

[33] Yang, H. J., K. B. Song, Application of lemongrass oil-containing polylactic acid films to the packaging of pork sausages, *Korean Journal for Food Science of Animal Resources* 36(3) (2016) 421-426.

[34] Yaghoubi, M., A. Ayaseh, K. Alirezalu, Z. Nemati, M. Pateiro, J. M. Lorenzo, Effect of Chitosan Coating Incorporated with Artemisia fragrans Essential Oil on Fresh Chicken Meat during Refrigerated Storage, *Polymers* 13(5) (2021) 716.

[35] Pelaes Vital, A. C., A. Guerrero, P. Guarnido, I. Cordeiro Severino, J. L. Olleta, M. Blasco, I. Nunes do Prado, F. Maggi, M. D. M. Campo, Effect of Active-Edible Coating and Essential Oils on Lamb Patties Oxidation during Display, *Foods* (Basel, Switzerland) 10(2) (2021).

[36] Smaoui, S., H. Ben Hlima, L. Tavares, K. Ennouri, O. Ben Braiek, L. Mellouli, S. Abdelkafi, A. Mousavi Khaneghah, Application of essential oils in meat packaging: A systemic review of recent literature, *Food Control* 132 (2022) 108566.

[37] Benbettaïeb, N., F. Debeaufort, T. Karbowiak, Bioactive edible films for food applications: mechanisms of antimicrobial and antioxidant activity, *Critical Reviews in Food Science and Nutrition* 59(21) (2019) 3431-3455.

[38] Herrera, N., H. Roch, A. M. Salaberria, M. A. Pino-Orellana, J. Labidi, S. C. M. Fernandes, D. Radic, A. Leiva, K. Oksman, Functionalized blown films of plasticized polylactic acid/chitin nanocomposite: Preparation and characterization, *Materials & Design* 92 (2016) 846-852.

[39] Tampau, A., C. González-Martínez, A. Chiralt, Polyvinyl alcohol-based materials encapsulating carvacrol obtained by solvent casting and electrospinning, *Reactive and Functional Polymers* 153 (2020) 104603.

[40] Mosquera, J. E., M. L. Goñi, R. E. Martini, N. A. Gañán, Supercritical carbon dioxide assisted impregnation of eugenol into polyamide fibers for application as a dental floss, *Journal of CO_2 Utilization* 32 (2019) 259-268.

[41] Liparoti, S., P. Franco, R. Pantani, I. De Marco, Supercritical CO2 impregnation of caffeine in biopolymer films to produce anti-cellulite devices, *Journal of Supercritical Fluids* 179 (2022).

[42] Topete, A., B. Saramago, A. P. Serro, Intraocular lenses as drug delivery devices, *International Journal of Pharmaceutics 602* (2021).

[43] Rosales, J. M., C. Cejudo, L. Verano, L. Casas, C. Mantell, E. J. Martínez de la Ossa, Supercritical impregnation of pla filaments with mango leaf extract to manufacture functionalized biomedical devices by 3d printing, *Polymers* 13(13) (2021).

[44] Fernández-Ponce, M. T., E. Gómez, C. Cejudo Bastante, L. Casas, A. Montes, C. Mantell, E. J. M. de la Ossa-Fernández, C. Pereyra, Development of functionalized alginate dressing with mango polyphenols by supercritical technique to be employed as an antidiabetic transdermal system, *Journal of Supercritical Fluids* 175 (2021).

[45] Chinnarasu, C., A. Montes, M. T. F. Ponce, L. Casas, C. Mantell, C. Pereyra, E. J. M. De La Ossa, Precipitation of antioxidant fine particles from Olea europaea leaves using supercritical antisolvent process, *Journal of Supercritical Fluids* 97 (2015) 125-132.

[46] Machado, A. P. D. F., M. Rueda, G. F. Barbero, Á. Martín, M. J. Cocero, J. Martínez, Co-precipitation of anthocyanins of the extract obtained from blackberry residues by pressurized antisolvent process, *The Journal of Supercritical Fluids* 137 (2018) 81-92.

[47] Frolov, A. I., M. G. Kiselev, Prediction of cosolvent effect on solvation free energies and solubilities of organic compounds in supercritical carbon dioxide based on fully atomistic molecular simulations, *Journal of Physical Chemistry B* 118(40) (2014) 11769-11780.

[48] Apicella, A., P. Scarfato, L. Di Maio, L. Incarnato, Transport properties of multilayer active PET films with different layers configuration, *Reactive and Functional Polymers* 127 (2018) 29-37.

[49] Bouledjouidja, A., Y. Masmoudi, M. Van Speybroeck, L. Schueller, E. Badens, Impregnation of Fenofibrate on mesoporous silica using supercritical carbon dioxide, *International Journal of Pharmaceutics* 499(1-2) (2016) 1-9.

[50] Li, X. K., H. Lu, G. P. Cao, Y. H. Qian, L. H. Chen, R. H. Zhang, H. L. Liu, Y. H. Shi, Experimental Study of the Synergistic Plasticizing Effect of Carbon Dioxide and Ibuprofen on the Glass Transition Temperature of Poly(methyl methacrylate), *Industrial & Engineering Chemistry Research* 53(14) (2014) 5873-5885.

[51] Goñi, M. L., N. A. Gañán, M. C. Strumia, R. E. Martini, Eugenol-loaded LLDPE films with antioxidant activity by supercritical carbon dioxide impregnation, *Journal of Supercritical Fluids* 111 (2016) 28-35.

[52] Champeau, M., J. M. Thomassin, C. Jerome, T. Tassaing, In situ investigation of supercritical CO_2 assisted impregnation of drugs into a polymer by high pressure FTIR micro-spectroscopy, *Analyst* 140(3) (2015) 869-879.

[53] Cejudo Bastante, C., M. J. Cran, L. Casas Cardoso, C. Mantell Serrano, E. J. Martínez de la Ossa, S. W. Bigger, Effect of supercritical CO_2 and olive leaf extract on the structural, thermal and mechanical properties of an impregnated food packaging film, The Journal of Supercritical Fluids 145 (2019) 181-191.

[54] Dias, A. M. A., M. E. M. Braga, I. J. Seabra, P. Ferreira, M. H. Gil, H. C. de Sousa, Development of natural-based wound dressings impregnated with bioactive compounds and using supercritical carbon dioxide, *International Journal of Pharmaceutics* 408(1) (2011) 9-19.

[55] Dias, A. M. A., A. Rey-Rico, R. A. Oliveira, S. Marceneiro, C. Alvarez-Lorenzo, A. Concheiro, R. N. C. Júnior, M. E. M. Braga, H. C. de Sousa, Wound dressings loaded with an anti-inflammatory jucá (Libidibia ferrea) extract using supercritical carbon dioxide technology, *The Journal of Supercritical Fluids* 74 (2013) 34-45.

[56] Champeau, M., J. M. Thomassin, T. Tassaing, C. Jerome, Drug Loading of Sutures by Supercritical CO_2 Impregnation: Effect of Polymer/Drug Interactions and Thermal Transitions, *Macromolecular Materials and Engineering* 300(6) (2015) 596-610.

[57] Cejudo Bastante, C., L. Casas Cardoso, C. Mantell Serrano, E. J. Martínez de la Ossa, Supercritical impregnation of food packaging films to provide antioxidant properties, *Journal of Supercritical Fluids* 128 (2017) 200-207.

[58] Zhao, Y., K. Jung, Y. Shimoyama, Y. Shimogaki, T. Momose, Kinetic effects of methanol addition on supercritical fluid deposition of TiO_2, *The Journal of Supercritical Fluids* 138 (2018) 63-72.

[59] Bouledjouidja, A., Y. Masmoudi, M. Sergent, V. Trivedi, A. Meniai, E. Badens, Drug loading of foldable commercial intraocular lenses using supercritical impregnation, *International Journal of Pharmaceutics* 500(1) (2016) 85-99.

[60] Milovanovic, S., M. Stamenic, D. Markovic, J. Ivanovic, I. Zizovic, Supercritical impregnation of cellulose acetate with thymol, *The Journal of Supercritical Fluids* 97 (2015) 107-115.

[61] Rojas, A., D. Cerro, A. Torres, M. J. Galotto, A. Guarda, J. Romero, Supercritical impregnation and kinetic release of 2-nonanone in LLDPE films used for active food packaging, *The Journal of Supercritical Fluids* 104 (2015) 76-84.

[62] Q. Q. Xu, J. Z. Yin, X. L. Zhou, G. Z. Yin, Y. F. Liu, P. Cai, A. Q. Wang, Impregnation of ionic liquids in mesoporous silica using supercritical carbon dioxide and co-solvent, *RSC Advances* 6(103) (2016) 101079-101086.

[63] Jacobs, L. J. M., M. F. Kemmere, J. T. F. Keurentjes, Sustainable polymer foaming using high pressure carbon dioxide: a review on fundamentals, processes and applications, *Green Chemistry* 10(7) (2008) 731-738.

[64] Champeau, M., J. M. Thomassin, C. Jérôme, T. Tassaing, In situ FTIR micro-spectroscopy to investigate polymeric fibers under supercritical carbon dioxide: CO_2 sorption and swelling measurements, *The Journal of Supercritical Fluids* 90 (2014) 44-52.

[65] Cejudo Bastante, C., M. J. Cran, L. Casas Cardoso, C. Mantell Serrano, E. J. M. de la Ossa, S. W. Bigger, Effect of Supercritical CO_2 and Olive Leaf Extract on the Structural, Thermal and Mechanical Properties of an Impregnated Food Packaging Film, *The Journal of Supercritical Fluids* (2018).

[66] Hirogaki, K., I. Tabata, K. Hisada, T. Hori, An investigation of the interaction of supercritical carbon dioxide with poly(ethylene terephthalate) and the effects of some additive modifiers on the interaction, *Journal of Supercritical Fluids* 36(2) (2005) 166-172.

[67] Manna, L., M. Banchero, D. Sola, A. Ferri, S. Ronchetti, S. Sicardi, Impregnation of PVP microparticles with ketoprofen in the presence of supercritical CO_2, The *Journal of Supercritical Fluids* 42(3) (2007) 378-384.

[68] Fanovich, M. A., J. Ivanovic, D. Misic, M. V. Alvarez, P. Jaeger, I. Zizovic, R. Eggers, Development of polycaprolactone scaffold with antibacterial activity by an integrated supercritical extraction and impregnation process, *The Journal of Supercritical Fluids* 78 (2013) 42-53.

[69] Milovanovic, S., G. Hollermann, C. Errenst, J. Pajnik, S. Frerich, S. Kroll, K. Rezwan, J. Ivanovic, Supercritical CO_2 impregnation of PLA/PCL films with natural substances for bacterial growth control in food packaging, *Food Research International* 107 (2018) 486-495.

[70] Maya, C., M. T. Fernández-Ponce, L. Casas, C. Mantell, E. J. Martínez de la Ossa, A comparative analysis on the impregnation efficiency of a natural insecticide into polypropylene films by means of batch against semi-continuous techniques using CO_2 as solvent, *Journal of Supercritical Fluids* 169 (2021).

[71] Valor, D., A. Montes, I. García-Casas, C. Pereyra, E. J. Martínez de la Ossa, Supercritical solvent impregnation of alginate wound dressings with mango leaves extract, *Journal of Supercritical Fluids* 178 (2021).

[72] Lopez de Dicastillo, C., C. Villegas, L. Garrido, K. Roa, A. Torres, M. Galotto, A. Rojas Sepúlveda, J. Romero, Modifying active compound release kinetic using supercritical impregnation process to incorporate an active agent into PLA electrospun mats, *Polymers* 10 (2018).

[73] Coutinho, I. T., L. P. Maia-Obi, M. Champeau, Aspirin-Loaded Polymeric Films for Drug Delivery Systems: Comparison between Soaking and Supercritical CO_2 Impregnation, *Pharmaceutics* 13(6) (2021).

[74] Rojas, A., A. Torres, A. Añazco, C. Villegas, M. J. Galotto, A. Guarda, J. Romero, Effect of pressure and time on scCO2-assisted incorporation of thymol into LDPE-based nanocomposites for active food packaging, *Journal of CO_2 Utilization 26* (2018) 434-444.

[75] Goñi, M. L., N. A. Gañán, R. E. Martini, A. E. Andreatta, Carvone-loaded LDPE films for active packaging: Effect of supercritical CO_2-assisted impregnation on loading, mechanical and transport properties of the films, *The Journal of Supercritical Fluids* 133 (2018) 278-290.

[76] Almeida, A. P., S. Rodríguez-Rojo, A. T. Serra, H. Vila-Real, A. L. Simplicio, I. Delgadilho, S. Beirão Da Costa, L. Beirão Da Costa, I. D. Nogueira, C. M. M. Duarte, Microencapsulation of oregano essential oil in starch-based materials using supercritical fluid technology, *Innovative Food Science and Emerging Technologies* 20 (2013) 140-145.

[77] Sharma, R., S. M. Jafari, S. Sharma, Antimicrobial bio-nanocomposites and their potential applications in food packaging, *Food Control* 112 (2020) 107086.

[78] Herek, L. C. S., R. C. Oliveira, A. F. Rubira, N. Pinheiro, Impregnation of pet films and PHB granules with curcumin in supercritical CO_2, *Brazilian Journal of Chemical Engineering* 23(2) (2006) 227-234.

[79] Franco, P., L. Incarnato, I. De Marco, Supercritical CO_2 impregnation of α-tocopherol into PET/PP films for active packaging applications, *Journal of CO_2 Utilization* 34 (2019) 266-273.

[80] Barbaro, G., M. R. Galdi, L. Di Maio, L. Incarnato, Effect of BOPET film surface treatments on adhesion performance of biodegradable coatings for packaging applications, *European Polymer Journal* 68 (2015) 80-89.

[81] Awaja, F. Autohesion of polymers, *Polymer* 97 (2016) 387-407.

[82] Gilciane Américo, A. Supercritical CO_2 Impregnation of Piper divaricatum Essential Oil in Fish (Cynoscion acoupa) Skin Gelatin Films, *Food and bioprocess technology* v. 13(no. 10) (2020) pp. 1765-1777-2020 v. 13 no. 10.

[83] Verano Naranjo, L., C. Cejudo Bastante, L. Casas Cardoso, C. Mantell Serrano, E. J. Martínez de la Ossa Fernández, Supercritical Impregnation of Ketoprofen into Polylactic Acid for Biomedical Application: Analysis and Modeling of the Release Kinetic, *Polymers* 13(12) (2021).

[84] Drago, E., P. Franco, R. Campardelli, I. De Marco, P. Perego, Zein electrospun fibers purification and vanillin impregnation in a one-step supercritical process to produce safe active packaging, *Food Hydrocolloids* 122 (2022) 107082.

[85] Pajnik, J., I. Lukić, J. Dikić, J. Asanin, M. Gordic, D. Misic, I. Zizović, M. Korzeniowska, Application of Supercritical Solvent Impregnation for Production of Zeolite Modified Starch-Chitosan Polymers with Antibacterial Properties, *Molecules* 25(20) (2020).

[86] Dikić, J., I. Lukić, J. Pajnik, J. Pavlović, J. Hrenović, N. Rajić, Antibacterial activity of thymol/carvacrol and clinoptilolite composites prepared by supercritical solvent impregnation, *Journal of Porous Materials* 28(5) (2021) 1577-1584.

[87] Darpentigny, C., P. R. Marcoux, M. Menneteau, B. Michel, F. Ricoul, B. Jean, J. Bras, G. Nonglaton, Antimicrobial Cellulose Nanofibril Porous Materials Obtained by Supercritical Impregnation of Thymol, *ACS Applied Bio Materials* 3(5) (2020) 2965-2975.

[88] Cejudo, C., M. Gallardo, M. T. Fernández-Ponce, L. Casas, C. Mantell, E. J. Martínez de la Ossa, Preservation of food with high oil content by supercritical impregnation techniques, *Chemical Engineering Transactions* 87 (2021) 541-546.

[89] Grüner-Richter, S., F. Otto, E. Weidner, Impregnation of oil containing fruits, *The Journal of Supercritical Fluids* 66 (2012) 321-327.

Chapter 9

Films Generated from Biowaste for Packaging Green Vegetables

Vijayalakshmi Rajarathinam[1,*], Vetriselvi Arumugam[1], Thilagavathy Duraisamy[2] and Vennilavan Thirumavalavan[2]

Post Graduate and Research Department of Botany, Queen Mary's College, Chennai, Food Technology Department, Anna University, Tamilnadu, India

Abstract

Packaging and processing are the two significant phases of operations in the food industry. The final phase is the packaging stage. Many automation strategies are frequently utilized in every phase of processing and packaging. The correct packaging provides processor to pack fresh cut fruits and vegetables and increase their shelf life. The important parameters for the shelf-life improvements are temperature moisture and modified atmosphere. If both packaging and temperature optimal ageing of fruits and vegetables can be reduced. In the last three chapters, my colleagues have already reviewed the possible use of postbiotics and probiotics, as well as the advances in supercritical solvent impregnation, in active food packaging. In this chapter and the next chapter, we and our colleagues will review the application of food packaging films in protecting fresh food products. To start, we will concentrate our discussion, with a special focus on preservation of green vegetables, in this chapter on different types of biowaste used in packaging.

Keywords: biowaste, vegetable, packaging, shelf-life, moisture content

---

* Corresponding Author's Email: rvijaya2799@gmail.com.

In: Food Packaging
Editor: Wing-Fu Lai
ISBN: 979-8-88697-249-8

1. Introduction

Prevention of microbial contamination and spread of infection through green vegetables is one of the major problems in the world. To retain freshness and increasing the self-life of green vegetables through biowaste in one of provision and developing field. Utilization of bio waste as a packaging material have high efficiency without any side effects. Modern packaging materials include metal, plastic, foil, clamshells, cardboard, paper board, bio plastic (or) combination of more than one material as a composition most of these enter municipal waste streams at the end of their service life. Million tone of packaging waste is generated manually in various countries [1].

In recent years, conversion of biowaste as a packaging material is a great challenge to typically most of the biowaste includes renewable raw materials such as starch or cellulose green vegetable packaging is a complete process which include packaging and bio waste management. Active biowaste packaging is an innovative technology to maintain internal freshness and extended shelf life of the vegetable in which the packaging product and the environment interact to maintain the quality of the product at the same time as prolong self-life and enhance safety. Active biowaste vegetables packaging includes microbial quality and temperature control, absorbing (or) scavenging properties removing features, releasing or emitting properties [2]. Bio based product such as corn starch, crustaceous waste, (chitosan) polysaccharide from red algae, furculum, agar agar, sugar cane fermentation products are commonly used as packaging products of green vegetable. Different types of plant extract, and various acids such as lactic acid are different types of biowaste product used as packaging material [3].

Fresh vegetable plays an important role in healthy nutrition and occupy a significant role in consumer priority list but most of the vegetables and cut vegetables have short life span leading to quit degradation a decomposition of the product. Vegetables when kept for a long period of time under goes various biological process such as decrease in respiration rate, cell death, ethylene production etc., It is essential to minimize physical damage of fresh produce in order to obtain optimal shelf live the use of suitable packaging is vital in this respect. Different type of packaging material such as coir packaging card board packaging tissue paper warps, trays, cups, pads are commonly used in vegetable packaging to reduce damage from abrasion. For every delicate green vegetable smaller poche with fever, layers are required to reduce from compression damage. Washed papers are commonly used to wrap a bunch of green vegetables, this imports the physical protection and also reduce the spread of disease origination within a pack. Good packaging is essential for long term transport of vegetables to avoid brushing and consequent decay of the raw material. The packaging of fresh vegetables provides the largest single use of printed poly ethyl bags.

The traditional polymers used for food packaging materials have several limitations like non-biodegradable, diminishing oil resources, environmental arrests for their disposal or incineration, and health or toxicity issues. These concerns and challenges motivate the development of a new environmentally-friendly field of packaging materials based on biodegradable polymers. Chitin and chitosan (CS) are polysaccharides from natural resources used for food contact applications that are nontoxic, biodegradable, and antimicrobial [4]. The role of a different kind of biowaste in active packaging properties in green vegetables are evaluated, with focus on the antimicrobial, their type of pathogens, their mechanism of action, and, finally, the antioxidant activity and its central mechanisms, as a potential to be used in active food packaging. About 9 million tons of plastic accumulate in soils decay. Whole food chain quality and has a high negative environmental impact. The aim of this chapter is to view the recent advances is the utilization of biowaste as packaging materials [5].

2. Types of Biowaste Used as Packaging Material for Vegetable Packaging

2.1. Corn Starch

Polystyrene packaging is non recycled and should be disposed only in in landfill method. Therefore, companies are looking forward for an efficient organic packaging material preferably a starch-based packing material. Here the product is biodegradable but not an inferior material when compared with synthetic polymer. Usage of packaging material made from synthetic polymer are replaced by materials developed using poly lactic acid. These polylactic acid are made from fermented sugar taken from corn starch. The material made from polylactic acid are eco-friendly and biodegradable [6].

The packaging material made from corn starch will break down in to carbon dioxide and water if disposed correctly. But if the material is not properly disposed it will take longer to decompose especially if there is no oxygen or light available. Most of the company prefer corn starch-based packaging material since the raw material corn is sustainable cheap and easy to produce whereas synthetic polymer material produced from petroleum are non-sustainable and expensive depending on the price of oil. Corn starch packaging has ability to reduce static electricity.

2.2. Chitosan

The second most abundant bio polymer present on earth is chitin. These are extracted from the crustaceous shell waste and are produced in as large amount as a bi product from seafood industry using protocol involving the demineralization, deproteinization, and decolouration. The deacylated derivative of chitin is chitosan. The presence of amino group in chitin are responsible for the diverse properties of this polymer. Chitosan also modifies the functional groups in chemicals its predominantly used in application such as agriculture water treatment, textile, biomedical, cosmetics, chromatography, electronic photography, paper industry, food industry, electronics. Chitosan is used as a bio degradable antimicrobial food packaging material. Many research specifies the chitosan based flexible food packaging and edible food coating to compete with non-biodegradable-based plastics-based packaging material to improve the property of chitosan. Many strategies have been used in corpora ting plasticizers embedding the polymers with fillers such as whiskers fibres and nanoparticles however still lot of research is needed in the field of food science [7].

2.3. Polysaccharide from Red Algae

Materials that come in contact of food products are called food contact material and they play an important role in food quality and safety. The major food packaging material such as plastics are very harmful to our environment as well as to the wild life since there are many research has been conducted to create alternative polymer which are similar to plastic like but also eco-friendly. In recent times the sea weed polysaccharide has been on the spotlight due to its non-toxicity biodegradability film formation ability and anti-oxidant capabilities but has series of disadvantages such as water solubility low tensile strength moderate anti-bacterial characteristics etc. [8].

2.4. Bamboo

Bamboo has become an important material in the sustainable material market. It is an environment friendly option than the plastics, metals, paper and cotton. Bamboo packaging gives a composable, renewable and durable option for the industry of food science. Bamboo, when used as a material for packaging, offers a high level of durability along with versatility. Bamboo has a rapid growth rate, making it an abundant and biodegradable renewable resource. Bamboo is a great replacement for materials such as plastic, and it can meet the sustainability goals of many businesses. As more and more customers prioritize purchases based on companies' sustainability goals, it's crucial that businesses are looking for ways to increase their environmentally friendly strategies.

2.5. Agar-Agar

Gel of the agar-agar is a prime component used as a packaging material for green vegetables. Agar absorbs 20 times its own weight and dissolves easily in boiling water but insoluble in cold water. Agar films were transparent and optically clear, showing good packaging properties. The potential application of the agar or glycerol solution to fresh vegetable preservation was tested. Model fruits and vegetables were coated with the biopolymer or plasticizer solution and compared with a control sample in terms of colour, firmness, weight loss and shelf life. Films of agar extracted from *G. vermiculophylla* coated with edible cherry tomato prevent dryness, weight loss and also reduce the moisture content [9, 10].

2.6. Sugar Cane Molasses

Supra pulp packaging material is commonly used to pack fresh vegetables. These are usually made from sugarcane waste that are fully compostable and yet durable to be used when the food is hot or wet. The product is made from the 100% renewable sugar cane fibres called bagasse. The dry pulpy fibrous matter which remains after the sugar cane and sorghum stalks are crushed. These supra pulp material can be frozen up to -40°C and reheated up to 270°C and this material can be disposed as organic wastes without affecting the environment [11].

2.7. Banana Peel

A new innovative packaging material made from banana plantation has been discovered by the researchers of University of South Wales. Many methods such as spray on edible peels and flexible films made from waste crustaceous shells has been explored by the packaging Europe over the past few months. Pseudo stem of banana consists of 90% of water and rest 10% being solid. The pseudo stem is bought into labs and chopped into pieces and dry it in very low temperature in drying oven, mill it into very fine powder. The powder under goes chemical treatment followed by isolation of nano particles. Which commonly used as high value package material. This material can be recycled for three times without changing its property. It also has no contamination risk.

2.8. Dry Leaves

Dry leaves such as grape and banana and pumpkin leaves are commonly used as a packaging material for vegetables. An earlier study discovered preservation technique of banana leaves. The technology enhances the cells, strengthen the cell wall of leaves and prevent pathogenic agents from destroying the cells. Increased in their stretchability, durability, crushability and these preserved leaves can resist extreme temperature and hold more weight their original state. The preservation capability of the leaves is up to one year and as extended shelf-life span of up to three years without its natural colour. These are human friendly, eco-friendly and are biodegradable [12]. The technology enhances cells, strengthens cell walls of leaves and prevents pathogenic agents from destroying the cells. Increased in their durability, stretchability and crushability, preserved leaves can resist extreme temperatures and hold more weight than in their original state. The preservation capability of leaves with natural green color is for a period of up to one year and an extended shelf lifespan of three years without its natural color. These processed biomaterials are biodegradable, healthy, pathogenic resistant, human-friendly and completely eco-friendly [13, 14].

Mint leaves are very popular in Mediterranean regions and represent a dominant part of the vegetation and they are mainly found only in summer. Drying is one of the traditional methods of preservation, which converts the vegetables into light weight, easily transportable and storable product. Advantage of this method is that the vegetable can be used throughout the year as it can be easily converted into fresh like form by rehydrating it. Blanching of Mint was done and it was checked by catalase test. Mint leaves (Mentha spicata L.) are a common name for members of the Labiatae Family. It is very difficult to preserve for other weather because they become spoil and its colour and flavour decreases. We want that Mint leaves are stored for whole year so dehydration was done. Dehydrated mint sample are packed in different packaging material such as LDPE, HDPE and Aluminium foil. Drying was completed under four different temperatures, 40ºC, 50ºC, 60ºC and Sundry. In Aluminium foil Moisture content and Ash content show best result under Sundry condition. The amount of β-carotene shows best result in Aluminium foil under 40ºC temperature and lower in Sundry condition. Chlorophyll a, Chlorophyll b and Total Chlorophyll were found best in Aluminium foil packaging bag under Sundry condition. For Mint sample Aluminium foil packaging material are the best packaging material for long time preservation. 40ºC and Sundry condition are the best for the drying of Mint because at this temperature the colour, flavour and nutritional quality of mint was maintained [15].

3. Use of Modified Atmosphere Packaging and Edible Coatings

Nitrogen (N_2) gas packaging for fresh-cut vegetables (lettuce and cabbage) has been examined as a means of modified atmosphere packing (MAP) for extending the shelf life of cut vegetables (Koseki and Itoh, 2002). Degradation of cut vegetables in terms of appearance was delayed by N_2 gas packaging. Because of this effect, the appearance of fresh-cut vegetables packaged with N_2 gas remained acceptable at temperatures below 5ºC after 5 days [16].

The application of edible coatings appears to be one of the most innovative approaches to extend the commercial shelf life of vegetables by, among other mechanisms, acting as a barrier against gas transport and showing similar effects to storage under controlled atmospheres [17]. Indeed, over the last two decades the development and use of bio-based packaging materials to prolong the shelf-life and improve the quality of fresh products has been receiving increased attention [18]. Edible coating formulations must wet and spread uniformly on the vegetable's surface and, after drying, a coating that has adequate adhesion, cohesion and durability to function properly must be formed [19]. A commercial edible coating formulation based on carboxymethylcellulose and sucrose fatty acid esters, has been applied to asparagus [20]. Proteins that can also be used in formulations of edible coatings for vegetables. These proteins can be obtained from animal sources, such as casein. Whey proteins are obtained from plant sources, like corn zein, wheat gluten and soy protein [21]. Whey protein-based coatings have been extensively used to extend the shelf life of vegetables [22].

4. Development of Nano Particle as Packaging Materials

The common nanocomposites used in the food packaging industry are polymer clay nanoclay, silica nanocomposites of nanosilver. The effects of nanoclay in polymers are increased stiffness, strength, nucleating agent in foams, smaller cell size, higher cell density, and flame retardant. Nanosilver is composed of de-ionized water with silver in suspension which has excellent antibacterial properties. Silver nanoparticles interact well with other particles as they have large surface area relative to volume which increases their antibacterial efficiency as a result of which they are extensively used in the food packaging industry [23]. One of the potential applications of nanotechnology in food packaging is polymer/clay nanocomposites; they have recently emerged due to their potential for improving properties of packaging materials such as increased mechanical, barrier and chemical properties with a small amount of nanoclays reinforcement. However, most work done on polymer/clay nanocomposites has focused mainly on synthetic polymers [24-26].

The packaging of fresh vegetables and fruit provides the largest single use of printed poly ethylene and polypropylene bags (Tables 1-2) [27-38]. The pre-packaging of potatoes, carrots, onions, parsnips, beets, radishes and so on have become a significant part of food distribution. The bags are sometime perforated to allow the product to breath. Permeation is the key to extending fresh produce shelf life. The key to extend fresh produce shelf life is permeation and for most of the commodity's light won't be a significant factor during their post-harvest handling. Polymeric films have been in prolonged use for packaging of fresh product for more than 35 years with many advantages such as protection from skin aberration, control of water loss and reduce contamination while handling. Polymeric film provides a protection from the spread of decay from one unit to another. The movement of respiratory gas will also be affected due to these films and lead to the development of raised carbon di oxide and decreased oxygen levels within a package. Active modification is defined as slight vacuum inside the pack and then replacing the atmosphere with preferable gas mixture. In passive modification the atmosphere is achieved by the respiration of the commodity within the pack. The final equilibrium atmosphere will more likely depend on the features of commodity and film. The main disadvantage in modified atmosphere packaging is reducing of oxygen levels and causing the production of unwanted off odour due to fermentation of tissues. It consists of closure of respiring produce in films where the gaseous environment is altered to slow the respiration moisture loss decay or to extend the shelf life of the product. [39] Most of the films are combined by the process co extrusion and laminations. There are several groupings in moderate atmosphere packing films hermaticoeal is provided by polythene and is a medium of control for characteristic like peeling ability anti fog ability and anti-fog ability and ability to seal through. The degree through which atmospheric modification happens depend upon variables such as film permeability to oxygen and carbon di oxygen. There has been a lot of commercial interest to develop a film with increase in gas transmission rates. This is obtained by altering the film manufacturing process so that gases such as oxygen, carbon di oxide and water vapour enter or exist the package in a controlled manner. For highly respiring product such as broccoli, beans, mushrooms, peas we use films like polyvinyl chloride low density polyethylene, vinyl acetate, oriented polypropylene and cellulose acetate which are not sufficiently permeable but the highly respiratory product are suitably packed in highly permeable micro perforated films [40].

Table 1. Classification of vegetables according to their respiration rates and maximum ethylene production rate

Commodities	**Ethylene production rate (μL/ kg. hr at 20°C)**	**Respiration Rate at 5°C (mL CO2 kg–1 h–1)**
Artichoke, cauliflower, asparagus, potato, leafy vegetables, root vegetables	Very low: 0.01-0.1	5–10
Cucumber, olive, sweet pepper, chili pepper, pumpkin.	Low: 0.1-1.0	10–20
Melon, mango, tomato	Moderate: 1.0-10.0	20–40
Brussels sprouts, green onion, snap bean	Moderate: 1.0-10.0	40–60
Asparagus, broccoli, mushroom, pea, spinach, sweet corn	Moderate: 1.0-10.0	>60

Table 2. Classification of vegetable/treatment, packaging materials/methods, and shelf-life

Vegetable/Treatment	Packaging materials/methods	Shelf life	Reference
Cauliflower, Truffle	Tray: PP; Cover: PE-LD/PET(40 μm), 0-14 microperforated package, all wrapped in PE	4 days at 4°C	[27]
Carrots,	Minimally processed 80% O_2 / 10% CO_2 / 10% N_2	7 days at 4°C	[28]
Cabbage, shredded	OPP (30 μm) Glass jar; PE (30 μm); PP (30 μm) in: Air, 100% N_2, MAP 1 (100% N_2), MAP 2 (5% O_2 / 95% N_2), MAP 3 (10% O_2 / 90%)	9-10 days at 3°C	[29]
Cabbage, shredded	MAP 4: 70% O_2 / 30% N_2 and 100% O_2	7 days at 0 and 10°C	[30]
Spinach Fresh-cut	PE-LD	8 days at 4°C and 90% HR	[31]
Spinach leaves	OPP	6 days at 4°C and 90% HR	
Lettuce	PE	10 days in the dark at 7°C.	[32]
Broccoli	PE-non-perforated	13 days at 4°C; 3 days at 20°C	[33]
Broccoli	PE-2 holes (750 μm in dia.)	23 days at 4°C	
Broccoli	PE-4 holes (8.8 mm in dia.)	5 days at 20°C	
Broccoli	unwrapped broccoli	10 days at 4°C,	
Fresh-cut vegetables, Fennels cut lettuce salads and shredded carrots	PA/PE (102 μm)		[34]
Carrots	Air	less than 5 days at 4°C	
Carrots	Vacuum	13.5 at 4°C	
Carrots	MAP: 70% N_2 / 30% CO_2	5.8 at 4°C	
Cherry tomatoes ozone treatment (5–30 mg/l ozone gas for 0–20 min)	CA 1: 5% CO_2 / 17% O_2 (7°C)	20 days at 7°C	[35]
Cherry tomatoes ozone treatment (5–30 mg/l ozone gas for 0–20 min)	CA 2: 5% CO_2 / 15% O_2 (22°C)	10 days at 22°C	
Cherry tomatoes ozone treatment (5–30 mg/l ozone gas for 0–20 min)	PE-LD (50 μm); MAP	10 days at 22°C	
Melon, minimally fresh processed	MAP passive in		[36]
Melon	PP	10 days at 5°C	
Melon	microperforated	10 days at 5°C	
Melon	BOPP OPP	10 days at 5°C	
Melon	PP-macroperforated (control)	10 days at 5°C	
Vegetable/Treatment	Packaging materials/methods	Shelf life	Reference
Tomato	PE-LD (50 mm) MAP + O2-absorber	8 days at 20°C	[37]
Lettuce (shredded)	OPP (30 μm)	8 days at 4°C	[38]

Several functions such as protection hygiene safety are covered by packaging. The important application of packaging is to prevent or minimise the product waste and to save the shelf life, hygiene is considered as very important. Packaging must ensure high product hygiene norms specifically for beverages and vegetables. Innovative sustainable packaging should meet all the requirements. In recent time special attention has been given to these kind of innovative packaging solutions. Without doubt innovative bio-based packaging can become a great substitute for traditional plastic packaging. But these friendly biobased biodegradable material does not properly meet producers and consumer's demand. Around 100 million tons of food in Europe are wasted per year. Food waste is attributed to bio waste category. Bio waste is called as bio degradable waste such as green garden or park waste which also includes left over kitchen waste and also biodegradable. Biodegradable waste category includes cardboard, textile, paper, processed wood, forest trees, agricultural residues, and sea weed sludge these are considered as bio degradable and not as bio waste. The amount of packaging waste is increasing drastically according to Eurostat data 31.21 million tons of packaging waste was produced in 2017 plastic packaging has reached 14.5million ton of waste in 2017 [41].

Raising packaging waste volume causes important environmental impact such as usage of primary raw material and disposal of huge amount of waste sustainable waste management and mitigation of negative environmental impact are advanced and interconnected process in the product film usage and disposal of food packaging plays an important role. As a sustainable waste management weigh composting has a better option in significant compound in waste management strategies. Materials of biological origin have a minimal environmental footprint compared to those made from food fuels the main advantage of replacing plastic packaging with compostable ones are conservation of fossil fuels mitigation of climate change caused by carbon dioxide and improvement of organic waste management the demand for biodegradable and composting packaging products are increasing. We must be aware the term that biodegradable and compostable are different and not the same. All compostable material naturally biodegradable but not all the materials that biodegradable used for composting. Composting is a process with specific conditions and environment where bio degradation occurs. The biodegradation process is called as environment and time frame. The time frame aspects are significant as most of the material are able to biodegradable in natural environment. But the duration can be more than 15 years also. So that composting process is framed with proper time frame and environmental condition. According to Muscat et al., (2012). The determination of composting criteria is also important because the material is not suitable for composting can reduce the final quality of compost. Biodegradation plays an important role in reducing the food packaging waste. Biodegradation is a natural disintegration process of organic matter into primary elements existing in nature by organism such as fungi, bacteria or yeast [42].

Biodegradation is called as events of action of enzyme and are chemical decomposition associated with living organisms. This process can occur with or without oxygen. In other words, it can be termed as aerobic or anaerobic. In the aerobic process the natural microbial colonizer consumes oxygen and used it as electron acceptor to convert the organic substance into water. The biproduct of aerobic process is in 3 mole of water, 5 mole of carbon di oxide and raised population of microorganism. In general, anaerobic treatment is a group of biological process where microorganism disintegrate organic biodegradable material during the absence of oxygen [43]. Oxygen which are used by the micro-organism in anaerobic biodegradation is replaced by sulphate, nitrogen di oxide, manganese carbon di oxide, iron. These compounds act as the electron acceptor and as by product which include hydrogen sulphide, nitrogen gas and methane. Composting is the process in which decomposition begins when micro-organism break down organic material into the final product. Biodegradation is the completely autonomous and long-lasting processes. But composting is time framed and require additional human labour. Both the composting and biodegradation can be aerobic and anaerobic. Aerobic treatment is called as degradation of material in the composting process while providing oxygen. Anaerobic process is called as degradation of material in the composting process without providing oxygen. Difference between the composting about the methane production. Methane is produced only under anaerobic condition. While the similarities between both processes are production of carbon di oxide and water. The compossibility process must be evaluated by four characteristics. Environmental impact and the quality of final product are given under chemical characteristics and ecotoxicity. Biodegradable and disintegration give degradation characteristics. Disintegration is the physical, material decomposition after the complete composting cycle. The biodegradation process ends with the complete breakdown of organic material into mineral and product. There is an official certificate called international standardisation and certain requirement that these materials have to meet. A compostable plastic should be biodegradable in the composting environment. It must be visually disintegrated up to 2mm and should not have any toxic foot prints which could negatively affect the soil quality [44].

5. Biosolids and Compost

The product should have less than 50% of maximum allowable concentration of heavy metals released by biosolids and the compost should support the germination of two different species of plants at the rate of at least 90% in a control sample. Many research indicate that the biodegradability is independent on the raw material but depends on the chemical structure of the polymer chain. It is important to note that compostable plastics can be plant based. But still they are petroleum based as well here negatively affect the environment. The term bio-based means that production material is obtained from bio mass. The most common biomass sources are sugarcane, corn, cellulose. The term biodegradable means the product can naturally biodegrade. It can be made from natural renewable resources or fossil based. Biodegradable packaging products are classified into fibre-based packaging and bioplastics.

Bioplastics are plastics made from biomass or biodegradable. Bioplastics when compared with fossil fuel when compared within fossil fuel based conventional plastics to neo based plastics are considered as more advantaged packaging material with less carbon foot prints and with more sustainable waste management option. An alternative bio packaging material are found which are called bio polymers. They are semi synthetic polymer made of renewable bio mass and are very advanced in nature which allow the packaging material to be biodegradable or compostable completely. Agricultural and Agri food residue can be termed into naturally biodegradable packaging material with a parent transparent eco efficiency assessment. Bio based plastic product made from food resource such as corn potato, rice, wheat, sugarcane, soy is facing different criticism both in a negative and positive aspects. And there is opinion that bio-based plastics contribute to an increase in food security concern and pressure or agriculture land [45].

Conclusion

One of the rapidly growing environmental concerns are food packaging waste and thus there is a demand for clear economical solution. Biodegradable composting package looks like a promising way which can be used to reduce waste volume and bring food packaging transformation from linear to circulate economy. The food packaging industry already has sum of the option for compostable packaging which does not impact an environmental contamination. Synthetic polymer such as poly lactic acid one of the most popular materials for packaging with bio compatibility and biodegradability. Broader reduction of renewable bio mass food packaging product seems to be the next big steps and will the strong growth in food packaging industry. However, there should be further research conducted to solve hygiene, productive shelf life, safety to evaluate the benefits and threats of the environment. In this chapter, we have presented how food packaging films can be designed and applied to package green vegetables. In the next chapter, our colleagues will have a more focused discussion on the development and applications of food packaging films for packaging mushrooms.

References

[1] Thompson, R. C., Moore, C. J., vom Saal, F. S, and Swan, S. H. (2009). Plastics, the environment and human health: current consensus and future trends. *Philosophical Transactions of the Royal Society B*. 364:2153-2166. https://doi.org/10.1098/rstb.2009.0053.

[2] Nile, S. H., Baskar, V., Selvaraj, D., Nile, A., Xiao, J., & Kai, G. (2020). Nanotechnologies in food science: Applications, recent trends, and future perspectives. *Nano-Micro Letters*, 12(1), 45. https://doi.org/10.1007/s40820-020-0383-9.

[3] Azevedo, V. M., Borges, S. V., Marconcini, J. M., Yoshida, M. I., Neto, A. R. S., Pereira, T. C., Pereira, C. F. G. Effect of replacement of corn starch by whey protein isolate in biodegradable film blends obtained by extrusion. *Carbohydr. Polym.* 2017, 157, 971–980. [CrossRef] [PubMed].

[4] Apjok, R., Mihaly Cozmuta, A., Peter, A., Mihaly Cozmuta, L., Nicula, C., Baia, M., & Vulpoi, A. (2019). Active packaging based on cellulose-chitosan-Ag/TiO2 nanocomposite for storage of clarified butter. *Cellulose*, 26(3), 1923–1946. https://doi.org/10.1007/s10570-018-02226-7.

[5] Guillard V., Gaucel S., Fornaciari C., Angellier-Coussy H., Buche P., Gontard N. (2018) The Next Generation of SustainableFood Packaging to Preserve Our

Environment in a Circular Economy Context. *Front. Nutr.*, 04 December 2018. https://doi.org/10.3389/fnut.2018.00121.

[6] *Corn starch packaging- The Balance Small Business* https://www.thebalancesmb.com › corn-starch-packagin.

[7] Ruchirpriyadarshi Jong WhanRhim 2020 Chitosan based biodegradable functional films for food pacaking applications do*Innovative Food Science & Emerging Technologies Vol 62*, 102346.

[8] Irina Elena Raschip, Nicusor Fifere, Maria Valentina Dinu 2021 Polysaccharide-Based Materials as Promising Alternatives to Synthetic-Based Plastics for Food Packaging Applications In book: *Bioplastics for Sustainable Development* (pp. 515-554) doi:10.1007/978-981-16-1823-9_19.

[9] Sousa, A. M. M., A. M. Sereno, L. Hilliou and M. P. Gonçalves 2010 *Biodegradable Agar extracted from Gracilaria Vermiculophylla: Film Properties and Application to Edible Coating Materials Science Forum Vols.* 636-637 (2010) pp 739-744 doi:10.4028/www.scientific.net/MSF.636-637.739.

[10] Agar-agar packaging | *wdkadesignforimpact* - WordPress.com. https://wdkadesignforimpact.wordpress.com › agar-agar...

[11] *Startup develops food packaging materials from sugarcane waste* https://packagingeurope.com.

[12] Tenith Adithyaa M *Banana leaf preservation technology- Tech to preserve.* https://bananaleaftechnology.com.

[13] Pablo Bordón, Rubén Paz, Carolina Peñalva, Gisela Vega, Mario Monzón and Lidia García *Biodegradable Polymer Compounds Reinforced with Banana Fiber for the Production of Protective Bags for Banana Fruits in the Context of Circular Economy Agronomy* 2021, 11, 242. https://doi.org/10.3390/agronomy.

[14] *Processed banana leaves, an eco-friendly packaging solution.* https://www.biobasedpress.eu › 2019/09.

[15] Ganesh Khule, Khupase Sp, Khandekar Sa, Ajay Kumar Singh 2019 Package and method for packaging and storing cut leafy vegetables *International Journal of Food Sciences and Nutr*ition 4(2):80-86.

[16] Koseki S., Itoh K. (2002) Effect of nitrogen gas packaging on the quality and microbial growth of fresh-cut vegetables under low temperatures, *Journal of Food Protection*, 65(2), pp. 326–332.

[17] Park, H. J. (1999) Development of advanced edible coatings for fruits, *Trends in Food Science & Technology*, 10, pp. 254–260.

[18] Galić K. (2009) Jestiva ambalaža u prehrambenoj industriji, Hrvatski časopis za prehrambenu tehnologiju, *biotehnologiju i nutricionizam*, 4 (1-2), pp. 23-31.

[19] Ribeiro C., Vicente A., Teixeira J. A., Miranda C., (2007) Optimization of edible coating composition to retard strawberry fruit senescence, *Postharvest Biology and Technology*, 44, pp. 63–70.

[20] Tzoumaki M. V., Biliaderis C. G., Vasilakakis M. (2009) Impact of edible coatings and packaging on quality of white asparagus (Asparagus offi cinalis, L.) during cold storage, *Food Chemistry*, 117 (1), pp. 55-63.

[21] Vargas M., Pastor C., Chiralt A., McClements D. J., Gonzalez-Martinez C. (2008) Recent Advances in Edible Coatings for Fresh and Minimally Processed Fruits, *Critical Reviews in Food Science and Nutrition*, 48 (6), pp. 496- 511(16).

[22] Cisneros-Zevallos L., Krochta J. M. (2003). Whey protein coatings for fresh fruits and relative humidity effects, *Journal of Food Science*, 68(1), pp. 176-181.

[23] Kader A. A. (2002) *Postharvest technology of horticultural crops. 3rd ed.*, Oacland, California, Publication no. 3311.

[24] Tonnie, A. O. (2007) A reference searching related to nanomaterials. Food Packaging and Su Brody A. L. (2007) Nanocomposite technology in food packaging, *Food Technology*, 61(10), pp. 80-83.

[25] Rhim J. W., Ng P. K. W. (2007) Natural biopolymer-based nanocomposite fi lms for packaging applications. *Critical Reviews in Food Science and Nutrition*, 47, pp. 411-433.

[26] Gonzalez-Buesa J., Ferrer-Mairal A., Oria R., L. Salvador M. (2009) A mathematical model for packaging with microperforated fi lms of fresh-cut fruits and vegetables, *Journal of Food Engineering*, 95 (1), pp. 158-165.

[27] Ayhan Z., Eştürk O., Taş E. (2008) Effect of modifi ed atmosphere packaging on the quality and shelf life of minimally processed carrots, *Turkish Journal of Agriculture and Forestry*, 32, pp. 57–64.

[28] Pirovani M. E., Piagentini A. M., Güemes D. R., Di Pentima J. H. (1998) Quality of minimally processed lettuce as infl uenced by packaging and chemical treatment, *Journal of Food Quality*, 21, pp. 475–484.

[29] Plestenjak A., Pozrl T., Hribar J., Unuk T., Vidrih R. (2008) Regulation of metabolic changes in shredded cabbage by modifi ed atmosphere packaging, *Food Technology and Biotechnology*, 46(4), pp. 427–433.

[30] Piagentini A. M., Güemes D. R. (2002) Shelf life of fresh-cut spinach as affected by chemical treatment and type of packaging fi lm, *Brazilian Journal of Chemical Engineering*, 19, pp. 383–389.

[31] Zhang FZ, Wagstaff C, Rae AM, et al. (2007) QTL for shelf life in lettuce co-locate with those for leaf biophysical properties but not for leaf developmental traits, *Journal of Experimental Botany*, 58, pp. 1433–1449.

[32] Jia C. G., Xu C. J., Wei J., Yuan J., Yuan G. F., Wang B. L. et al., (2009) Effect of modifi ed atmosphere packaging on visual quality and glucosinolates of broccoli fl orets, *Food Chemistry*, 114, pp. 28–37.

[33] Corbo M. R., Del Nobile A., Sinigaglia M. (2006) A novel approach for calculating shelf life of minimally processed vegetables, *International Journal of Food Microbiology*, 106, pp. 69–73.

[34] Das E., Candan G. G., Bayındırlı A. (2006) Effect of controlled atmosphere storage, modifi ed atmosphere packaging and gaseous ozone treatment on the survival of Salmonella Enteritidis on cherry tomatoes, *Food Microbiology*. 23(5), pp. 430-438.

[35] Aguayo E., Allende A., Artes F. (2003) Keeping quality and safety of minimally fresh processed melon, *European Food Research and Technology*, 216, pp. 494–499.

[36] Charles F., Sanchez J., Gontard N. (2003) Active modifi ed atmosphere packaging of fresh fruits and vegetables: modeling with tomatoes and oxygen absorber, *Journal of Food Science*, 68(5), pp. 1736–1742.

[37] Pirovani M. E., Piagentini A. M., Güemes D. R., Di Pentima J. H. (1998) Quality of minimally processed lettuce as infl uenced by packaging and chemical treatment, *Journal of Food Quality*, 21, pp. 475–484.

[38] Kader A. A. (1980) Prevention of ripening in fruits by use of controlled atmospheres, *Food Technology*, 34 (3), pp. 45.

[39] Lange D. L. (2000). *New fi lm technologies for horticultural products, Horticultural Technology*, 10, pp.

[40] Eurostat. *Packaging waste statistics* (2020). Available online: https://ec.europa.eu/eurostat/statistics-explained/pdfscache/10547.pdf (Accessed 31 March, 2020).

[41] Muscat D., Adhikari B, Adhikari R, Chaudhary. DS: Comparative study of film forming behaviour of low and high amylose starches using glycerol and xylitol as plasticizers, *Journal of Food Engineering*, Vol. 109, (2012) Issue 2, 189-201. https://doi. org/10.1016/j.jfoodeng.2011.10.019.

[42] Lee D. S., Haggar P. E., Lee J., Yam K. L. (2006) Model for Fresh Produce Respiration in Modifi ed Atmospheres Based on Principles of Enzyme Kinetics, *Journal of Food Science*, 56(6), pp. 1580 – 1585.

[43] Simon Angelo Cichello 2015 Oxygen absorbers in food preservation: a review *J Food Sci Tech*nol.; 52(4): 1889–1895. doi: 10.1007/s13197-014-1265-2.

[44] Naik K. R. and A. K. Rokhade. 1996. Effect of polyethylene packaging on the storage of different ber cultivars. *Karnataka J. Agri. Sci*. 9: 85—88.

[45] Siddiqui S. and O. P. Gupta. 1997. Effect of individual fruit wrapping by different materials on the shelf-life of guava cv Allahabad Safeda. *Haryana J. Hort. Sci*. 26: 102—104.

Chapter 10

Applications of Edible Packaging for the Preservation of Mushrooms

Nishant Kumar[1], Pratibha[2], Neeraj[1] and Anka Trajkovska Petkoska[3,*]

[1]National Institute of Food Technology Entrepreneurship and Management, Sonipat, India
[2]National Institute of Technology, Delhi, India
[3]Faculty of Technology and Technical Sciences, Dimitar Vlahov - Veles, St. Kliment Ohridski University - Bitola, R. North Macedonia

Abstract

In the previous chapter, our colleagues have discussed the use of food packaging films for vegetable packaging. In this chapter, we will focus our discussion on mushroom packaging. The mushrooms are edible and known for their savory flavor worldwide. They have been used for medicinal purpose from the ancient time, due to their nutritional and medicinal values such as antioxidant, antimicrobial, anti-inflammatory, anti-tumor and other phytochemical properties. They also play an important role in prevention and treatment against various chronic diseases. The non-enzymatic and enzymatic compounds such as ascorbate peroxidase, glutathione S-transferase, superoxide dismutase, catalase, glutathione peroxidase, glutathione reductase, glutathione and vitamins (C and E) play an important role against reactive oxygen species. Apart from this, mushrooms have very short shelf life due to their high perishability, transpiration and respiration rate, enzymatic activation, lack of physical protection, high tendency to turn brown and microbial load. Different packaging options and treatments for mushroom have been proposed. One of them is edible packaging as an

* Corresponding Author's Email: anka.trajkovska@uklo.edu.mk.

In: Food Packaging
Editor: Wing-Fu Lai
ISBN: 979-8-88697-249-8

eco-friendly and sustainable way to extend the shelf life and quality of mushrooms for a longer period of time by maintaining nutritional and organoleptic properties. The present review describes the nutritional and functional properties of edible mushroom, physiological disorder and shelf life extension strategy of mushroom using different types of biopolymer-based edible packaging. Variety of bio-based edible materials as potentials for active edible coatings on quality and shelf life of mushrooms are also reviewed here.

Keywords: mushrooms, nutritional properties, antioxidant activity, edible packaging, shelf life

1. Introduction

Mushroom is a macro fungus, widely used as a food and folk medicinal purpose. It is considered as a prime and renewable source of nutrients in the food processing sector to develop functional food. Mushroom has been considered as a "super food" and becoming a vital component for human health due to its medicinal and nutritional properties; it is a rich source of good quality of proteins with tryptophan and lysine, contains 4.5-5.0% of carbohydrates in form of glycogen, starch (chitin and hemicellulose), 0.3% of fat content, but also it is rich in essential fat (linoleic acid), dietary fibers, vitamins (B and C) i.e., thiamine, riboflavin, biotin and niacin, and minerals (iron and potassium) (Kunzek et al. 2002; Cheung 2008; Aida et al. 2009; Ifuku et al. 2011; Mehta et al. 2011). The role of mushroom has been described from the ancient times, e.g., in the Bible as "Bread from heaven" and "manna of the Israelities"; the mushroom was found in the nature as fossil food as 300 million years old (Pegler 2002). Mushrooms have a long tradition to be used as a human food for thousands of years due to its functional and therapeutic properties (Mortimer al. 2021). The common types of mushrooms have umbrella shape with a pilus (cap), stipe (stem), other species of mushrooms have cup (volva) known as *Volvariella volvacea* and annulus (ring) known as *Agaricus campestris*. Most of cultivated gourmet mushrooms are saprophytic fungi. Some are mycorrhizal i.e., *Tuber melanosporum* (Perigold black truffle) and *Tricholoma matsutake* (matsutake); these types of mushroom have symbiotic relationship with vegetation and can be grown in dung, bark, live trees and soils (Cheung 2008).

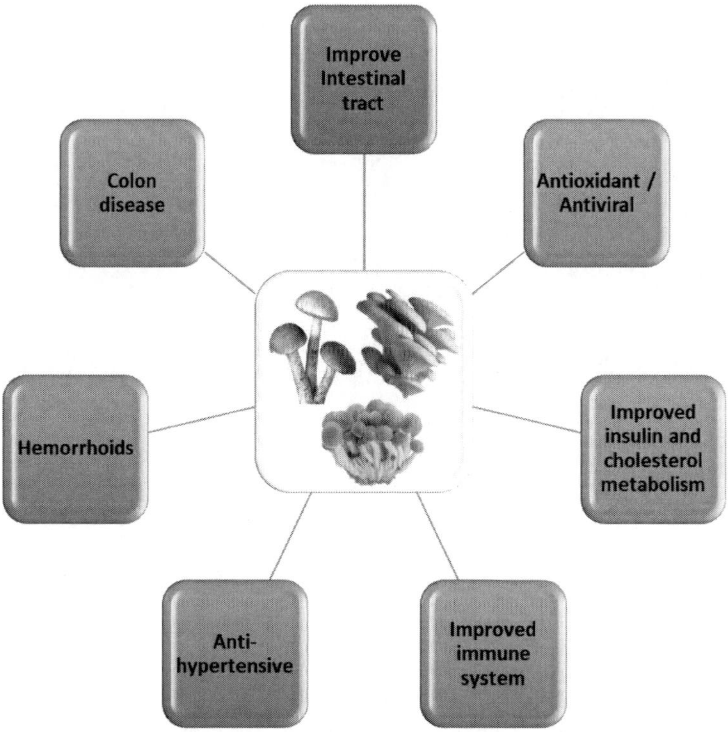

Figure 1. Functional activities and health benefits of mushrooms.
(Chang and Buswell 1996; Cheung 2013; Fernandes et al. 2015; Jayachandran et al. 2017; Sridhar 2019; Cateni et al. 2021; Niazi and Ghafoo 2021).

Around the world, the most popular mushroom is *Agaricus bisporus* also known as common mushroom followed by shiitake, oyster and enoki mushrooms (Feeney et al. 2014; Valverde et al. 2015; Samsudin and Abdullah 2019). Species of edible and medicinal mushroom such as *Agaricus bisporus*, *Pleurotus ostreatus*, *Flammulina velutipes*, *Auricularia auricular*, *Poria cocos*, *Cordyceps sinensis*, and *Ganoderma lucidum* are the most cultivated mushroom around the world (Ma et al. 2018). Worldwide 14,000 species of mushrooms are identified and approx. 1800 of species of mushrooms are known which are potential for food and medical purposes (Chang 2006; Cheung 2008; Islam et al. 2019; Rathore et al. 2019; Zhou et al. 2020; Martinez-Medina et al. 2021). During the last 10 years mushroom production has been increased from 6.9 to 10.24 million metric tons (MMT) due to a higher demand and consumption (FAOSTAT 2019); namely, the consumption of mushroom has risen 21-fold over the last 56 years (FAOSTAT 2017). As per report of Food Agriculture Organization (FAO), China is the largest producer of mushroom around the world which covered approx. 77% of the total globally production followed by other countries including US (0.42 MMT), Netherlands (0.3 MMT), Poland (0.3 MMT) and Spain (0.16 MMT). Mushrooms are considering as a cheap pharmacology due to its biological activities, such as anti-angiogenic (Jana and Acharya 2020), antidiabetic, anti-obesity, anticancer (Liu et al. 2015; Friedman 2016; Hereher et al. 2018), antioxidants (Garrab et al. 2019; Živković et al. 2019), antiviral, anti-metastatic, immunoregulatory, cardiovascular disease and anti-inflammatory (Liu et al. 2015; Bunbamrung et al. 2017; Muszyńska et al. 2018; Smiderle et al. 2013; Friedman 2016; Kothari et al. 2018). They also play an important role in prevention and treatment against chronic diseases. Figure 1 presents the functional and health benefits of mushrooms.

In addition, mushrooms are used in functional food to incorporate bioactive substances such as β-D-glucans polysaccharides, heteroglucans, chitinous, peptidoglucans, proteoglucans, lectins, RNA components, dietary fiber, terpenoids, steroids, and phenolic contents (Zhang et al. 2007; Perera and Li 2011; Erjavec et al. 2012; Kosanic et al. 2013; Chowdhury et al. 2015; Meng et al. 2016; Khan et al. 2018). The non-enzymatic and enzymatic compounds such as ascorbate peroxidase, glutathione S-transferase, superoxide dismutase, catalase, glutathione peroxidase, glutathione reductase, glutathione and vitamins (C and E) play an important role against reactive oxygen species (ROS) (Islam et al. 2019). There are various direct and indirect applications reported of mushrooms in the food processing sector; they are used as an additive, flavoring agent, tastemaker, color stabilizers in the bakery, beverages, fermented and other food industries. The presence of high nutritional components in mushrooms like, digestible and non-digestible carbohydrates, proteins, minerals, vitamins, dietary fibers and phenolic fractions are responsible to refer edible mushrooms as functional foods. Besides that, mushroom losses their commercial value and consumer acceptance due to their high perishability and short shelf life due to the high transpiration, high respiration rate, browning, enzymatic activation, lack of physical protection, higher tendency to turn brown and microbial load (Mohapatra et al. 2010; Jiang 2013; Jiang et al. 2013; Nasiri et al. 2017; Salehi 2020). There are numerous ways of postharvest management that have been applied to prolong the shelf life of mushrooms by maintain their nutritional, physico-chemical and sensorial attributes. The application of biopolymer-based semipermeable edible coatings and films has been shown to improving the storage life of different types of mushrooms. The edible coatings and films are biodegradable and eco-friendly in nature; they are produced from bio-based polymers i.e., hydrocolloids (polysaccharide, proteins), lipids (wax, oil) and their composites. They also can act as carriers of bioactive compounds and nutraceuticals (active form of packaging) as well as to show suitable gas and water vapor barrier properties which resulting in the minimizing of respiration rate, ethylene production, and lipid oxidation of products. The current study describes the mushrooms in details, their nutritional and physico-chemical properties, health benefits, postharvest decay and effects of different types of biopolymers and nano-based edible coatings on the shelf life of mushroom during different storage conditions.

2. Mushroom as a Functional Food

In past decade, the functional and nutraceutical food have higher demand due to their functional, nutritional and organoleptic properties as well as they are beneficial in reducing the risk against various chronic diseases (Shamtsyan 2016; Raghavendra et al. 2017; Niego et al. 2021). According to Silva et al. (2010) mushrooms as functional food are not only source of nutrients but also are recommended as therapeutic foods that help in the prevention against diseases; they can be used as whole and fortified in the form of powder to other foods (Hasler 2002). Mushrooms also have wide application in food sector (ready to eat, ready to drink, dietary supplement with probiotic), agriculture sector (bio-char, bio-sorbent, media) and product development sector (edible packaging, nanoparticles and skin care formulations) using raw mushroom, mushroom stalk, mushroom extract, and powder form of mushroom (Kumar et al., 2021a). Mushrooms are rich in proteins, water, carbohydrates, fiber, vitamins and other phytochemicals with delicious taste and organoleptic characteristics (Chang and Miles 2008; Valverde et al. 2015).

Mushrooms have potential applications in food sector due to their antioxidant, anti-inflammatory, anticancer, antibacterial properties (Heleno et al. 2010; Alves et al. 2012; Reis et al. 2012; Sevindik 2019; Waktola and Temesgen 2021). The functional and nutritional properties of mushroom are owned to presence of phytochemicals and bioactive compounds such as phenolic, flavonoid, terpenoids, β-glucans, etc. The presence of β-glucans in mushrooms is a major constituent responsible for antioxidant, antibacterial, anticancer, hepatoprotective and natural preservative properties; it provides opportunities for the origination of novel nutraceutical and therapeutic agents

(Khan et al. 2018; Fatima et al. 2021; Martinez-Medina et al. 2021). The consumption of mushroom also improved the gut microbiota pattern of *Bifidobacterium* and *Lactobacillus* genera in human body due to their pharmalogical and functional properties (Vamanu and Pelinescu 2017; Vamanu et al. 2021).

Many researchers and food engineers have developed different functional foods like biscuits, soup powder, nuggets, papad, pickle, murraba, jam, candy, pakoda, chips by incorporating mushroom as an active ingredient, then mushroom fortified corn (extruded), cake, vegetarian sausage, instant noodles, and tomato soup (Wakchaure et al. 2010; Wakchaure et al. 2011). On other hand, Ribeiro et al. (2015) have improved the antioxidant activity, nutritional profiles, and color of cottage cheese by incorporating extract of two species of mushrooms *Suillus luteus* and *Coprinopsis atramentaria* while Sabaratnam et al. (2013) have reported the functional food prepared using edible mushroom (*H. erinaceus)* effective against neurological disorders, nerve growth factor and Alzheimer diseases. Ma et al. (2018) have critically reviewed and reported the ability of mushrooms for correlation between host health problems and gut microbiota, immune system, inflammation, neurological disease and colon cancer. Other researchers have incorporated different types of mushroom to improve the physico-chemical properties of food products, for example, Lu et al. (2016) have prepared pasta using white, porcini and shitake mushroom as partial substitute of durum wheat semolina and reported improvement of the mechanical and textural properties of pasta with reduced cooking loss. Furthermore, the 10% of wheat was replaced with powder of edible mushroom (Desert truffle) as reported by Gadallah and Ashoush (2016); actually they found that the incorporation of mushroom powder in biscuit, improved the nutritional, antioxidants and organoleptic properties. Salehi (2019) has characterized different mushroom powders and their incorporation in bakery products for improving their nutritional and functional properties.

2.1. Nutritional Properties of Mushrooms

It is well known that the mushroom is considered as a functional food due to its properties and health benefits. Generally, mushrooms possess excellent nutritional values due to their content of proteins, carbohydrates, minerals, vitamins and dietary fibers and many researchers have reported the nutritional profiles of different types of mushrooms (Kayode et al. 2015; Kozarski, et al. 2015; Han et al. 2016 Wani et al. 2010; Nicolcioiu et al. 2018; USDA 2019; Samsudin and Abdullah 2019; Fijałkowska et al. 2020). Table 1 summarizes the nutritional profiles of different mushrooms such as common mushrooms, shiitake, oyster and enoki mushrooms as per report of USDA (2019). In this context, Han et al. (2016) have reported 60.47 g carbohydrates with 451.60 cal/g; 22.4 g of protein, 56.99 g of dietary fibers, 7.79 g of ash and 2.3 g of crude fat per 100g in *Pleurotus sajor-caju* variety of mushrooms, while Aremu et al. (2009) have reported energy value of 1476.7 kJ/100g in *Ganoderma spp.* and 1513.5 kJ/100g in *Hebeloma mesophaeum* varieties of mushrooms. Moreover, Chatterjee et al. (2021) have also conducted study on nutritional compositions (proteins, carbohydrates, lipids, fibers, moisture, ash and minerals) of different types of mushroom species (Table 2).

Table 3 presents the nutritional compositions of different species of mushroom (in % of dry weights). Many researchers have elaborated about nutritional status of variety of wild and edible mushrooms, for example, Barros et al. (2008) have investigated the proximate analysis of wild and commercial mushrooms (*Agaricus bisporus, Agaricus silvicola, Boletus edulis, Calocybe gambosa, Cantharellus cibarius, Craterellus cornucopioides, Marasmius oreades*) using HPLC refractive index detector. They reported the range of fat content between 0.92-4.49 g/100g, protein 17.18-80.93 g/100g, ash content 7.07-16.48 g/100g, carbohydrate 8.25-71.15 g/100g, reducing sugar 1.26-3.39g/100g and energy 1460.87-1671.83 g/100g of tested samples. As sugar constituents, they reported mannitol (0.27-19.57 g/100g), trehalose (0.11-10.49 g/100g) and total sugar (3.72-10.87 g/100g), respectively. The maltose (0.44-0.57) was reported only in two species of mushrooms (*Agaricus silvicola* and *Agaricus silvaticus)* whereas *Cantharellus cibarius* and *Craterellus cornucopioides* species of mushroom do not showed melezitose content. On the other hand, Fogarasi et al. (2018) have evaluated proximate analysis of some species of mushrooms (*Pleurotus ostreatus, Agaricus bisporus, Cantharellus cibarius* and *Boletus*) and reported the range of 2.17-14.20 g/100g of moisture content, 1.26-2.69 g/100g fat content, 17.92-36.24 g/100g protein, 8.30-11.11 g/100g ash, 43.01-62.45g/100g carbohydrates and 347.5-361.70 kcal/100g of energy, respectively.

Table 1. Nutritional profile of mushrooms

Macro nutrients (per 100g of edible portion)	**Type of mushrooms**			
	Common	**Shiitake**	**Oyster**	**Enoki**
Proximate composition (g/100g)				
Moisture	92.45	89.74	89.18	88.34
Protein	3.09	2.24	3.31	2.66
Ash	0.85	0.73	1.01	0.91
Fat	0.34	0.49	0.41	0.29
Carbohydrates	3.26	6.79	6.09	7.81
Dietary fibers	1.0	2.5	2.3	2.7
Minerals (mg/100g)				
Calcium	3	2	3	0
Copper	0.32	0.14	0.24	0.11
Iron	0.5	0.41	1.33	1.15
Magnesium	9	20	18	16
Manganese	0.05	0.23	0.11	0.08
Phosphorus	86	112	120	105
Potassium	318	304	420	359
Selenium (µg/100 g)	9.3	5.7	2.6	2.2
Sodium	5	9	18	3
Zinc	0.52	1.03	0.77	0.65
Vitamins (mg/100g)				
Thiamin (B1)	0.081	0.015	0.125	0.225
Riboflavin (B2)	0.40	0.22	0.35	0.20
Niacin (B3)	3.61	3.88	4.96	7.03
Pantothenic acid (B5)	1.50	1.50	1.29	1.35
Pyridoxine (B6)	0.10	0.29	0.11	0.10
Ergosterol (D)	56	85	64	36

Source: USDA (2019).

Table 2. Nutritional and mineral compositions of different mushroom species

Mushroom spp.	**Proteins**	**Carbohydrates**	**Lipids**	**Fibres**	**Moisture**	**Ash**	**Minerals**
Pleurotus ostreatus	17-42	37-48	0.5-5	24-31	85-87	5.7	4-10
Pleurotus sajor-caju	16-38	37-40	1-5	22-31	85-87	6.3	5-9
Pleurotus florida	15-21	40-43	1-5	23-27	87-88	-	8-12
Pleurotus cystidiosus	17-18	43-45	5-6	25-26	86-87	-	7-8
Pleurotus highking 51	20-21	36-37	5-6	30-31	85-86	-	5-6
Pleurotus eryngii	11-12	39-40	7-8	28-29	85-90	6.2	4-5
Pleurotus tuberregium	13-17	53-54	0.2-2	15-16	88-89	-	4-10
Pleurotus flabellatus	21-22	59-60	1-2	10-12	91-94	-	6-7
Pleurotus giganteus	17.7	78.0	4.3	-	-	-	-
Pleurotus geestaranus	19-20	45-46	3-4	26-27	85-86	-	5-6
Agaricus bisporus	14.1	74	2.2	-	-	9.7	-
Agaricus blazei (dry powder formulations)	31.3	59.	4 1.8	-	-	7.5	-
Lentinula edodes	4.5	87.1	1.73	-	-	6.7	-

Values as per (dry weight/weight) g/kg.

Reference: (Groves 1981; Fasidi and Ekuere 1993; Dillard and German 2000; Mattila et al. 2001; Miles and Chang 2004; Pinelo et al. 2005; Aida et al. 2009; Krishnamoorthy et al. 2010; Heleno et al. 2012; Patel and Goyal 2012; Valverde et al. 2015).

Table 3. Nutrient compositions of selected edible and wild mushrooms (% dry weight)

Species	Moisture	Proteins	Carbohydrates	Fatty acids	Ash	Crude fiber	Reference
Amanita ceciliae, Volvariella bommbycina, Collybia dryophila, C. ellipsoidia, Flammulina velutipes, Laccaria globosia, Marasmiellus najmii, Marasmius oreades, Oudemansiella radicata, Russula aeruginea and R. chamaeleontina	10.17-20.17	15.65- 30.56	24.78-55.61		5.99-20.3	5.44-17.93	Sabir et al. (2003)
Agaricus arvensis	-	32.87	32.91		0.18	-	Kumar et al. (2013)
Agaricus langei	-	35.14	34.83	-	14.1	-	
Aricularia auricula-judae	-	36.3	33.23	-	7.07	6.9	
Boletus aestivalis	-	32.76	52.07	-	14.97	-	
Cantharellus cibarius	-	34.17	47.00	-	7.78	11.2	
Heimeiella retispora	-	21.1	44.5	6	11.6	-	
Lepiota lilacea	-	28.12	49.33	-	8.09	-	
Lepiota magnispora	-	27.55	35.00	-	7.07	-	
Pleurotus pulmonaritus	-	37.63	43.4	-	10.17	-	
Lactarius hygrophoroides	-	44.93	42.00	-	2	-	
Russula alboareolata	-	21.2	41.6	9.5	17.6	-	
Russula lepida	-	18.3	60.1	5.6	7.6	-	
Russula nigricans	-	22.6	56.3	4.8	6.7	-	
Russla virences	-	20	54.7	4.3	11.3	-	
Russla xerampelina	-	22.4	55.83	4.5	6.7	-	
Schizophyllum commune	-	22.5	32.4	-	10.1	-	
Lepista irina	-	26.12	50.2	-	3.16	-	
Melanoleuca grammopodia	-	36.27	33.04	-	4.13	-	

Species	Moisture	Proteins	Carbohydrates	Fatty acids	Ash	Crude fiber	Reference
Wild mushrooms [*Amanita calyptroderma*, *Amanita princeps* A., *Astraeus odoratus*, *Heimiella retispora* Boedijn., *Mycoamaranthus cambodgensis* Trappe, *Russula alboareolata* Hongo, *Russula cyanoxantha* Schaeff.ex.Fr., *Russula emetic* (Schaeff. ex Fr.) S.F.Gray., *Russula virescens* (Schaeff.) fr., *Termitomyces clypeatus* Heim]; Cultivated mushrooms [*Auricularia auricula*-judae, *Lentinus polychrous* Lev., *Lentinus squarrosulus* Mont., *Pleurotus] sajor-caju* (Fr.) Sing, *Volvariella vovacea* (Bull. Ex.Fr.) Sing.	84.15-90.12	11.16–50.29	2.11- 59.73	-	2.56–13.96	1.43–21.94	Srikram and Supapvanich, (2016)
Russula	-	38.08-38.53	39.29-41.64	4.06-5.70	12.7-13.80	9.59-19.78	Appolinaire et al. (2016)
Ganoderma lucidum		15.04-24.8	65.34-82.47	0.53-2.02	2.0-9.99	6.11-54.12	Sharif et al. (2016)
P. ostreatus	92.33	18.08	51.31		12.125	25.036	Dilfy et al. (2020)
A. bisporus	89.025	29.64	34.54		7.425	20.995	

The values of proximate composition range in ten (cultivated and wild) mushroom species have been reported by Srikram and Supapvanich (2016). The mineral content of zinc, iron, copper, manganese, calcium, magnesium, phosphorus, sodium and pottasium has been investigated by Zahid et al. (2010) in edible species of mushrooms (*Pleurotus ostreatus, Pleurotus sajor-caju, Pleurotus florida, Pleurotus HK-51 Calocybe indica*). The range for zinc (0.78-1.24 mg/100g), copper (0.14-0.91 mg/100g), iron (0.81-1.81 mg/100g), manganese (0.17-0.53 mg/100g), calcium (0.12-0.58 mg/100g), magnesium (27.26-51.21 mg/100g), phosphorus (22.20-62.10 mg/100g), sodium (3.18-37.23 mg/100g) and potassium (19.83-197.24 mg/100g), respectively. Sabir et al. (2003) have been dedicated to proximate composition of 11 different species of *Azad Kashmir* mushrooms. Adejumo and Awosanya (2005) have investigated nutritional and mineral composition of four edible mushroom species of south western Nigeria such as *Lactarius triviralis, Lentinus tigrinus, Termitomyces mammiformis and Russula vesca.* They have reported the higher proximate composition like, protein (37%), fiber (7%) and minerals - calcium (216 g/kg) and manganese (136 mg/kg), respectively. The higher carbohydrates content (71%) and magnesium (14 g/kg) has been reported in *Russula vesca.* The moisture content (37%), carbohydrates (64%), calcium (210 g/kg), manganese (120 mg/kg), iron (1230 mg/kg) and copper (8 mg/kg) were also reported in *Lactarius triviralis mushroom.* Protein content range (21.23-28.23 g/100g) of four different commercial species of mushroom has been evaluated by Naknaen et al. (2015); they also reported mannitol (73.27-75.53 mg/g) and trehalose (12.97-42.57 mg/g). Moreover, the proximate and mineral composition of four south Indian mushroom species such as *Auricularia polytricha, Macrolepiota rhacodes, Clitocybe Sp., Lentinus tigrinus* etc. were analyzed by Manjunathan et al. (2011); they found that the mushroom species *Auricularia polytrichais* are an excellent source for proteins (37%), fiber (21.97%), ash (6.87%), calcium and manganese (607 mg/g and 136 mg/kg). *Macrolepiota rhodocus* contained higher carbohydrate content (48%) and magnesium (250 mg/g) while iron (1230mg/g) and copper (9.0mg/g) have been reported higher in *Clitocybe sp* mushroom. The lipids, sodium and phosphorus content were reported very low in all selected mushroom species.

Similarly, Muthu and Shanmugasundaram (2016) have also reported proximate composition - moisture content (92.36%), fiber (8.05mg/100g), ash (1.22 mg/100g) and fat (0.20mg/100g) and mineral contents i.e., sodium (0.293mg/g), potassium (0.085 mg/g), calcium (0.273 mg/g), magnesium (0.096 mg/g), manganese (0.058 mg/g), iron (3.85 mg/g), zinc (0.153 mg/g), copper (0.096 mg/g), selenium (0.007 mg/g) and phosphorus (0.951 mg/g) in Agrocybe aegerita mushroom. On the other side, other researchers have been investigated vitamins profiles of different types of wild and edible mushrooms; for example, Bano and Rajarathnam (1986) have reported vitamin C (82-144 mg/100g), vitamin B1 (0.16-2.23 mg/100g), vitamin B2 (0.48-8.97 mg/100g), vitamin B3 (4.10-73.3 mg/100g), vitamin B5 (21.1-33.3 mg/100g) and vitamin B9 (933-1412 mg/100g), respectively, in four different species of mushroom. A white button mushroom vitamins profile was investigated by Furlani and Godoy (2008) using HPLC. They reported the range between 0.004-0.08mg/100g for vitamin B1 (thiamine) and 0.04-0.3mg/100g for vitamin B2 (riboflavin), respectively. Similarly, vitamin B3 (154-362 mg/100g) and B2 (1.57-5.06 mg/100g) were reported in white button mushrooms by Bernaś and Jaworska (2016), and vitamins B2 (1.8-5.1 mg/100g), B3 (31-65 mg/100g) and B9 (0.30-0.64 mg/100g) were reported in white, brown, Lentinus edodes, and Pleurotus ostreatus mushrooms by Mattila et al. (2001). Dilfy et al. (2020) was investigated the nutritional profiles of two edible mushroom like P. ostreatus and A. bisporus; the reported mineral content was: sodium (2.341 mg/kg or 681.215 mg/kg) and calcium (135.761 mg/kg or 569.047 mg/kg); vitamins B1 (0.061 mg/100g or 0.048 mg/100g), vitamins B2 (0.153 mg/100g or 0.342 mg/100g), vitamins B3 (3.244 mg/100g or 37.226 mg/100g), vitamins B12 (0.281 µg/100g or 0.054 µg/100g), vitamin C (1.405 mg/100g or 1.271 mg/100g) and vitamin D (0.129 µg/100g or 0.177 µg/100g), respectively. Amino acids such as lysine, arginine, serine, methionine, cysteine, alanine, phenylalanine, threonine, valine, tyrosine, asparagine acid, glycine, glutamic acid, histidine and isoleucine are reported in both A. bisporus and P. ostreatus species of mushrooms. Based on the previous evidences, it can be concluded that the mushroom possesses an excellent balance of nutritional, vitamins and minerals composition which can help in the functional attributes to provide health benefits; overall most of the evidence showed lower fat value in edible and wild species of mushroom that could help to control the cholesterol, diabetes and other types of human diseases.

2.2. Bioactive Compounds and Pharmalogical Activity of Mushrooms

Natural bioactive compounds and their antioxidant properties found in mushrooms act as free radical scavenger, which eliminate the free radicals (ROS) by reducing oxidative stress. Oxidative stress and ROS could lead to health disorders for humans, but the antioxidant compounds are effective in improving defense mechanism against various types of human disorders (Kozarski et al. 2015). Mushrooms have been considered as an important source for bioactive compounds and natural antioxidants agents like phenolic compounds, β-glucans, lipids, terpenoids, polysaccharide etc. (Cateni et al. 2021) which can be utilized in food and other industries at commercial scale and β-glucans as the major component of mushrooms are a key constituent in pharmaceuticals (Khan et al. 2018; Gupta and Pandey 2019). Mushrooms contain phenolic acids that are categorized in two groups: hydroxybenzoic and hydroxycinnamic acid; hydrolysable tannins, sugar, organic compounds and lignins - they are considered as derivatives of hydroxybenzoic acid in mushrooms. Furthermore, cellulose, lignins, organic acids, gallic acid, tannic acid, tartaric acids, caffeic acid, sinapic, and ferulic acid are derivatives of hydroxycinnamic acid (Manach et al. 2004; Ferreira et al. 2009). Apart from that the different types of flavonoid compounds, catechin, quercetin, pyrogallol, rutin, kaempferol, chrysin, myricetin, biochanin, formometin, naringin, naringenin and hesperetin are reported in edible mushrooms (Ferreira et al. 2009; Gil-Ramirez et al. 2016). The demand of mushrooms has been increasing recently due to their pharmalogical properties - antioxidant, antimicrobial, antiviral, antifungal, anticancer, antidiabetic, anti-inflammatory, etc. (Wu and Xu 2015; Jayachandran et al. 2017; Bhatt et al. 2018; Adeoyo and Oluborode 2020; Krakowska et al. 2020; Al-Faqeeh et al. 2021; Eraslan et al. 2021; Żukowska et al. 2021); all of them are good for human health by improving intestinal, microflora gut, boosting the immunity, liver protection and elimination of pathogens. Fogarasi et al. (2018) have confirmed the presence of phenolic acid and flavonoid compounds in different species of mushrooms by HPLC-MS methods. Phenolic compounds (8.74 and 2.21 mg/g) and flavonoid compounds (2.83 and 1.02 mg/g) have been reported in *R. virescens* mushroom by Hasnat et al. (2014) with water and ethanol as extractive solvents. The quinic acid (14.6 mg/100g), quercetin (3.01 mg/100g), isoflavonoids (0.023 mg/100g) and daidzein (0.061mg/100g) of dry weight were reported by Nowakowski et al. (2020). Mushroom also contains terpenoids including volatile fractions. Terpenoids like α- and β-pinene, β-phellandrene, β-myrcene and D-limonene were reported by Fogarasi et al. (2018) in *A. bisporus* and *B. edulis* mushrooms.

The properties as free radical scavenging activity, antioxidant capacity, metal chelating activity and super-oxide anion scavenging activity of mushrooms can be resulting from present phenolic compounds, flavonoid compounds, β-carotene and α-tocopherol. The physico-chemical properties, amount of polysaccharide and degree of branching β-glucans are dependent on the type of mushrooms (Su et al. 2016). Numerous researchers have reported the phenolic activity, antioxidant, free radical and other potential activities of different species of mushrooms. For example, Elmastas et al. (2007) has been studied the mushroom as an excellent source for bioactive compounds which are responsible for antioxidant and super-oxide anion scavenging activity. Mohamed et al. (2011) have investigated antioxidant activity, total phenolic content and total flavonoid content in two edible mushrooms (*Pleurotus florida* and *Pleurotus eous*) and reported the phenolic content activity (3.125 mg/g and 2.725 mg/g) and flavonoid content (4.4 mg/g & 3.0 mg/g), respectively. Similarly, Puttaraju et al. (2006) was investigated the phenolic profiling, and antioxidant potential of 23 different species of mushrooms under the extraction in methanol and water solvents. HPLC method was used for identification of phenolic acids; they reported the higher range of phenolic content in water extract (2-37 mg/g) compared to methanolic extract (0.7-11.2 mg/g); HPLC analysis indicated the presence of gallic acid, tannic acid, gentisic acid and protocatacheuic acid in *Termitomyces heimii* and *Termitomyces mummiformis* species of mushroom and suggested to use for developing nutraceutical. On other hand, Cheung et al. (2003) have reported positive correlation between the phenolic content and antioxidant activity of shitake and straw mushrooms extract prepared using water, methanol, petroleum ether and ethyl acetate as extractive solvents; they also reported water as a potential solvent to extraction of phenolic compounds (1.33 and 1.34 mg/g) as compared to other solvents in case of *L. edodes* and *V. volvacea* mushrooms. Furthermore, Keleş et al. (2011) have studied higher antioxidant activity (97.96%) in *Leccinum Scabrum* species of edible mushroom compared to other 23 species in methanolic extraction. The higher half maximal effective concentration (EC^{50}) was reported in *Pleurotous dryinus* (24.71) and *Lactarius piperatus* (24.12) species of edible mushrooms.

The antioxidant activity of *Phellinus torulosus* species of mushroom are also reported in methanolic extract by Hokmollahi et al. (2012). Many scientists have reported the functional and medicinal properties like anti-angiogenic (Jana and Acharya 2020), antidiabetic, anti-obesity, anticancer (Liu et al. 2015; Friedman 2016; Hereher et al. 2018), antiviral, anti-metastatic, immune regulatory, cardiovascular disease and anti-inflammatory activity (Liu et al. 2015; Bunbamrung et al. 2017; Muszyńska et al. 2018; Smiderle et al. 2013; Friedman 2016; Kothari et al. 2018). In this context, Dilfy et al. (2020) also have reported the total phenolic content in two species of mushrooms such as *P. ostreatus* and *A. bisporus*. The higher phenolic content (0.205 mg/kg) caffeic acid (5.541 mg/kg), ρ-Coumaric acid (14.995 mg/kg), ferulic acid (6.42 mg/kg) and cinnamic acid (19.207 mg/kg) in *A. bisporus* mushroom as compared to *P. ostreatus* mushroom species. Thus, mushroom is considered as an excellent source of phytochemicals, secondary metabolites compounds which improve their medicinal and functional properties (Thu et al. 2020; Zhou et al. 2020; Kushwaha et al. 2021; Kumla et al. 2021).

3. Mushroom as a Source of Chitin and Chitosan Biopolymers

Mushrooms are considered as an alternatives source of chitin and chitosan biopolymers (nanofibers); their cellular cell wall is an excellent source of chitin which can transform into chitosan form using deacetylation process (Bowman et al. 2006; Ifuku et al. 2011; Ifuku et al. 2012). Chitin and chitosan biopolymers have versatile applications in agriculture, biomedical and food processing sectors due to their natural properties like biocompatibility, biodegradable, eco-friendly materials, non-toxic with antibacterial activity (Ospina et al. 2014; Abo et al. 2019). Chitin is considered as the second most abundant biopolymer after cellulose in nature with biosynthesis of 100 billion tons annually (Tharanathan and Kittur 2003; Kumar & Neeraj 2019). It was first isolated from the cell wall of mushroom in by Braconnot (1811) and named as "*fungine,*" but it is renamed as chitin by Odier (1823) (Elsoud and Kady 2019). The shell waste of crustaceans and exo-skeletion of mollusks, fungi and insects are the main sources for production of chitin (Bhuiyan et al. 2013; Yeul et al. 2013). For example, Fazli et al. (2019) have extracted 42 g of nanofibers from 3 kg of whole mushrooms. Different methods for extraction are known - with hot water and alkaline as well as low energy methods that can be used to produce chitosan nanofibers from mushrooms by the removal of glucans and proteins in the matrix. Ivshina et al. (2006) have isolated and extracted chitin from honey mushroom (*Armillariella mellea*) and yellow morel mushroom (*Morchella esculenta*) under acidic conditions including de-proteinization (2% NaOH with 0.1% sodium stearate at 83-85°C for 2 h), de-mineralization (1% HCl at 55-60°C for 2 h) and de-glucanization (2% NaOH at 83-85°C for 2 h). The content of chitin in final products was reported 70% in honey mushroom (*A. mellea*) and 50% in morel mushroom (*M. esculenta*), respectively. Erdogan et al. (2017) was reported 11.4% and 7.9% of chitin extracted from *Lactarius vellereus* and *Phyllophora ribis* mushroom species

under acidic condition (60% NaOH) at 130°C for 4 h; the chitosan yield from chitin was reported as 73.1% (*Lactarius vellereus*) and 75.1% *(P.ribis)*, respectively. Kulshreshtha and Thakur (2020) have studied the mushroom waste as an important source for bioactive or nutritional components; also it can be used to extract a nutritional compound - vitamin D_2.

4. Shelf life and Physiological Disorder in Mushrooms

Mushrooms are popular food due to its nutritional, medicinal and organoleptic properties; they contain important nutrients and considered as low calories food with free of fat, cholesterol and gluten. FAO has recommended consumption of edible mushrooms as replacement of cereals for protein source (Rahman et al. 2020). The fresh mushrooms belong to highly perishable commodity with short shelf life of 3-4 days that is due to the fast spoilage depending on the poor infrastructure facilities, and also due to thin epidermal structure of mushroom, higher respiration rate, color browning, water loss, polyphenol oxidase (PPO), wilting, and microbial contamination (Kamal et al. 2015; Rahman et al. 2020; Samadpour 2020). The process of degradation has been started after maturation of fruiting body of mushroom and become inconsumable due to loss of nutritional and organoleptic properties. The respiration rate and other biochemical or enzymatic activities are responsible for the postharvest browning of mushrooms which reduce the shelf life and quality deterioration of mushrooms (Singh et al. 2020). Other studies have reported also the relationship between respiration rate of mushrooms and their reduced shelf life (Rai and Arumuganathan 2008). The postharvest management technologies such as Modified atmosphere packaging (MAP), Controlled atmosphere (CA), edible packaging and others, can be used to extend the shelf life of mushrooms during the storage at ambient and cold conditions by maintaining their nutritional and organoleptic quality attributes. The physical (packaging, irradiation, pulsed electric field), chemical (washing, coatings - edible/nano, ozone, electrolyzed water) and thermal (drying, cooling, packaging) methods can be used for prevention of quality attributes of mushrooms for a longer period of time (Zhang et al. 2018). Recently, the trends of application of biopolymers (polysaccharides, proteins, and lipids) in form of edible coatings and films, composites and active packaging have been used to extend the shelf-life of fruits, vegetables and other food categories by maintaining their nutritional, organoleptic properties and food safety during the storage time (Suhag et al. 2020; Petkoska et al. 2021; Kumar et al. 2020).

The preparation of edible packaging using biopolymers (chitin) extracted from mushroom is also as an emerging and very attractive option for this purpose.

5. Application of Edible Packaging on Mushrooms

Edible mushroom has commercial importance due to its health benefits, nutrients, flavor, taste and functional properties. Approximately 45% of mushroom are consumed as fresh, 50% are processed and 5% in dehydrate form (powder). It is considered as a highly perishable with very short shelf life at room temperature due to the high respiration rate, water transpiration, browning, shrinkage, reduced flavor and texture. *Tyrosinase* activity is responsible for browning of mushrooms by *melanine* synthesis (Singh et al. 2010; Nasiri et al. 2017). The microbial load is increased very fast in fresh mushrooms due to its perishability and other properties. Previous studies done by Doores et al. (1987); Halami et al. (1999); Chikthimmah and Beelman (2005) were reported that the fresh mushroom has high populations of microbes (bacteria and molds), 6.3 to 7.2 *log cfu g-1* bacterial (*Pseudomonas* and *Flavobacteria*) count in fresh mushrooms tissues. Halami et al. (1999) have reported bacterial i.e., *Chryseobacetrium* (5.5 *log cfu g-1* and *Coryneform* (5.6 *log cfu g-1*) of bacterial genus, moulds (3 *log cfu*) and yeast (6 *log cfu*) per gram of fresh tissue of mushrooms. Various postharvest technologies such as MAP, CA, edible and nano-composite coatings can be used to extending the shelf life of mushroom by maintained their quality attributes (Singh et al. 2010; Sami et al. 2020). The application of edible coatings is developed by using different types of biopolymers and their composites; this is the most effective strategy to extending the shelf life of mushrooms by reducing undesirable effects and microbial loads.

Edible packaging in form of coatings and films is promising as an excellent technology to extending the shelf life of fruits and vegetables by reducing undesirable effects, providing microbial safety, reducing physiological disorder, barrier against moisture, oxidation, color and sensorial attributes. They act as an alternative to synthetic plastic materials; they are biodegradable, biocompatible and provide suitable mechanical strength, good water and gas resistance, suitable carrier of bioactive and antioxidant agents, and economically affordable (Kumar and Neeraj 2019; Kumar et al. 2020; Kumar et al. 2021b; Kumar et al. 2021c; Kumar et al. 2021d). The edible biopolymers like polysaccharides, proteins, lipids and waxes, are mainly used to form edible packaging (Suhag et al. 2020). Table 4 summarizes different types of edible coatings and films on the shelf life and quality of mushrooms during storage at different conditions.

Table 4. Application of different types of edible coatings/films and their effects on mushrooms

Edible coating (materials)	Application/Effects	References
Nano-coating (Chitosan +nisin)	Nano formulation maintained the freshness and quality of white button mushrooms by retarded weight loss, maintained texture property, reducing enzymatic activity (PPO), respiration rate and microbial counts (yeast & molds).	Sami et al. (2021)
Chitosan + titanium dioxide nanoparticles	The nanocoating was applied to extend the shelf life of button mushroom to prolong the shelf life up to 12 days. This coating showed higher superoxide dismutase and lower peroxidase activity and respiration rate. The antioxidant and phenolic activity was higher in mushroom treated with nanocoating.	Khoja et al. (2021)
Apple peel powder, CMC, tartaric acid and glycerol monostearate	The application of apple peel powder (1.17wt/v) and CMC (1.8%) coating maintained higher quality attributes of button mushroom at 75 s of dipping time as compared to other treatments and dipping times. The coating helps to extending shelf life of mushroom up to 5 and 9 days at ambient and refrigerated condition, respectively.	Thakur et al. (2021)
Chitosan nano-packaging	The application of chitosan nano-packaging + MAPs help to extending the shelf life of button mushrooms by controlling respiration rate, improving physical and mechanical properties.	Gholami et al. (2020)
Lipid based edible coating (glycerol monoestearate + thyme extract)	The coating treatment (GMS+EO150) significantly preserved the quality attributes i.e., texture, color, weight loss and others of oyster mushrooms during storage period.	Samadpour and Beheshti (2020)
Chitosan and guar gum composite	Composite edible coating was effective to maintained firmness, delaying changes in ascorbic acid and total soluble solid and extending shelf life of shiitake mushroom during storage at 4°C condition for 16 days.	Huang et al. (2019)
Chitosan, CMC, sodium alginate and modified starch	All the treatments were found effective to preserve the quality of paddy straw mushroom but the treatment with chitosan based coating (0.20%) was more potential to retarded the weight loss, maintained TSS, acidity, ascorbic acid, sensory characteristics, reduced microbial count and deterioration effects at 20°C and shelf life of mushrooms extending by 21 days.	Minh and Hang (2019)

Table 4. (Continued)

Edible coating (materials)	**Application/Effects**	**References**
Sodium alginate	The composite edible coating prepared with soduim alginate with thyme essential oil and nisin was potential to prevent the quality and sensory attributes of *P. nameko* mushrooms during storage. The applied composite material significantly controlled the weight loss, browning, MDA contents, PPO/POD, and cellulose activity, maintained sugar content, ascorbic acid and inhibit microbial activity.	Zhu et al. (2019)
Carboxy-methyl cellulose + ascorbic acid	Treatment of CMC+ascorbic acid based coating effectively help to extending shelf life 8-15 days of mushroom at 4°C. The coating treatment delayed the changes in TSS, firmness and reduced the physiological loss in weight, microbial counts (yeast & molds).	Vaziri et al. (2019)
Chitosan/tripolyphosphate nano-aggregates	Chitosan loaded with tripolyphosphate nano-aggregates control the enzymatic browning, maintained firmness, phenolic compounds, ascorbic acid, antioxidant capacity, inhibited polyphenol oxidase (PPO) of mushrooms at 5°C for 16 days of storage.	Ojeda et al. (2019)
Arabic gum/CMC	Both treatments were potential to inhibite weight loss, color browning and texture (firmness) properties of button mushrooms up to 12 days of storage at 4°C.	Srivastava and Bala (2016)
Polysaccharide (Chitosan, sodium alginate and guar gum)	In this study, the selected formulations [(i) 0.5% (w/v) of chitosan, 0.125% (v/v) of glycerol and 0.15% (v/v) of Tween 80 + eugenol and carvacrol (1:1) and; (ii) 0.5% (w/v) of alginate, 0.125% (v/v) of glycerol and 0.05% (v/v) of Tween 80 + eugenol and carvacrol (1:1)] were applied on *Agaricus bisporus* mushrooms (white button) to extend the shelf life at 6-8°C throughout 10 days of storage period, but the mushroom treated with chitosan based formulation degraded their quality and showed negative impacts, so the formulation was excluded for further application. The sodium alginate based edible formulation was selected and applied on white button mushroom; it was potential to prolong the shelf life of mushrooms by reducing weight loss, maintained color, TSS, acidity, pH, reducing bacterial counts (mesophilic, psychrophilic) during storage at refrigerated conditions.	de Sousa (2015)

Edible coating (materials)	Application/Effects	References
Chitosan	The chitosan (0.2%) based coating functionalized with 0.5% of acetic acid solution and 2% of calcium chloride was applied on whole and sliced mushrooms during modified atmosphere packaging (PVE/PPE-1 & PPE-2) for 7 days of storage period at 12°C. The combination of MAP and edible coating treatment was effective to prolong the shelf life of mushroom with maintaining the consumer preferences by reducing weight loss and lowering the maturity index.	Ban et al. (2014)
Alginate	Alginate based coating extended postharvest shelf life of button mushrooms up to 16 days. With maintained quality attributes i.e., firmness, control browning, changes in soluble solid, ascorbic acid, inhibit enzymatic activity (PPO/POD).	Jiang (2013)
Alginate/Nano coating (alginate + Ag)	Application of alginate/nano coating (alginate + Ag) was found effective to maintained physicochemical and sensory attributes of shiitake mushroom at 4°C storage condition. The coatings were significantly inhibited the weight loss, firmness and microbial counts of mushroom after 16 days of storage. The sugar content, TSS, was maintained by reducing electrolytic leakage rate.	Jiang et al. (2013)
Chitosan	The treatment of chitosan based edible coating functionalized with thyme oil help to extend the postharvest shelf life of shiitake mushrooms at 4°C for 16 days by reducing loss of firmness, weight loss, reducing respiration rate and microbial growth (molds, yeast and pseudomona), controlled MDA and electrolytic leakage rate and maintained higher phenolic, flavonoid content, ascorbic acid and organoleptic properties.	Jiang et al. (2012)
Gum arabic/CMC	The edible coating was found effective to prevent the quality attributes of mushrooms as compared to acidic washing. Edible treatment helps to significantly reducing weight loss, maintained firmness, color attributes and other postharvest characteristics of mushrooms.	Sedaghat and Zahedi (2012)
Aloe Vera and Gum Tragacanth	The edible coating based on aloe vera and gum tragacanth and their composites were effective to extend the shelf life of button mushrooms at different storage temperatures (4°C, 10°C and 15°C) up to 13 days of storage. The edible treatments significantly delayed the weight loss, color browning and maintained other quality parameters of mushrooms at different storage temperatures as compared to control samples.	Mohebbi et al. (2012)

Table 4. (Continued)

Edible coating (materials)	Application/Effects	References
Methoxy Pectin (HMP), starch and CMC	The formulation of edible coating was potential to maintain moisture content, reduced weight loss, maintain $CaCl_2$ (firming agent), color and sugar content at refrigerator (4°C) condition for 14 days of storage.	Niazmand et al. (2009)
Chitosan	Treatment of chitosan based coating significantly control the reduction of overall quality and extended the shelf life of fresh-cut mushrooms by reducing enzymatic activities (PPO/POD/PAL/Laccase), cellulose, total amalyase and alpha amylase, etc. The microbial load of fresh cut mushroom was also inhibited by chitosan coating treatment. Results of this study have showed that the increased concentration of chitosan coating is more beneficial to extend shelf life of fresh cut mushrooms.	Eissa (2007)
Chitosan and $CaCl_2$	The packed slices and whole mushrooms' shelf life was extended using chitosan and $CaCl_2$ based coating by reducing weight loss, lowering the maturity index but chitosan coating was more potential to extend the shelf life of mushrooms (whole & slices) during storage period at 12°C.	Kim et al. (2006)
Alginate and Alginate-Ergosterol	The applied coating treatment was found to lead better wettability on surface of mushrooms with better color retention, appearance and lower reduction of weight. The alginate coating functionalized with ergosterol and emulsifier (Tween) was found moitable to maintain the size and shape of treated mushrooms as compared to control samples.	Hershko and Nussinovitch (1998)
Alginate	The different concentration of alginate coatings were investigated on mushroom and found effective alginate based coating to prolong the shelf life of mushrooms during storage at 20°C and 4°C. Both (1% and 2%) alginate coatings were effective to protect the color and appearance of the treated mushroom during storage as compared to control samples.	Nussinovitch and Kampf (1993)

Many researchers have been focused on the extending shelf life of mushrooms with maintaining the quality attributes during the storage conditions. For example; Kim et al. (2006) have investigated chitosan edible coating under MAP conditions on the whole and sliced mushrooms stored at 12°C; they reported that mushrooms treated with chitosan coating were in better condition with lower maturity index, lower weight loss, and color change as compared to wrap packaging mushrooms due to barrier properties of chitosan coating. Chitosan coating also was applied on different fruits and mushrooms by Tamer and Copur (2009); the coating treatment improved the

storability of the all treated fruits and mushrooms by reducing the activity of polyphenol oxidase (PPO), peroxidase (POD), catalase, phenylalanine ammonia lyase (PAL), laccase, cellulose, alpha amylase, total amylase and maintained phenolic activity. Wrona et al. (2015) have developed a sulphur dioxide based active packaging functionalized with citric acid, cinnamon essential oil, green tea extract and purple carrot extract; it was applied on mushrooms to evaluate their effects on shelf life and other quality attributes such as reducing microbial growth, and maintain color properties. Furthermore, incorporation of green tea extract in packaging materials was more effective to reduce the color loss of mushrooms and showed potential antioxidant activity.

On the other hand, addition of cinnamon essential oil and purple carrot extract did not showed good antioxidant activity to reduce the spoilage and prolong the shelf life of mushrooms (Wrona et al. (2015). The button mushrooms shelf life was prolonged by Mohebbi et al. (2012) by using aloe vera and gum tragacanth and their composite coatings at different temperature conditions (4°C, 10°C and 15°C) over the storage period of 13 days. The rapid loss of mass, color, texture was found in untreated mushrooms as compared to treated. All the formulations of edible coating were effective in prevention of quality of mushroom, but the composite coatings of *Aloe vera* and gum tragacanth were more potential as compared to other formulations. Luttmann (2011) have studied the effects of three different formulations of calcium alginate based coating which include: i) alginate; ii) alginate + bee wax; and iii) alginate + bee wax + nisin + Na_2EDTA (ABN). The optimized formulation of coating (alginate - 2.49%; beeswax-0.82%; Na_2EDTA (8.18 mg/mL) and nisin (4000 IU/mL) was prolonged the shelf life of button mushrooms up to 7 days at 12°C. Jiang et al. (2012) have reported the effects of chitosan coating functionalized with thyme oil on shiitake mushroom during 16 days of storage at 4°C. These results showed that the application of thyme oil rich edible coating was effective to extend the shelf life of mushroom during storage condition by decreasing respiration rate, maintaining high phenolic and flavonoid contents, inhibiting the growth of microbes (*Pseudomonas*), yeast and molds, and maintaining sensory attributes as compared to untreated shiitake mushrooms. Similarly, the shitake mushrooms shelf life was extended by using alginate (1%, 2% and 3%) based edible coating under high oxygen modified atmosphere (Jiang 2013). These results indicated that the application of 2% alginate coating under high oxygen modified atmosphere was effective to extending the shelf life of shiitake mushrooms up to 16 days by retaining high textural property, delaying respiration rate, reducing mass loss, maintaining color, TSS and ascorbic acid. In another study done by Jiang et al. (2013) was reported potential of arabic gum and natamycin coating to prolong the shelf life of shiitake mushroom up to 16 days by reduction in changes of quality attributes and physicochemical properties.

Recently, Huang et al. (2019) have also prolonged the shelf life of shiitake mushrooms using chitosan (1%) and guar gum (5%, 15% and 25%) composite coating. Namely, formulation prepared with 1% chitosan with 15% guar gum was found the most effective to maintain the textural property of tissue, reduced the loss of ascorbic acid and soluble protein, maintained TSS, malondialdehyde and reduced electrolyte leakage as compared to other formulations. They also suggested the best formulation for commercial applications to prolong the shelf life of shiitake mushrooms. The lipid based edible coating incorporated with different doses of glycerol monoestearate and thyme extract was applied on mushroom by Samadpour and Dastgheib (2020) and evaluated their effects; these results showed that the treatment of lipid based coating significantly retained the quality of oyster mushroom for longer time of storage by retained texture quality; color attributes, inhibited microbial growths, reduced weight loss and retained sensory characteristics. On the other hand, the treatment of chitosan and chitosan + nano titanium dioxide (TiO_2) coating was found potential to significantly lower the respiration rate, reduced weight loss, antimicrobial efficiency against microbial growth (4.27 *log cfu/g*) and total yeast and molds (5.93 *log cfu/g*), maintained acidity and other quality parameters of button mushrooms by Sami et al. (2020); treatment of chitosan nano coating with TiO_2 delayed the aggeing process of mushroom and was suggested as novel packaging for other food products to extend their shelf life with an appropriate food safety and quality. Similarly, the button mushroom shelf life was also prolonged by Rad et al. (2020) by using CMC and sodium metabisulphite based coating at 4°C storage temperature. These results showed that the treatment of 2% CMC and 2% sodium metabisulphite coating extended the shelf life of button mushroom by maintaining the physiochemical properties, physiological properties, organoleptic properties and microbial properties. In another study, the different formulations of apple peel powder and CMC coating were applied on button mushroom to investigate their effects on shelf life and other quality attributes by Thakur et al. (2020). They found the best formulation loaded with moderate concentration (1.2% w/v) of apple peel powder and high concentration (1.8%w/v) of CMC to prolong the shelf life of mushrooms up to 5 days by retained higher quality i.e., reducing water loss, maintain color properties, reduced microbial growth and antioxidant activity as compared to other formulations.

Conclusion

Mushrooms are known for thousands of years, but in the last few decades they attract a big attention worldwide due to rising awareness and recognition of many health benefits that they provide. Namely, mushrooms have been shown that are important food for humans full of nutrition and bioactive compounds; most of them have preventive role against many diseases and contribute to better human health. In general, mushrooms are rich source of vitamins and minerals, proteins, dietary fibers, β-glucans; they possess an immuno-modulatory effect, but also they could provide anti-tumor, antioxidant, anti-hypertensive, anti-inflammatory, anti-diabetic, anti-microbial, cardiovascular benefits, cholesterol-lowering, and anti-obesity effects. They also have various enzymes that could be used in food industry. All these attributes lead to increased interest for edible mushrooms; moreover, they could be an alternative source for income especially in less developed countries.

Novel technologies for safe storage of mushrooms for a longer period of time with good quality are very important; edible packaging is one of the suitable method for this purpose. There are many proposed bio-based materials that could form edible films or coatings for mushroom packaging, protection and extending their shelf life. However, beside the current utilization of mushrooms in everyday life, there are some gaps that should be filled up with better understanding of different types of mushrooms all over the globe, their safety and edibility, bioactive compounds and effects on humans, or medical supplements based on natural plant origin. Inventing, designing and optimization study for novel cross-cultural functional foods based on mushrooms for everyone's taste, texture and smell are not less important and should be a strategic plan for food engineers and scientists in the coming years. Therefore, mushrooms are unlocking potential that could be added value to a variety of food items or medical supplements.

References

Abo EMM, El Kady EM (2019) Current trends in fungal biosynthesis of chitin and chitosan. *Bull Natl Res Cent* 43(1):59.

Adejumo TO, Awosanya OB (2005) Proximate and mineral composition of four edible mushroom species from South Western Nigeria. *Afr J Biotechnol* 4(10):1084-1088.

Adeoyo OR, Oluborode OO (2020) Antimicrobial properties of some Nigerian edible mushrooms. *J Phytopharmacol,* 9(2): 110-114.

Aida FMNA, ShuhaimI M, Yazid M, Maaruf AG (2009) Mushroom as a potential source of prebiotics: a review. *Trends Food Sci Technol* 20:567-575.

Al-Faqeeh LAS, Naser R, Kagne SR, SUBUR WK (2021) Activity of mushrooms against diabetic and inflammation: A review. *GSC Biol Pharm Sci* 14(2):37-44.

Alves M, Ferreira IFR, Dias J, Teixeira V, Martins A, Pintado M. (2012) A review on antimicrobial activity of mushroom (*Basidiomycetes*) extracts and isolated compounds. *Planta Med* 78(16):1707-1718.

Appolinaire KK, Hubert KK, Eugène KJP, Ahipo DE, Lucien KP (2016) Proximate composition, minerals and amino acids profiles of selected wild edible Russula species from Côte d'Ivoire. *TURJAF* 4(10):882-886.

Aremu MO, Basu SK, Gyar SD, Goyal A, Bhowmik P, Banik SD (2009) Proximate composition and functional properties of mushroom flours from Ganoderma spp., Omphalotus olearius (DC.) sing. and Hebeloma mesophaeum (Pers.) Quél. used in Nasarawa State. *Nigeria Malays J Nutr* 15(2):233-241.

Ban Z, Li L, Guan J, Feng J, Wu M, Xu X, Li J (2014) Modified atmosphere packaging (MAP) and coating for improving preservation of whole and sliced Agaricus bisporus. *J Food Sci Technol* 51(12):3894-3901.

Bano Z, Rajarathnam S (1986) Vitamin values of Pleurotus mushrooms. *Qual Plant Plant Foods Hum Nutr* 36(1):11-15.

Barros L, Cruz T, Baptista P, Estevinho L, Ferreira I (2008) Wild and commercial mushrooms as source of nutrients and nutraceuticals. *Food Chem Toxicol* 46(8):2742–2747.

Bernaś E, Jaworska G (2016) Vitamins profile as an indicator of the quality of frozen Agaricus bisporus mushrooms. *J Food Compos Anal* 49:1-8.

Bhatt RP, Singh U, Uniyal P (2018) Healing mushrooms of Uttarakhand Himalaya, India. *Curr Res Environ Appl Mycol,* 8(1): 1-23.

Bhuiyan MR, Shaid A, Bashar MM, Haque P, Hannan MA (2013) A novel approach of dyeing jute fiber with reactive dye after treating with chitosan. *Open J Org Polym* Mater, 3(4):87-91.

Bowman SM, Free SJ (2006) The structure and synthesis of the fungal cell wall. *BioEssays: news and reviews in molecular, cellular and developmental biology* 28(8):799-808.

Bunbamrung N, Intaraudom C, Dramae A, Boonyuen N, Veeranondha S, Rachtawee P, Pittayakhajonwut P (2017) Antimicrobial activity of illudalane and alliacane sesquiterpenes from the mushroom Gloeostereum incarnatum BCC41461. *Phytochem Lett* 20:274-281.

Cateni F, Gargano ML, Procida G, Venturella G, Cirlincione F, Ferraro V (2021) Mycochemicals in wild and cultivated mushrooms: nutrition and health. *Phytochem Rev* 1-45.

Chang ST (2006) The world mushroom industry: trends and technological development. *Int J Med Mushrooms* 8(4).

Chang ST, Buswell JA (1996) Mushroom nutriceuticals. *World J Microbiol Biotechnol* 12:473-476.

Chang ST, Miles PG (2008) *Mushrooms: Cultivation, Nutritional Value, Medicinal Effect, and Environmental Impact.* 2nd. Boca Raton, Fla, USA: CRC Press.

Chatterjee D, Halder D, Das S (2021) Varieties of Mushrooms and their Nutraceutical Importance: A Systematic Review. *J clin diagn* 15(3).

Cheung LM, Cheung PCK, Ooi VEC (2003) Antioxidant activity and total phenolics of edible mushroom extracts. *Food Chem* 81(2):249-255.

Cheung PCK (2008) *Mushrooms as Functional Foods.* John Wiley & Sons, Inc 3:23-24.

Cheung PCK (2013) Mini-review on edible mushrooms as source of dietary fiber: preparation and health benefits. *Food Sci Hum Well* 2:162-166.

Chikthimmah N, Beelman RB (2005) Microbial spoilage of fresh mushrooms, in *Microbiology of Fruits and Vegetables,* ed. by Sapers M, Gorny JR and Yousef AE. CRC Press, Boca Raton, FL, 138-146.

Chowdhury MMH, Kubra K, Ahmed SR (2015) Screening of antimicrobial, antioxidant properties and bioactive compounds of some edible mushrooms cultivated in Bangladesh. *Ann Clin Microbiol Antimicrob* 14(1):1-6.

de Sousa MF (2015) *Edible coatings as a means to improve the quality and shelf-life of Agaricus bisporus (white button) mushroom.*

Dilfy SH, Hanawi MJ, Al-bideri AW, Jalil AT (2020) Determination of Chemical Composition of Cultivated Mushrooms in Iraq with Spectrophotometrically and High Performance Liquid Chromatographic. *J Green Eng* 10:6200-6216.

Dillard CJ, German JB (2000) Phytochemicals: Nutraceuticals and human health. *J. Sci Food Agric* 80(12):1744-1756.

Doores S, Kramer M, Beelman R (1987) Evaluation and bacterial populations associated with fresh mushrooms (Agaricus bisporus), in *Cultivating Edible Fungi: Developments in Crop Science,* ed. by Wuest PJ, Royce DL and Beelman RB. Elsevier, Amsterdam, 283-294.

Eissa HA (2007) Effect of chitosan coating on shelf life and quality of fresh-cut mushroom. *J Food Qual* 30(5): 623-645.

Elmastas M, Isildak O, Turkekul I, Temur N (2007) Determination of antioxidant activity and antioxidant compounds in wild edible mushrooms. *J Food Compos Anal* 20(3-4): 337-345.

Elsoud MMA, El Kady EM (2019) Current trends in fungal biosynthesis of chitin and chitosan. *Bulletin of the National Research Centre* 43(1):59.

Eraslan EC, Çırçırlı B, Özkan A, Akgul H (2021) Anticancer Mechanisms of Action of Macrofungus Extracts. *Eurasian J Med Biol Sci* 1-12.

Erdogan S, Kaya M, Akata I (2017) Chitin extraction and chitosan production from cell wall of two mushroom species (Lactarius vellereus and Phyllophora ribis). In *AIP Conference Proceedings,* 1809(1):020012. AIP Publishing LLC.

Erjavec J, Kos J, Ravnikar M, Dreo T, Sabotic J (2012) Proteins of higher fungi from forest to application. *Trends Biotechnol* 30(5):259–273.

FAOSTAT (2019) *Mushrooms and Truffles* [Internet]. Available from: http://www.fao.org/faostat/en/#data/QC [Accessed 12 August 2019].

Fasidi IO, Ekuere UU (1993) Studies on Pleurotus tuber-regium (Fries) Singer: cultivation, proximate composition and mineral contents of sclerotia. *Food Chem* 48(3):255-258.

Fatima N, Upadhyay T, Ahmad F, Arshad M, Kamal MA, Sharma D, Sharma R (2021) Particulate β-glucan activates early and delayed phagosomal maturation and autophagy within macrophage in a NOX-2 dependent manner. *Life Sci* 266: 118851.

Fazli WNWM, Lee KY, Kontturi E, Murphy RJ, Bismarck A (2019) Chitin nanopaper from mushroom extract: natural composite of nanofibers and glucan from a single biobased source. *ACS Sustain Chem Eng* 7(7):6492-6496.

Feeney MJ, Dwyer J, Hasler-Lewis CM, Milner JA, Noakes M, Rowe S, Wach M, Beelman RB, Caldwell J, Cantorna MT, Castlebury LA, Chang ST, Cheskin LJ, Clemens R, Drescher G, Fulgoni VL 3rd, Haytowitz DB, Hubbard VS, Law D, Myrdal Miller A, Minor B, Percival SS, Riscuta G, Schneeman B, Thornsbury S, Toner CD, Woteki CE, Wu D. (2014) *Mushrooms and health summit proceedings.* JN 144(1):1128S-1136S.

Fernandes A, Barreira JCM, Antonio AL, Morales P, Férnandez-Ruiz V, Martins A, Oliveira MBPP, Ferreira ICFR (2015) Exquisite wild mushrooms as a source of dietary fiber: analysis in electron-beam irradiated samples. *LWT-Food Sci Technol* 60:855-859.

Ferreira ICFR, Barros LL, Abreu RMV (2009) Antioxidants in wild mushrooms. *Curr Med Chem* 16:1543-1560.

Fijałkowska A, Muszyńska B, Sułkowska-Ziaja K, Kala K, Pawlik A, Stefaniuk D, Matuszewska A, Piska K, Pekala E, Kaczmarczyk P, Pietka J, Jaszek M (2020) Medicinal potential of mycelium and fruiting bodies of an arboreal mushroom Fomitopsis officinalis in therapy of lifestyle diseases. *Sci Rep* 10(1): 1-12.

Fogarasi M, Socaci SA, Dulf FV, Diaconeasa ZM, Fărcaş AC, Tofană M, Semeniuc CA (2018) Bioactive Compounds and Volatile Profiles of Five Transylvanian Wild Edible Mushrooms. *Molecules* 23(12):3272.

Food and Agriculture Organization Statistical (2017) FAOSTAT— Data-crops-visualized-mushrooms and truffles.

Friedman M (2016) Mushroom polysaccharides: Chemistry and antiobesity, antidiabetes, anticancer, and antibiotic properties in cells, rodents and humans. *Foods* 5(4):80.

Furlani RPZ, Godoy HT (2008) Vitamins B1 and B2 contents in cultivated mushrooms. *Food Chem* 106(2):816-819.

Gadallah MG, Ashoush IS (2016) Value addition on nutritional and sensory properties of biscuit using desert truffle (*Terfezia claveryi*) powder. *Food Nutr Sci* 7(12):1171-1181.

Garrab M, Edziri H, El Mokni R, Mastouri M, Mabrouk H, Douki W (2019) Phenolic composition, antioxidant and anticholinesterase properties of the three mushrooms Agaricus silvaticus Schaeff., Hydnum rufescens Pers. and Meripilus giganteus (Pers.) Karst. in Tunisia. *S Afr J Bo*t 124:359-363.

Gholami R, Ahmadi E, Ahmadi S (2020) Investigating the effect of chitosan, nanopackaging, and modified atmosphere packaging on physical, chemical, and mechanical properties of button mushroom during storage. *Food Sci Nutr* 8(1):224-236.

Gil-Ramirez A, Pavo-Caballero C, Baeza E, Baenas N, Cristina GV, Martin FRM, Rivas CS (2016) Mushrooms do not contain flavonoids. *J Funct Foods* 25:1-13.

Groves JW (1981) *Edible and Poisonous mushrooms of the Canada*. Canadian Government Publishing Centre Supply and Services Canada 1-346.

Gupta VG, Pandey A (2019) *New and Future Developments in Microbial Biotechnology and Bioengineering: Microbial Secondary Metabolites Biochemistry and Applications.* Elsevier.

Halami PM, Chandrashekar A, Joseph R (1999) Characterization of bacteriocinogenic strains of lactic acid bacteria in fowl and fish intestine and mushroom. *Food Biotechnol* 13:121-136.

Han NS, Ahmad WANW, Ishak WRW (2016) Quality characteristics of Pleurotus sajor-caju powder: Study on nutritional compositions, functional properties and storage stability. *Sains Malays* 45(11):1617-1623.

Hasler CM (2002) Functional Foods: Benefits, Concerns and Challenges- A Position Paper from the American Council on Science and Health. *The Journal of Nutrition,* 132(12):3772-3781.

Hasnat MA, Pervin M, Debnath T, Lim BO (2014) DNA protection, total phenolics and antioxidant potential of the mushroom Russula virescens. *J Food Biochem* 38:6-17.

Heleno SA, Barros L, Martins A, Queiroz MJ, Santos-Buelga C, Ferreira IC (2012) Phenolic, polysaccharidic, and lipidic fractions of mushrooms from Northeastern Portugal: Chemical compounds with antioxidant properties. *J Agric Food Chem* 60(18):4634-4460.

Heleno SA, Barros L, Sousa MJ, Martins ARL, Ferreira ICFR (2010) Tocopherols composition of Portuguese wild mushrooms with antioxidant capacity. *Food Chem* 119(4):1443-1450.

Hereher F, ElFallal A, Toson E, Abou-Dobara M, Abdelaziz M (2018) Pilot study: Tumor suppressive effect of crude polysaccharide substances extracted from some selected mushroom. *Beni-Suef Univ J Basic Appl Sci* 7(4):767-775.

Hershko V, Nussinovitch A (1998) Relationships between hydrocolloid coating and mushroom structure. *J Agric Food Chem* 46(8):2988-2997.

Hokmollahi F, Rafati H, Riahi H, Hakimi MH, Heydari H, Haghirosadat F, Azimzade M, Frotan A, Mossazade SA (2012) Investigation of the antioxidant activities and determination of the phenol content of fractional extracts of Iranian medicinal mushroom *Phellinustorulosus*. *J Rafsanjan Univ Med Sci* 11(1):55-66.

Huang Q, Qian X, Jiang T, Xiaolin Z (2019) Effect of chitosan and guar gum based composite edible coating on quality of mushroom (*Lentinus edodes*) during postharvest storage. *Sci Hortic* 253:382-389.

Ifuku S, Nomura R, Morimoto M, Saimoto H (2011) Preparation of chitin nanofibers from mushrooms. *Mater* 4(8):1417-1425.

Ifuku S, Saimoto H (2012) Chitin nanofibers: preparations, modifications, and applications. *Nanoscale* 4:3308-3318.

Islam T, Ganesan K, Xu B (2019) New insight into mycochemical profiles and antioxidant potential of edible and medicinal mushrooms: A review. *Int J Med Mushrooms* 21:237-251.

Ivshina TN, Artamonova SD, Ivshin VP, Sharnina FF (2009) Isolation of the chitin-glucan complex from the fruiting bodies of mycothallus. *Appl Biochem Microbiol* 45(3):313-318.

Jana P, Acharya K (2020) Mushroom: A New Resource for Anti-Angiogenic Therapeutics. *Food Rev Int* 1-22.

Jayachandran M, Xiao J, Xu B (2017) A critical review on health promoting benefits of edible mushrooms through gut microbiota. *Int J Mol Sci* 18(9):1934.

Jiang T (2013) Effect of alginate coating on physicochemical and sensory qualities of button mushrooms (*Agaricus bisporus*) under a high oxygen modified atmosphere. *Postharvest Biol Technol* 76:91-97.

Jiang T, Feng L, Wang Y (2013) Effect of alginate/nano-Ag coating on microbial and physicochemical characteristics of shiitake mushroom (*Lentinus edodes*) during cold storage. *Food Chem* 141(2):954-960.

Jiang T, Feng L, Zheng X (2012) Effect of chitosan coating enriched with thyme oil on postharvest quality and shelf life of shiitake mushroom (Lentinus edodes). *J Agric Food Chem* 60(1):188-196.

Kała K, Kryczyk-Poprawa A, Rzewińska A, Muszynska B (2020) Fruiting bodies of selected edible mushrooms as a potential source of lovastatin. *Eur Food Res Technol* 246(4):713-722.

Kamal ASM, Khair A, Begum F, Chowdhury K, Karim R, (2015) Effect of respiratory gases (O_2; CO_2) on shelf-life of fresh oyster mushrooms packaged with different sealable polymeric materials. *Bangladesh J Sci Ind Res* 50(3):205-210.

Kayode RMO, Olakulehin TF, Adedeji BS, Ahmed O, Aliyu TH, Badmos AHA (2015) Evaluation of amino acid and fatty acid profiles of commercially cultivated oyster mushroom (Pleurotus sajor-caju) grown on gmelina wood waste. *Niger Food J* 33(1):18-21.

Keleş A, Koca I, Gençcelep H (2011) Antioxidant properties of wild edible mushrooms. *Int J Food Process Technol* 2(6):2-6.

Khan AA, Gani A, Khanday FA, Masoodi FA (2018) Biological and pharmaceutical activities of mushroom β-glucan discussed as a potential functional food ingredient. *Bioact Carbohydr Diet Fibre* 16:1-13.

Khojah E, Sami R, Helal M, Elhakem A, Benajiba N, Alharbi M, Alkaltham MS (2021) Effect of Coatings Using Titanium Dioxide Nanoparticles and Chitosan Films on Oxidation during Storage on White Button Mushroom. *Crystals* 11(6): 603.

Kim KM, Ko JA, Lee JS, Park H, Hanna M (2006) Effect of modified atmosphere packaging on the shelf-life of coated, whole and sliced mushrooms. *LWT- Food Sci Technol* 39(4):365-372.

Kosanic M, Rankovic B, Dasic M (2013) Antioxidant and antimicrobial properties of mushrooms. *Bulg J Agric Sci* 19:1040-1046.

Kothari D, Patel S, Kim SK (2018) Anticancer and other therapeutic relevance of mushroom polysaccharides: A holistic appraisal. *Biomed Pharmacother* 105:377-394.

Kozarski M, Klaus A, Jakovljevic D, Todorovic N, Vunduk J, Petrović P, Niksic M, Vrvic MM, van Griensven L (2015) Antioxidants of edible mushrooms. *Molecules* 20(10):19489-19525.

Kozarski M, Klaus A, Vunduk J, Zizak Z, Niksic M, Jakovljevic D, Vrvic MM, Van Griensven LJ (2015) Nutraceutical properties of the methanolic extract of edible mushroom Cantharellus cibarius (Fries): primary mechanisms. *Food Funct* 6(6): 1875-1886.

Krakowska A, Zięba P, Włodarczyk A, Kała K, Sułkowska-Ziaja K, Bernaś E, Sękara A, Ostachowicz B, Muszyńska B (2020) Selected edible medicinal mushrooms from Pleurotus genus as an answer for human civilization diseases. *Food Chem* 327: 127084.

Krishnamoorthy AS, Muthuswamy MT, Nakkeeran S (2000) Technique for commercial production of milky mushroom Calocybe indica P&C. *Indian J Mush* 18:19-23.

Kulshreshtha S, Thakur M (2020) Mushroom: A Potential Tool for Food Industry Waste, In book: *Sustainable Food Waste Management Concepts and Innovations.* Thakur, M., Modi, V. K., Khedkar, R., & Singh, K. (Eds.). 17: 307-331, Springer.

Kumar H. Bhardwaj K, Sharma R, Nepovimova E, Cruz-Martins N, Dhanjal DS, Singh R, Chopra C, Verma R, Abd-Elsalam KA, Tapwal A, Musilek K, Kumar D, Kuˇca K (2021a) Potential Usage of Edible Mushrooms and Their Residues to Retrieve Valuable Supplies for Industrial Applications. *J. Fungi* 7: 427. https://doi.org/10.3390/jof7060427.

Kumar N, Neeraj (2019) Polysaccharide-based component and their relevance in edible film/coating: A review. *Nutr. Food Sci* 49(5):793-823.

Kumar N, Neeraj, Pratibha, Singla M (2020) Enhancement of Storage Life and Quality Maintenance of Litchi (*Litchi Chinensis Sonn.*) Fruit Using Chitosan: pullulan Blend Antimicrobial Edible Coating. *Int J Fruit Sci* 20(3):S1662-S1680.

Kumar N, Neeraj, Pratibha, Trajkovska Petkoska A (2021b) Improved Shelf Life and Quality of Tomato (*Solanum lycopersicum L.*) by Using Chitosan-Pullulan Composite Edible Coating Enriched with Pomegranate Peel Extract. *ACS Food Sci Technol* https://doi.org/10.1021/acsfoodscitech.0c00076.

Kumar N, Pratibha, Neeraj, Ojha A, Upadhaya A, Singh R, Kumar S (2021c) Effect of active chitosan-pullulan composite edible coating enrich with pomegranate peel extract on the storage quality of green Bell pepper. *LWT- Food Sci Technol* 138:110435.

Kumar N, Pratibha, Neeraj, Trajkovska Petkoska A, Ali S, Fawole OA (2021d) Effect of Chitosan–Pullulan Composite Edible Coating Functionalized with Pomegranate Peel Extract on the Shelf Life of Mango (Mangifera indica). *Coatings* 11: 764 https://doi.org/10.3390/coatings11070764.

Kumar R (2013) The politics of medical practice license and its impact on primary care workforce: International developments and Indian perspective. *J Med Prim Care* 2(1): 1.

Kumla J, Suwannarach N, Tanruean K, Lumyong S (2021) Comparative Evaluation of Chemical Composition, Phenolic Compounds, and Antioxidant and Antimicrobial Activities of Tropical Black Bolete Mushroom Using Different Preservation Methods. *Foods* 10(4): 781.

Kunzek H, Müller S, Vetter S, Godeck R (2002) The significance of physico chemical properties of plant cell wall materials for the development of innovative food products. *Eur Food Res Technol* 214:361-376.

Kushwaha A, Chaturvedi VK, Rai SN, Masih SC, Singh MP (2021) Therapeutic Efficacy of Mushroom in Neurodegenerative Diseases. *Recent Advances in the Treatment of Neurodegenerative Disorders,* 120, 120.

Liu RM, Zhang XJ, Liang GY, Yang YF, Zhong JJ, Xiao JH (2015) Antitumor and antimetastatic activities of chloroform extract of medicinal mushroom Cordyceps taii in mouse models. *BMC Complement Altern Med* 15(1):1-13.

Lu X, Brennan MA, Serventi L, Mason S, Brennan CS (2016) How the inclusion of mushroom powder can affect the physicochemical characteristics of pasta. *Int J Food Sci* 51(11):2433-2439.

Luttmann A (2011) Optimization of an alginate based edible coating with beeswax, nisin and EDTA to maximize shelf life of fresh mushrooms (*Agaricus bisporus*). *Master of Science thesis.*

Ma G, Yang W, Zhao L, Pei F, Fang D, Hu Q (2018) A critical review on the health promoting effects of mushrooms nutraceuticals. F*ood Sci Hum Well* 7(2):125-133.

Manach C, Scalbert A, Morand C, Rémésy C, Jiménez L. (2004) Polyphenols: Food sources and bioavailability. *Am J Clin Nutr* 79:727-747.

Manjunathan J, Subbulakshmi N, Shanmugapriya R, Kaviyarasan V (2011) Proximate and mineral composition of four edible mushroom species from South India. *Int J Biodivers Conserv* 3(8):386-388.

Martinez-Medina GA, Chávez-González ML, Verma DK, Prado LA, Martinez-Hernandez JL, Flores C, Thakur M, Srivastav PP, Aguilar CN (2021) Bio-funcional components in mushrooms, a health opportunity: Ergothionine and huitlacohe as recent trends. *J Funct Foods* 77:104326.

Mattila P, Könkö K, Eurola M, Pihlava JM, Astola J, Vahteristo L, Hietaniemi V, Kumpulainen J, Valtonen M, Piironen V (2001) Contents of vitamins, mineral elements, and some phenolic compounds in cultivated mushrooms. *J Agric Food* Chem 49(5):2343-2348.

Mehta BK, Jain SK, Sharma GP, Doshi A, Jain HK (2011) Cultivation of button mushroom and its processing: an techno-economic feasibility. *Int J Adv Biotechnol Res* 2(1):201-207.

Meng X, Liang H, Luo L (2016) Antitumor polysaccharides from mushrooms: a review on the structural characteristics, antitumor mechanisms and immunomodulating activities. *Carbohydr Res* 424:30-41.

Miles PG, Chang ST (2004) *Mushrooms: Cultivation, Nutritional Value, Medicinal Effect, and Environmental Impact.* CRC Press.

Minh NP, Hang LP (2019) Several factors affecting to shelf-life of paddy straw mushroom (*Volvariella spp.)* in preservation. *Plant Arch* 19(2):444-448.

Mohamed IM, Mohamed MRM, Abdul BJ, Basith JA, Asarudeen A (2011) Determination of total phenol, flavonoid and antioxidant activity of edible mushrooms Pleurotus florida and Pleurotus eous. *Int Food Res J* 18(2):579-582.

Mohapatra D, Bira ZM, Kerry JP (2010) Postharvest hardness and color evolution of white button mushrooms (*Agaricus bisporus*). *J Food Sci* 75(3):E146-E152.

Mohebbi M, Ansarifar E, Hasanpour N, Amiryousefi MR (2012) Suitability of aloe vera and gum tragacanth as edible coatings for extending the shelf life of button mushroom. *Food Bioproc Tech* 5(8):3193-3202.

Mortimer PE, Boa E, Hyde KD, Li H (2021) Macrofungi as Food : Encyclopedia of Mycology. In book: *Reference Module in Life Sciences.* doi:10.1016/B978-0-12-819990-9.00071-8.

Muszyńska B, Grzywacz-Kisielewska A, Kała K, Gdula-Argasińska J (2018) Anti-inflammatory properties of edible mushrooms: A review. *Food Chem* 243:373-381.

Muthu N, Shanmugasundaram K (2016) Proximate and mineral compositions of edible mushroom Agrocybe aegerita. *J Pharmacogn Phytochem* 5(1):116.

Naknaen P, Itthisoponkul T, Charoenthaikij P (2015) Proximate compositions, nonvolatile taste components and antioxidant capacities of some dried edible mushrooms collected from Thailand. *J Food Meas Charact* 9(3):259-268.

Nasiri M, Barzegar M, Sahari MA (2017) Tragacanth gum containing Zataria multiflora Boiss. essential oil as a natural preservative for storage of button mushrooms (*Agaricus bisporus*). *Food Hydrocoll* 72:202-209.

Niazi AR, Ghafoor A (2021) Different ways to exploit mushrooms: A review. *All Life,* 14(1): 450-460.

Niazmand A, Ghodusi HB, Shahidi F, Niazmand R (2009) Effects of three polysaccharide coatings on physicochemical and organoleptic properties of mushroom (Agaricus bisporus). Am Eurasian *J Agric Environ Sci* 5:740-744.

Nichols R (1985) Postharvest physiology and storage. In: *The Biology and Technology of the Cultivated Mushroom* (Eds. P. B. Flegg, D. M. Spencer and D. A. Wood). John Wiley and Sons Ltd, New York. 195-210.

Nicolcioiu MB, Popa G, Matei F (2018) Biochemical Investigations of Different Mushroom Species for Their Biotechnological Potential. In *"Agriculture for Life, Life for Agriculture" Conference Proceedings,* 1(1): p. 562-567, Sciendo.

Niego AG, Rapior S, Thongklang N, Raspé O, Jaidee W, Lumyong S, Hyde KD (2021) Macrofungi as a Nutraceutical Source: Promising Bioactive Compounds and Market Value. *J. Fungi* 7(5): 397.

Nowakowski P, Markiewicz-Żukowska R, Gromkowska-Kępka K, Naliwajko SK, Moskwa J, Bielecka J, Grabia M, Borawska M, Socha K (2021) Mushrooms as potential therapeutic agents in the treatment of cancer: Evaluation of anti-glioma effects of Coprinus comatus, Cantharellus cibarius, Lycoperdon perlatum and Lactarius deliciosus extracts. *Biomed. Pharmacother* 133: 111090.

Nowakowski P, Naliwajko SK, Markiewicz-Zukowska R, Borawska MH, Socha K (2020) The two faces of Coprinus comatus—functional properties and potential hazards. *Phytother Res* 34(11):2932-2944.

Nussinovitch A, Kampf N (1993) Shelf-life extension and conserved texture of alginate-coated mushrooms (Agaricus bisporus). *LWT- Food Sci Technol* 26(5):469-475.

Odier A (1823) Mémoir sur la composition chimique des parties cornées des insectes. *Mémoirs de la Societé d'Histoire Naturelle* [Memory on the chemical composition of the horny parts of insects. *Memories of the Natural History Society*]. 1:29-42.

Ojeda GA, Sgroppo SC, Martín-Belloso O, Soliva-Fortuny R (2019). Chitosan/tripolyphosphate nanoaggregates enhance the antibrowning effect of ascorbic acid on mushroom slices. *Postharvest Biol Tech* 156:110934.

Ospina Álvarez, S. P., Ramírez Cadavid, D. A., Escobar Sierra, D. M., Ossa Orozco, C. P., Rojas Vahos, D. F., Zapata Ocampo, P., & Atehortúa, L. (2014) Comparison of extraction methods of chitin from Ganoderma lucidum mushroom obtained in submerged culture. *BioMed Res Int* 169071.

Patel S, Goyal A (2012) Recent developments in mushrooms as anti-cancer therapeutics: A review. 3 *Biotech* 2(1):1-5.

Pegler DN (2002) Useful Fungi of the World: the 'Poor man's tuffles of Arabia' and 'Manna of the Israelites'. *Mycologist,* 16(1):8-9.

Perera PK, Li Y (2011) Mushrooms as a functional food mediator in preventing and ameliorating diabetes. *Funct Food Health Dis* 1(4):161-171.

Petkoska AT, Daniloski D, D'Cunha NM, Naumovski N, Broach AT (2021) Edible packaging: Sustainable solutions and novel trends in food packaging. *Food Res Int* 140: 109981.

Pinelo M, Rubilar M, Jerez M, Sineiro J, Nunez MJ (2005) Effect of solvent, temperature, and solvent-to-solid ratio on the total phenolic content and antiradical activity of extracts from different components of grape pomace. *J Agric Food Chem* 53(6):2111-2117.

Puttaraju NG, Venkateshaiah SU, Dharmesh SM, Urs SM, Somasundaram R (2006) Antioxidant Activity of Indigenous Edible Mushrooms. *J Agric Food Chem* 54(26):9764-9772.

Rad M, Ghafori H, Gholami Z (2020) Effect of edible coating containing carboxymethyl cellulose and sodium metabisulfite on the shelf life of the button mushroom. *Iran J Sci Technol Trans A Sci* 16(5):581-605.

Raghavendra VB, Venkitasamy C, Pan Z, Nayak C (2017) Functional foods from mushroom. In *Microbial Functional Foods and Nutraceuticals;* John Wiley & Sons, Ltd.: Chichester, UK; 65-91.

Rahman MS, Hassan MK, Talukder FU, Rahman MS (2020) Effect of Low Temperature on Postharvest Behaviors of Oyster Mushroom (Pleurotus spp.). *Int J Hortic Sci Technol* 7(1):213-225.

Rai RD, Arumuganathan T (2008) Post harvest technology of mushrooms. Chambaghat, India: National Research Centre for Mushroom, Indian Council of Agricultural Research. *Technical Bulletin.*

Rathore H, Prasad S, Kapri M, Abhay T, Satyawati S (2019) Medicinal importance of mushroom mycelium: Mechanisms and applications. *J Funct Foods* 56:182-193.

Reis FS, Barros L, Martins A, Ferreira IC (2012) Chemical composition and nutritional value of the most widely appreciated cultivated mushrooms: an inter-species comparative study. *Food Chem Toxicol* 50(2):191-197.

Ribeiro A, Ruphuy G, Lopes JC, Dias MM, Barros L, Barreiro F, Ferreira IC (2015) Spray-drying microencapsulation of synergistic antioxidant mushroom extracts and their use as functional food ingredients. *Food Chem* 188:612-618.

Sabaratnam V, Kah-Hui W, Naidu M, David PR (2013) Neuronal health–can culinary and medicinal mushrooms help? *J Tradit Complement Med* 3(1):62-68.

Sabir SM, Hayat I, Hussain I, Gardezi SRA (2003) Proximate analysis of mushrooms of Azad Kashmir. *Plant Pathol J* 2(2): 97-10.1.

Salehi F (2019) Characterization of different mushrooms powder and its application in bakery products: A review, *International J. Food Prop* 22(1): 1375-1385.

Salehi F (2020) Edible Coating of Fruits and Vegetables Using Natural Gums: A Review. *Int J Fruit Sci* 20(2):S570-S589.

Samadpour R (2020) The effect of edible coating with combined Thymus Vulgaris extract and glycerol monoestearate on oyster mushroom's shelf life. *Future of Food: Journal on Food, Agriculture and Society* 8(2):1-13.

Samadpour R, Dastgheib BMK (2020) The effect of edible coating with combined Thymus Vulgaris extract and glycerol monoestearate on oyster mushroom's shelf life. *Future of Food: J Food, Agric Soc* 8 (2): 1-13.

Sami R, Elhakem A, Alharbi M, Benajiba N, Almatrafi M, Jing J, Helal M (2020) Effect of Titanium Dioxide Nanocomposite Material and Antimicrobial Agents on Mushrooms Shelf-Life Preservation. *Processes* 8(12):1632.

Sami R, Elhakem A, Almushhin A, Alharbi M, Benajiba N, Fikry M, Helal M (2021) Enhancement in physicochemical parameters and microbial populations of mushrooms as influenced by nano-coating treatments. *Sci Rep* 11(1):1-9.

Samsudin NIP, Abdullah N (2019) Edible mushrooms from Malaysia; a literature review on their nutritional and medicinal properties. *Int. Food Res. J* 26(1):11-31.

Sedaghat N, Zahedi Y (2012) Application of edible coating and acidic washing for extending the storage life of mushrooms (*Agaricus bisporus*). *Food Sci Technol Int* 18(6):523-530.

Sevindik M (2019) Wild edible mushroom Cantharellus cibarius as a natural antioxidant food. *Turk J Food Agric Sci* 7(9):1377-1381.

Shamtsyan M (2016) Potential to develop functional food products from mushroom bioactive compounds. *J Hyg Eng Des* 15: 51-59.

Sharif S, Mustafa G, Munir H, Weaver CM, Jamil Y, Shahid M (2016) Proximate Composition and Micronutrient Mineral Profile of wild Ganoderma lucidum and Four Commercial Exotic Mushrooms by ICP-OES and LIBS. *J Food Nutr Res* 4(11):703-708.

Silva MCS, Nozuka J, Oliveira PV, Vanetti MC, Bazzolli DM, Costa NM, Kasuya MC (2010) In vivo bioavailablity of selenium in enriched *Pleurotus ostreatus* mushrooms. Metallomics 2:162-166.

Singh P, Langowski HC, Wani AA, Saengerlaub S (2010) Recent advances in extending the shelf life of fresh Agaricus mushrooms: a review. *J Sci Food Agric* 90(9):1393-1402.

Smiderle FR, Alquini G, Tadra-Sfeir MZ, Iacomini M, Wichers HJ, Van Griensven LJ (2013) Agaricus bisporus and Agaricus brasiliensis (1→ 6)-β-d-glucans show immunostimulatory activity on human THP-1 derived macrophages. *Carbohyd Polym* 94(1):91-99.

Sridhar KR (2019) Expedition with micro-and macro-fungi: New perspectives to bridge the gaps. *Kavaka* 52: 1-19.

Srikram A, Supapvanich S (2016) Proximate compositions and bioactive compounds of edible wild and cultivated mushrooms from Northeast Thailand. *Agric Nat Resour* 50(6):432-436.

Srivastava S, Bala KL (2016) Effect of arabic gum-carboxymethylcellulose edible coatings on shelf life of button mushroom (Agaricus bisporus). *Int J Res Eng Technol* 5(6):484-494.

Su CH, Lai MN, Lin CC, Ng LT (2016) Comparative characterization of physicochemical properties and bioactivities of polysaccharides from selected medicinal mushrooms. *Appl Microbiol Biotechnol* 100(10):4385-4393.

Suhag R, Kumar N, Petkoska AT, Upadhyay A (2020) Film formation and deposition methods of edible coating on food products: A review. *Food Res Int* 136:109582.

Tamer CE, Çopur OU (2009) Chitosan: an edible coating for fresh-cut fruits and vegetables. In *VI International Postharvest Symposium* 877:619-624.

Thakur RR, Shahi NC, Mangaraj S, Lohani UC, Chand K (2020) Effect of apple peel based edible coating material on physicochemical properties of button mushrooms (*Agaricus bisporus*) under ambient condition. *Int J Chem Stud* 8(1):2362-2370.

Thakur RR, Shahi NC, Mangaraj S, Lohani UC, Chand K (2021) Development of an organic coating powder and optimization of process parameters for shelf life enhancement of button mushrooms (*Agaricus bisporus*). *J Food Process Preserv* 45(3):e15306.

Tharanathan RN, Kittur FS (2003) Chitin-the undisputed biomolecular of great potential. *Crit Rev Food Sci Nutr* 43:61-87.

Thu ZM, Myo KK, Aung HT, Clericuzio M, Armijos C, Vidari G (2020) Bioactive phytochemical constituents of wild edible mushrooms from Southeast Asia. *Molecules* 25(8): 1972.

USDA (2019) *Food Data Central* [Internet]. Available from: https://fdc.nal.usda.gov [Accessed: 12 August 2019].

Valverde ME, Hernández-Pérez T, Paredes-López O (2015) Edible mushrooms: Improving human health and promoting quality life. *Int J Microbiol* 2015:1-14.

Vamanu E, Dinu LD, Pelinescu DR, Gatea F (2021) Therapeutic Properties of Edible Mushrooms and Herbal Teas in Gut Microbiota Modulation. *Microorganisms* 9: 1262.

Vamanu E, Pelinescu D (2017) Effects of mushroom consumption on the microbiota of different target groups—Impact of polyphenolic composition and mitigation on the microbiome fingerprint. *LWT Food Sci. Technol* 85: 262–268.

Vaziri A, Safekordi AA, Shekarabi A, Shamlou S (2019) Extending the shelf life of edible button mushroom (Agaricus bisporus) by edible coatings on the basis of natural polymers. *Food Sci Tech* 16(91):243-256.

Wakchaure GC, Shirur M, Manikandan K, Rana L (2010) Development and evaluation of oyster mushroom value added products. *Mushroom Res* 19(1):40-44.

Wakchaure GC, Singh M, Vijay B, Wakchaure GC (2011) *Mushrooms: cultivation, marketing and consumption*. Directorate of Mushroom Research (Indian Council of Agricultural Research), New Delhi, India.

Waktola G, Temesgen T (2018) Application of mushroom as food and medicine. *Adv Biotech & Micro* 113: 1-4.

Wani BA, Bodha RH, Wani AH (2010) Nutritional and medicinal importance of mushrooms. *J. Med. Plant Res* 4(24): 2598-2604.

Wrona M, Bentayeb K, Nerín C (2015) A novel active packaging for extending the shelf-life of fresh mushrooms (*Agaricus bisporus*). *Food Control* 54:200-207.

Wu T, Xu B (2015) Antidiabetic and Antioxidant Activities of Eight Medicinal Mushroom Species from China. *Int. J. Med. Mushrooms* 17(2): 129–140.

Yeul VS, Rayalu SS (2013) Unprecedented chitin and chitosan: a chemical overview. *J Polym Environ* 21(2):606-614.

Zahid M, Barua S, Huq SM (2010) Proximate composition and mineral content of selected edible mushroom varieties of Bangladesh. *Br J Nutr* 22-23: 61-68.

Zhang K, Pu YY, Sun DW (2018) Recent advances in quality preservation of postharvest mushrooms (Agaricus bisporus): A review. *Trends Food Sci Technol* 78: 72-82.

Zhang M, Cui SW, Cheung PCK, Wang Q (2007) Antitumor polysaccharides from mushrooms: a review on their isolation process, structural characteristics and antitumor activity. *Trends Food Sci. Technol* 18(1):4-19.

Zhou LW, Ghobad-Nejhad M, Tian XM, Wang YF, Wu F (2020) Current Status of 'Sanghuang' as a Group of Medicinal Mushrooms and Their Perspective in Industry Development, *Food Rev. Int* 1-19.

Zhu D, Guo R, Li W, Song J, Cheng F (2019) Improved Postharvest Preservation Effects of Pholiota nameko Mushroom by Sodium Alginate–Based Edible Composite Coating. *Food Bioproc Tec* 12:587-598.

Živković L, Bajić V, Bruić M, Borozan S, Popić K, Topalović D, Santibanez J, Spremo-Potparević B (2019) Antigenotoxic and antioxidant potential of medicinal mushrooms (Immune Assist) against DNA damage induced by free radicals-an in vitro study. *Mutat Res* 845:403078. doi: 10.1016/j.mrgentox.2019.06.008. Epub 2019 Aug 1. PMID: 31561902.

Section III: Quality Control and Assurance in Food Packaging

Chapter 11

Quality Control and Safety Considerations of Active Food Packaging

Viviane Patrícia Romani[1*], Vilásia Guimarães Martins[2], Meritaine da Rocha[3] and Roberta Letícia Krüger[1]

[1] Department of Food Engineering, State University of Midwest, Guarapuava, PR, Brazil

[2] School of Chemistry and Food, Federal University of Rio Grande, Rio Grande, RS, Brazil

[3] School of Chemistry and Food, Federal University of Rio Grande, Santo Antônio da Patrulha, RS, Brazil

Abstract

The need to increase food safety and reduce food waste has been driving the development of new food packaging technologies. In this sense, the interest in active packaging is growing especially among researchers and industries. In Section II, our colleagues have comprehensively reviewed the latest advances and techniques involved in active food packaging. Based on the contents in Section II, we can see that the active packaging technology is predominantly based on the interaction of the material with the food and/or its direct environment. Among the different mechanisms of active packaging, food contact materials have been designed to release or absorb a wide variety of natural or synthetic substances into or from the packaged food or from the environment surrounding the food. Their performance in the food quality, as well as the safety regarding the effects that they may have on the consumer's health if migration to the food

[*] Corresponding Author's Email: vivi.patricia@hotmail.com.

In: Food Packaging
Editor: Wing-Fu Lai
ISBN: 979-8-88697-249-8

occurs, are of great importance. Thus, in this chapter, which is also the first chapter of this Section, we will focus our discussion on the main mechanisms of active packaging with a special emphasis on the quality and safety concerns of such materials for food packaging. In addition, the global market and future trends will be addressed.

Keywords: active packaging mechanisms, food packaging, food quality, food safety

1. Introduction

There is a growing concern about the nature and composition of the food being consumed. During food processing, different types of additives are added to guarantee product quality and safety. The problem is that most of the additives used are of synthetic origin, that is why the benefits and harms to the consumer's health are still scientifically discussed. The point is that the consumer is looking for products containing the smallest quantity of synthetic additives possible. In this case, active packaging can make an important contribution, because instead of adding the synthetic and/or natural compounds to the food, it is possible for the incorporation of the package material to have a positive interaction with the food during its shelf life. In the literature, it is possible to find a lot of different definitions and points of view of what active packaging is. Ahvenainen (2003) defined it as a packaging system that changes the condition of the packed food to extend the shelf life or to improve food safety or sensory properties while maintaining the quality of the product.

The technology of active packaging can involve physical, chemical, and/or biological actions, promoting the interaction between the packaging material and the food product through releasing or absorbing different compounds. The absorbing systems are generally used to absorb oxygen, carbon dioxide, ethylene, and moisture. Releasing systems are often used to emit carbon dioxide, ethanol, antimicrobials, antioxidants, enzymes, flavors, and nutraceuticals. Both releasing and absorbing mechanisms can be used in the form of films, sachets, or adhesives in packaging materials. Such systems can be applied to preserve the quality of food during its shelf life, increase food safety, extend the shelf life of the product, decrease food loss, reduce the use of food additives, and as a marketing strategy.

Another important point to be highlighted is regarding the environmental problems caused by the incorrect disposal of synthetic packaging. Considering this issue, active packaging can be produced using biodegradable raw materials, such as proteins, polysaccharides, and lipids. Then, biodegradable and/or sustainable materials can bring in addition to the fast degradation in the environment, advantages as bioactivities, such as antimicrobial and antioxidant activity. Various natural compounds can be added to packaging to make them active, including plant extracts and essential oils, mainly used in antimicrobial and antioxidant active packaging. Regarding oxygen scavengers, some examples of the most used are iron or organic compounds such as ascorbic acid and polyunsaturated fatty acids. Moisture absorbers are generally made of silica gel, calcium chloride, and calcium oxide. Besides that, there are substances that can absorb different compounds, such as zeolite, which can absorb CO_2, water, and ethylene.

Food packaging that remains in direct contact and interacts with the foods must fulfill additional requirements compared to traditional materials, such as biochemical properties, microbial stability, absence of toxins, acceptable sensory characteristics, and safety (Debeaufort et al. 1998). In this sense, it is important to note that the compounds added to active packaging, whether in a sachet, adhesive or film must be food grade and non-toxic. In this way, if the compound migrates to food, the safety of the consumer is guaranteed. Otherwise, if non-food grade compounds are used, it is necessary to ensure that they will not migrate to the food product. Thus, this chapter will be focused on the main mechanisms of active packaging with a special emphasis on the quality and safety concerns of such materials for food packaging, including compounds migration aspects. In addition, the global market and future trends will be addressed.

2. Active Packaging Mechanisms

Traditionally, food packaging is used as a mechanical and barrier protection for products, that contributes to their quality and safety. More recently, packaging has also been contributing to extending the shelf life of products and/or acting as a quality indicator. The technology that interacts with the product to increase its shelf life is called active packaging (Dobrucka and Cierpiszewski, 2014). The industrial interest in these technologies has been growing mainly due to two aspects: protecting consumers from diseases caused by spoilage foods and preventing food loss due to deterioration (Firouz et al. 2021).

In active packaging, intentionally added materials (absorbent or emitters) interact with the package's internal environment, promoting an increase in product conservation, either by microbial inhibition, preventing oxidation reactions, or ethylene release. Active packaging can be classified into two types of systems: absorber and releasing systems. The absorber systems aim to remove unwanted compounds from the free space of the packaging or around the food, such as: oxygen, ethylene, carbon dioxide, water, and other specific compounds, which accelerate the degradation of food products (Braga and Silva, 2017). Releasing systems, on the other hand, incorporate substances into the packaging material (e.g., plastic, paper, etc.), sachets, or adhesives which are gradually released into the food, such as carbon dioxide, ethanol, antioxidants, antimicrobials, and other preservatives (Muriel-Galet et al. 2015). Some of the main mechanisms used in active packaging technologies will be discussed below.

2.1. Absorber Systems

Moisture Absorbers

In food products such as fresh vegetables, bakery products, dairy and meat products, a high relative humidity inside the package during transport, storage and marketing can cause fast microbial multiplication. It promotes product degradation and nutritional losses, consequently decreasing the quality and safety of the food product being marketed. Besides that, the attractiveness of a product is affected when consumers perceive liquid in the packaging (in sliced ham, for example). A strategy is to control the fraction of water present inside the packaging (Day, 2008; Droval et al. 2012). Thus, in addition to the packaging material itself, which requires resistance to gas exchange and moisture absorption, it is possible to add, in the form of sachets, absorbent films or trays, desiccant materials that have a high affinity with water molecules. They sequester the humidity of the environment and prevent its collaboration in microbial development or other biochemical reactions, such as darkening and discoloration (Yildirim et al. 2018).

Different substances have the characteristic of absorbing moisture, such as traditional silica gel, calcium oxide, potassium chloride, carbonate potassium, calcium chloride, zeolite, cellulose, and bentonite (Dainelli et al. 2008; Yildirim et al. 2018). Alternatively, studies are aiming to optimize the use and efficiency of organic materials for this purpose, such as sorbitol, fructose, and cellulose, which have a higher cost when compared to inorganic materials because they are organic (Maboonchuay et al. 2016; Gaikwad et al. 2019; Abral et al. 2020; Ghanbarzadeh et al. 2010; Wang et al. 2018).

The main parameters that need to be studied in the design of moisture absorbers include storage temperatures, equilibrium absorption, and the suitable range of relative humidity for the packaged product to minimize the activity of microorganisms. The materials used to produce sachets, for example, must be semi-permeable to water and resistant to breakage (Gaikwad et al. 2018).

Oxygen Absorbers

The presence of oxygen in packaged foods contributes to the rapid loss of the product quality by oxidizing lipids or vitamins, or by promoting the growth of aerobic or facultative microorganisms. Consequently, there is a change in the color, smell, and/or taste of the product (Jacobsson et al. 2004). An alternative to prevent these changes is to reduce the oxygen inside the packaging through the application of a vacuum. It consists of the removal of atmospheric air from inside the packaging, therefore, making the environment free of oxygen. Another important alternative is the removal of oxygen from the packaging headspace by modifying the gaseous atmosphere present (Dey and Neogi, 2019). The gas composition of the headspace can be optimized for each type of product, varying the concentrations of different gases, such as oxygen, carbon dioxide, and nitrogen. It will delay biochemical and microbiological reactions that lead to the degradation of the food product. However, even with the application of a vacuum or modified atmosphere, it is known that the residual oxygen in the headspace remains between 0.5 to 2% (by volume). This concentration may be enough to allow the occurrence of some degradation reactions (Gibis and Rieblinger, 2011). In this sense, oxygen absorbers are more effective in reducing the oxygen concentration since they can decrease it to levels lower than 0.1% (by volume) leading to a longer shelf-life (Mills et al. 2006).

Oxygen absorbers can be found in different formats (sachets, stickers, labels, cards, etc.) and have been commercially available in Japan since the late 1970s (Braga and Silva, 2017). Structurally, they may contain different chemical elements, such as those with a metallic base (metallic iron, platinum, palladium, zinc, copper), of antioxidant nature [butyl hydroxyanisole (BHA), butyl hydroxytoluene (BHT)], enzymes (glucose, oxidase, lactase), or based on unsaturated fatty acids (oleic, linoleic), among others (Firouz et al. 2021).

Some examples of oxygen scavenger applications are ground coffee, tea, roasted nuts, potato chips, chocolate, fat powdered milk, powdered drinks, bread, tortillas, pizza, pizza crust, refrigerated fresh pasta, fruit tortes, cakes, cookies, beer, deli meats, smoked and cured meats, fish, cheese (Restuccia et al. 2010).

Metallic iron is one of the main active components of O_2 absorbers because it has the advantages of being low cost, easy to oxidize in the presence of oxygen, being approved by the Food and Drug Administration (FDA) (Hernandez and Giacin, 1998). The process of oxygen absorption involves the oxidation of iron in the presence of water and oxygen. This process can occur in multiple steps, generating $Fe(OH)_3$. The end of the process is obtained by the reaction of two $Fe(OH)_3$ molecules to form the non-toxic and stable oxide $Fe_2O_3.3H_2O$. This process is dependent on the storage temperature and the moisture arising from the product (water activity of the food) (Robertson, 2006).

Nano-iron has a better scavenging capacity compared to iron particles of superior size. It can be used both in a hydrous or anhydrous environment and reduces oxygen levels to 9.45% v/v. In addition to the substances already mentioned, platinum group metals, such as palladium, can be used as the base for scavenging systems. The role of palladium is as a catalyst that produces water from oxygen in the presence of hydrogen presenting a high scavenging rate. However, this system must be used in a modified atmosphere environment with the presence of hydrogen (Dey and Neogi, 2019).

Ethylene Absorbers

Ethylene is a hormone released from fruits and vegetables that is associated with growth, ripening, and germination processes (Sun et al. 2019). The accumulation of ethylene inside the packaging during transport and storage can accelerate the maturation and deterioration process, reducing the shelf life of these products and causing losses. Thus, the presence of inhibitors of ethylene synthesis or ethylene absorbers can significantly contribute to slowing down the ripening and preserving the global quality of vegetables in general (Pathak et al. 2017). According to Janjarasskul and Suppakul (2018), strategies that control the presence of ethylene in the post-harvest of vegetables are the key to reducing losses and ensuring product quality. The double bond that exists in the molecular structure of ethylene gives it a high level of reactivity, which provides many commercial possibilities to neutralize or eliminate it.

One of the most used mechanisms for the elimination of the ethylene hormone is based on its oxidation by potassium permanganate ($KMnO_4$). In some cases, potassium permanganate is immobilized on alumina or silica gel. Alumina acts as an absorption surface for ethylene molecules and as a vehicle for permanganate (Firouz et al. 2021). Other ethylene absorber options can be clays, zeolites, vermiculite, and activated carbon. These materials can be used in the form of sachets, incorporated into the packaging material, or added to the medium in the form of films (Pathak et al. 2017; Firouz et al. 2021). Catalysts can also be added to increase the efficiency of ethylene absorbers, such as palladium and titanium dioxide (TiO_2), together with activated carbon. Photocatalytic systems based on titanium dioxide nanoparticles can be used to eliminate ethylene gas, reducing it to water and carbon dioxide (Gaikwad et al. 2020).

2.2. Releasing or Emitter Systems

Ethanol Emitters

Ethanol-emitting sachets with controlled release are a method to extend the shelf life of bakery and confectionery products. Such products have a short shelf life due to their high-water activity that promotes the growth of fungi, and even spoilage bacteria. The sachets present in the packaging, upon absorbing the product moisture, diffuse ethanol vapor into the free space, inhibiting microbial growth and consequently increasing the preservation of the product during storage and commercialization (Sarantópoulos and Cofcewicz, 2016). Ethanol emitters can be associated with other mechanisms of active packaging, as in the study of Latou et al. (2010), which used an oxygen absorber. These authors studied the shelf life extension of sliced wheat bread using either an ethanol emitter or an ethanol emitter combined with an oxygen absorber as alternatives to chemical preservatives. Based on the results of bread texture and microbiological data, shelf life was around 4 days for samples without preservatives, 6 days for bread containing commercial preservatives, 24 days for bread containing the ethanol emitter, and at least 30 days for samples containing the ethanol emitter combined with an oxygen absorber.

Sulfur Dioxide Emitters

Sulfur dioxide (SO_2) emitters have an antimicrobial function to control the deterioration caused by fungi in foods like fruits, vegetables, and seeds. As the moisture from the fruit's respiration, for example, is absorbed by the active system, there is a controlled emission of gaseous sulfur dioxide with antimicrobial action. SO_2 has the advantage of inactivating even fungal spores. The SO_2 emission can be performed in two different phases: a fast phase for immediate control of product quality; and a slow phase for disease control during storage (Sarantópoulos and Cofcewicz, 2016). Briefly, SO_2 emitters' mechanism is based on metabisulfite hydrolysis and the reaction of calcium sulfite with moisture (Kuswandi and Jumina, 2020). Besides microbial inactivation, other sulfites can be used to prolong the shelf life of products through the inhibition of enzymatic and nonenzymatic browning reactions during processing and storage. These substances include sulfur dioxide, sodium sulfite, sodium bisulfite, sodium metabisulfite, potassium metabisulfite, and potassium sulfite (Firouz et al. 2021).

Carbon Dioxide Emitters

Carbon dioxide (CO_2) has been extensively applied in the food industry to extend shelf life and preserve the product quality due to its antimicrobial properties. Generally, CO_2 is used for products such as meat, fish, poultry, and fresh fruits (Kuswandi and Jumina, 2020). Sodium bicarbonate ($NaHCO_3$) and citric are two commonly used substances in CO_2 releasing systems. CO_2 is soluble both in the aqueous and fatty phases of food products, that is why it can be applied to a wide variety of products. The antimicrobial effect of CO_2 emitters is very dependent on the quantity of CO_2 dissolved in the food product (Yildirim et al. 2018).

The releasing systems containing CO_2 are activated when in contact with the liquid from the food product, which mechanism works according to Le Chatelier's principle. This mechanism is based on the pH decrease with the liquid entrance into the system, followed by the production of carbon dioxide. Such systems are flexible to have a pH compatible with the pH of the food product, which has the capacity to change the ratio between the active materials. The capacity of CO_2 releasing systems can be optimized according to the variable requirements of the food product, including size, packaging volume, gas compositions, and physicochemical properties (Vilela et al. 2018).

Antioxidant and Antimicrobial Materials

Oxidation is one of the main causes of food spoilage. For example, oil oxidation produces low molar mass substances with off-flavors, as well as toxic compounds. It also causes other negative effects such as vitamin degradation and loss of essential fatty acids, which decrease the nutritional value of foods (Umaraw et al. 2020). Besides oxidation, microbial growth is also an important cause of food deterioration and might cause health problems for consumers (Martins et al. 2018).

Early forms of antioxidant and antimicrobial packaging did not depend on the shape of the packaging devices and can be in the form of sachets, pads, or labels. This form is still widely used in the food industry, where the most common active ingredients are iron and ferrous oxide (Mills, 2005). In recent years, extensive studies have been conducted on active packaging materials with antioxidants and antimicrobials incorporated in the polymer matrix or on its surface (Kuai et al. 2021).

Among the most used antioxidants, butylated hydroxyanisole (BHA) and butylated hydroxytoluene (BHT) are common synthetic antioxidants used for food preservation. However, the US Department of Health and Human Services (HSS) reported their carcinogenic potential and thus they should be replaced by safe antioxidants (Kuswandi and Jumina, 2020). To reduce the use of synthetic additives in the food industry, natural additives have been explored, as well as alternatives to control the release of these additives in food through functional packaging (Ganiari et al. 2017). Examples of natural additives applied in active packaging include extracts of ε-polylysine and astaxanthin derived from shrimp by-products (*Litopenaeus vannamei*) (Xu et al. 2020), antioxidants of rice straw (ferulic acid *P*-coumaric acid and protocatechuic acid) (Menzel et al. 2020), ferulic acid or ferulated arabinoxylan-oligosaccharides from beer processing waste (Moreirinha et al. 2020), eugenol (Garrido-Miranda et al. 2018), yerba mate extract (Reis et al. 2015), essential oils (Sanches-Silva et al. 2014), among others.

Besides antioxidants, antimicrobials have also been widely studied in the development of active materials. Silver, silver zeolite, triclosan, glucose oxidase, chlorine dioxide, natamycin, and allyl isothiocyanate are widely used as active compounds in commercially available antimicrobial active packaging systems (Fang et al. 2017). Gold, copper, titanium dioxide, and zinc oxide are other metal-based materials that present important antimicrobial activity. Compared to the organic antimicrobials, metallic-based nanoparticles have superior antimicrobial performance (Firouz et al. 2021). Silver, for example, has stability against high temperature and low volatility. At the nanoscale, it is known for its antifungal and antimicrobial efficiency against several microorganisms (Sarantópoulos and Cofcewicz, 2016). On the other hand, studies incorporating natural antimicrobial agents in active packaging are of great interest in the scientific community. Essential oils extracted from plants (e.g., coriander, clove, rosemary, oregano, lemongrass, tea tree, basil, and fennel) have a great impact on inhibiting the growth of microorganisms. This creates a strong potential for their use as antimicrobial additives for food packaging applications (Maisanaba et al. 2017).

The main issue in utilizing essential oils for antioxidant and antimicrobial applications is that they must be used in high concentrations to achieve the significant impact required, which could undesirably affect the organoleptic properties of the packaged food (Vilela et al. 2018). To solve this and other problems, scientists have focused on the incorporation of natural active materials, including ascorbic acid, essential oils, polyphenols, etc. into active matrices (e.g., chitosan) (Bi et al. 2019; Sobeh et al. 2018; Wang et al. 2015).

3. Migration from Packaging to Food

Materials based on petroleum, such as polypropylene (PP), polyvinyl chloride (PVC), polystyrene (PS), polyethylene terephthalate (PET), low-density polyethylene (LDPE), high-density polyethylene (HDPE), ethylene vinyl alcohol copolymer (EVOH) and others, are widely used for food packaging. These types of packaging have been used to facilitate the transport and storage, and for the overall protection of products from the outside environment. However, they don't fully protect food against spoilage and don't ensure product quality due to the absence of substances to avoid degradation reactions (Kuai et al. 2021; Nikolic et al. 2021; Silva et al. 2021).

In the last decades, other food packaging types, such as active and intelligent packaging, have been developed. These packaging technologies can release or scavenge active compounds or additives from packaging to food or can respond to the type of reactions occurring with the food product and/or the environment surrounding the packaging (Robertson, 2006; Alamri et al. 2021). In active packaging, different compounds can be added to the packaging material to have a positive interaction with the food (migrating or not) and protect it from microbial spoilage, UV radiation effects, oxygen, and others (Vilela et al. 2018; Nikolic et al. 2021).

Generally, there is a direct contact between packaging materials and foods (Lee, 2014; Kuai et al. 2021) and migratory or non-migratory mechanisms of action can be involved. In the non-migratory active packaging, substances are supposed to act without an intentional migration, while migratory active packaging has a controlled migration of non-volatile agents or an emission of volatile compounds (Dainelli et al. 2008). In the case of non-migratory materials design, the active substances might be incorporated into polymer structures by immobilization techniques. However, their efficacy, stability, and non-migratory nature must have been evaluated and proven by the packaging manufacturers (Werner et al. 2017).

For the development of active packaging material, many factors must be observed related to the food, such as water activity, nutritional components, pH, and others. Also, characteristics of the active compounds, such as bioactivity and stability, migration, and toxicity, must be taken into account. These characteristics are crucial to improving the safety, quality, and shelf life of foods (Werner et al. 2017; Vilela et al., 2018).

According to Arvanitoyannis and Kotsanopoulos (2014) migration is recognized as a phenomenon whereby packaging materials can act as a source of active compounds. However, it is important to highlight that the migration of components between food and packaging can possess a beneficial effect (as the migration of bioactive compounds) or the migration of undesirable compounds might occur. In order to improve the physical and chemical properties of the materials, many additives are added to resins during the development of plastic (polymeric) food packaging. These additives are antioxidants, plasticizers, glidants, stabilizers, lubricants, UV absorbers, and others (Sarantópoulos et al. 2002; Coltro and Machado, 2011), which are not supposed to migrate. According to Alamri et al. (2021), migration can result in detrimental alterations of food characteristics, such as nutrients, flavor, and color, impairing food safety and quality. This is possible due to the transfer of undesirable components from the packaging material to food. In this sense, it is crucial to consider the toxicity of compounds used in the material formulation and their behavior under the different environmental conditions to ensure the safety of consumers, according to the legislative parameters.

The migration of substances can be observed in two ways: from packaging to food and from food to packaging (Mousavi et al. 1998; Arvanitoyannis and Kotsanopoulos, 2014). In the following section, the migration mechanisms will be discussed focused in the migration from packaging to food.

3.1. Migration Mechanisms

The migration of compounds from packaging materials to foods might be mainly influenced by the food characteristics, such as chemical composition and nature (Arvanitoyannis and Stratakos, 2011). Indeed, the migration mechanisms are considered a complex phenomenon. The diffusion is the most important and frequent mechanism due to the macroscopic mass movement of molecules from higher to lower concentration gradients until the equilibrium (Alamri et al. 2021; Miltz et al. 1997). The migration content improves when the packaging material is in direct contact with food. This occurs due to the increase in the mass-transfer rate (Alamri et al. 2021; Anderson and Castle, 2003; Dainelli et al. 2008).

Active packaging made of plastic films with the incorporation of antioxidant and antimicrobial compounds to protect the foods against degradation reactions have been developed (Shin and Lee, 2003; Tovar et al. 2005; Wessling et al. 2000). The active materials produced must follow the established limits of overall (OM) and the specific migrations (SML). The OM is the total mass of compounds transferred from the polymer packaging to food. On the other hand, the SML is the mass of each individual compound transferred to the food in contact with the packaging. The migration tests need a fixed period of time and temperature to quantify the OM and SML (Alamri et al. 2021; Tovar et al. 2005). Though, migration determination in real conditions is difficult, and simulating methods can be used to determine migration using different conditions (Arvanitoyannis and Stratakos 2011). These conditions simulate the food's main characteristics, using, for example, distilled H_2O, sunflower oil or rectified olive oil, aqueous acetic acid, and aqueous ethanol (Alamri et al. 2021; Rossi, 2000).

The migration of many bioactive compounds, like antioxidants and antimicrobials, occurs by diffusion from the polymer structure to the food surface and distribution between interfaces. When the compound is incorporated in a carrier, the compound diffuses from the carrier followed by distribution between the carrier and the polymer matrix (Kuai et al. 2021; Lee, 2014). The migration rate of substances is not constant and varies with time, that is why the compound migration is a complex phenomenon. Some natural compounds require a migration where the concentration range of antioxidants or antimicrobials is constant with time to ensure their effectiveness. Thus, to optimize the mass transfer to ensure effectiveness, some mathematical models of mass transfer of compounds from active packaging are studied (Kuai et al. 2021; Ozer et al. 2016).

According to Karen et al. (2006) and Alamri et al. (2021), there are different types of migration of compounds from packaging to food:

- *According to the number of migrants*: related to the overall migration and specific migration. Overall migration represents all releasing substances from a packaging material (the sum of the mass transfer), while the specific migration indicates the migration of specific chemical species.

- *Related to foods nature*: divided into three categories, non-migrating, volatiles and leaching systems. A non-migrating type has no migration from the material to the food; the volatile type has minor volatile compounds being transferred even without the direct contact between the food and the packaging; and the leaching type consist in the contact between the food and the packaging to promote the transfer of migrants.
- *Based on the coefficient of diffusion*: depending on the packaging contact with the food (low coefficient) or not (constant coefficient). However, the precise diffusion estimation is difficult when the food and package are directly in contact because the diffusion rate in the packaging material is altered.
- *Contact migration*: this type occurs only when there is contact between the packaging material and the food. One important parameter to consider in this type of migration is the nature and extent of contact. For example, solid foods have generally limited contact with the packaging material while liquids have a more extensive contact.
- *Gas-phase migration*: the active compound permeates from the outer layer to the inner layer (the mass transfer is through the gas). It is important to consider that the mass transfer through the gas medium is slower in comparison to direct contact.
- *Penetration migration*: based on the migration of a compound through the inner layer or contacting side of the packaging through the packaging material. After reaching the inner side of the package, it migrates to the food by direct contact or gas-phase migration.
- *Set-off migration*: related to the mass transfer from the outer printed side to the inner side of the packaging films by stacking or during reeling (for inks, varnishes, and coatings, for example). Depending on the substance, it can be visible or invisible. From the inner surface, it can be transferred to the food by gas-phase migration or direct contact.
- *Condensation/distillation migration*: the transfer of the volatile components from the packaging material, as well as from distillation of moisture from the steam released from aqueous foods migrates from package to food and vice versa in boiling or sterilization processes of packaged food products.

Besides the nature of foods and composition, as previously mentioned, other factors affect the migration phenomenon. Such factors include the type of contact (if it is direct or indirect), the duration of contact, the temperature of contact (migration increases at higher temperatures), nature of packaging material (thickness and plasticization), migrant characteristics (spherical configurations and volatile substances exhibit higher rates of migration), and migrant concentration in the packaging (migration increases with higher concentrations) (Alamri et al. 2021; Anderson and Castle, 2003; Triantafyllou et al. 2005). Thus, in the development of new active packaging technologies, such factors are generally studied. For example, Braga et al. (2018) studied the antimicrobial effect of active packaging by the incorporation of silver ions into polyvinyl chloride (PVC). They verified that even PVC films containing the highest silver concentration didn't show antimicrobial action due to the low migration through the PVC matrix. Marini et al. (2007), in the evaluation of the antibacterial activity of plastics coated with silver-doped organic-inorganic hybrid coatings prepared by sol-gel processes, observed that incorporation of inorganic salts in the structure of polymers results in poor migration.

The migration limits in food packaging systems established by the European Food Safety Authority (EFSA) are stipulated by the risks of the migration of substances from food contact materials. For this reason, many tests are realized, such as migration tests, which assess if the compounds pose health risks to consumers. The compound which has authorization for application in food packaging must be compliant with applicable food legislation (Alamri et al. 2021; Cushen et al. 2012; Cushen et al. 2013). Cushen et al. (2013) evaluated the migration of silver from plasticized PVC nanocomposites to chicken meat according to parameters such as time and temperature of storage. They verified a migration ranging from 0.03 to 8.4 mg/kg. These results are below European Union (EU) legislation limits, that allow substances to be no more than 60 mg/kg or 10 mg/dm^2 (Commission Regulation EU, 2011; Cushen et al. 2013). Then, the PVC nanocomposites containing silver are safe for consumers.

4. Safety and Legal Aspects

The main parameters evaluated in traditional packaging are the migration rate of packaging components (e.g., polymers) for the food and the capacity of the material to protect the product from the external environment. The safety assessment of active packaging is more complex than traditional food packaging. In active packaging materials, other characteristics must be evaluated. When active components are incorporated into the traditional packaging or new materials are synthesized, factors such as mechanical and barrier properties, optical characteristics, and sealing ability are important. If the packaging material performance is not enough to protect the food, the consequent loss of safety will occur (Pereira et al. 2020). Besides that, diverse substances can be used to promote the active function of the packaging, and especially when synthetic components are used, their migration profile is important to be known. In fact, global and specific migration is demanded to assess the safety of conventional and active packaging materials (Alamri et al. 2021).

According to Dainelli et al. (2008), three main pillars are the base for the safety of active packaging: the first is labelling to avoid problems such as misuse and misunderstanding by the consumers; the second is the careful consideration of migration and breakdown of products, because of the potential toxicity (and releasing materials must act according to the legislation); and the third is the efficacy of the packaging, related to the ability of the packaging to perform the claimed function.

As the studies in this area are expanding, innovative materials are being developed and other aspects need to be considered. Pereira et al. (2020) mentioned that the legislation, for example, must be re-evaluated with the inclusion of new guidelines considering new substances used and technologies developed. Regarding the labeling and reliability, detailed information needs to be added by the packaging manufacturer about substances used and conformity. Also, environmental impacts and economic performance should be taken into account, considering the production, use, and disposal of the materials to prevent their accumulation in nature. A wide variety of biodegradable alternatives are available in the market and can be used to produce such new technologies.

The benchmark for conventional and active packaging materials selection involves compliance with valid legislation and regulations (Alamri et al. 2021). The regulation is one of the challenges that new technology transfer faces, impacting the growth of active packaging. The active packaging technologies are considered food contact materials for regulation terms. The Federal Food, Drug, and Cosmetic Act (FFDCA) defines a food contact substance (FCS) as any substance that is intended for use as a component of materials used in manufacturing, packing, packaging, transporting, or holding food if the use is not intended to have any technical effect in such food (section 409(h)(6) of the FD&C Act) (Food and Drug Administration – FDA, 2021).

In the US, the Food and Drug Administration (FDA) is the regulatory agency for food ingredients and food packaging from the FFDCA. If the components of active packaging materials migrate to the food, then expected to become components of food, they are subject to the premarket approval requirements of the FFDCA. For these types of components (FCS), the FDA's premarket regulatory programs include the food contact substance notification (FCN), generally recognized as safe (GRAS) notification and threshold of regulation exemption (TOR). Based on such programs, the industry is responsible to establish that the proposed use of the material is safe (Song and Hepp, 2005).

Werner et al. (2017) mentioned that for the material considered safe through the mentioned programs, the criteria are the same for traditional packaging or active packaging. The manufacturer must give information regarding the chemical, environmental, and toxicological data, to ensure that the food contact material is safe under the intended use. Also, it must provide data to determine the minimum quantity of the substance needed to perform the intended technical effect. It is important to mention that for releasing systems, the active substance needs to be declared in the ingredients list in the food packaging labels. In the case of active components immobilized in the packaging material, which are not expected to migrate, they are considered indirect additives. The FCN is also required, except if the substance is considered GRAS.

In the European Union (EU), the Regulation 1935/2004 (European Parliament, Council of the European Union, 2004). was created to enable the use of packaging with components that could migrate to food. This regulation referred to all materials aimed at coming into contact with food. Regulation 450/2009 (European Commission, 2009) was set up in 2009 to expand upon the Framework Regulation 1935/2004. While Regulation 1935/2004 provided general conditions on the safety of active packaging, Regulation 450/2009 established specific rules for active materials and objects. The last EU regulation brings specific aspects such as the use of nanotechnology and labeling, for example. Nanotechnology is prohibited to be used without advanced evaluation, even when not directly in contact with the food. To avoid ingestion of inedible parts of active materials, the recommendation is to label them "inedible" (Dainelli et al. 2008; Pereira et al. 2020).

In MERCOSUR legislation, that includes countries such as Brazil, Argentina, Uruguay and Paraguay, there is no specific regulation for active packaging, because the concept of active packaging was not yet introduced. The use of active packaging is limited to the substances described in the Positive Lists, which have materials that can come into contact with food, according to Resolution RDC No. 17, March 17, 2008 (ANVISA, 2009), and Resolution RDC No. 56, November 2012 (ANVISA, 2012). Currently, it is not necessary to declare the active component on the ingredients list on the label of the product, and neither is the conformity declaration provided by the packaging manufacturer. The specific regulation of active packaging by MERCOSUR is expected in the near future.

5. Global Market and Future Trends

Packaging has been an essential component of the market for industries since many times the consumers are positively influenced by the packaging to buy the product, besides the protection role. Each product needs a specific form of packaging for its protection during transportation, handling, and storage. Around 99.8% of all food and beverage items have some type of packaging. It makes the food industry involved in the development of new technologies for packaging, improving safety and quality, extending the shelf life, and reducing waste and deterioration of food products (Müller and Schmid, 2019).

According to the Food and Agriculture Organization of the United Nations (2021), every year, approximately one-third of the food produced globally is wasted. People throw away more than 1.3 billion tons of food every 365 days. It represents more than USD 900 billion in losses, and it is more than twice in developed countries compared to developing nations. These numbers provide the reasons for the permanent development of active and intelligent packaging in the food industry to keep its freshness as intact as possible. Food loss and waste have indeed become an issue of great public concern. The 2030 Agenda for Sustainable Development reflects the increased global awareness of the problem.

Active packaging can contribute to the reduction of food waste and losses. Besides the main functions of traditional packaging, such as protection and containment of the product, it has other several functions associated with food preservation, such as absorbing or emitting compounds that can enhance the safety, quality, and shelf life of the food products (Vilela et al. 2018). Moisture and oxygen-absorbent packaging for meat products had been among the first active packaging to be successfully developed and applied commercially to maintain food quality, safety, and increase its expiration date (Dainelli et al.

2008). On the internet, it is possible to find a lot of different sites that give provisions of the growing market and future trends for active and intelligent packaging. Below we bring some views from different sources about those markets.

The active and intelligent packaging market was valued at USD 17.5 billion in 2020. It is expected to reach a value of USD 25.16 billion by 2026, registering an increase of almost 50% during 2021-2026. For food, pharmaceutical, and other industries that deal with products with limited expiration dates and specific environmental requirements, it is critical and urgent to improve the efficiency of packaging to keep those products of high quality and safety as long as possible. There is a clear interest in packaging that can maintain food freshness, and also in packaging that shows to the consumer, through a color change in the packaging, if the packaged product is still safe to be consumed (intelligent packaging) (Mordor Intelligence, 2021).

According to Data Bridge Market Research (2021), the active and intelligent packaging market is segmented into six segments which are based on type, function, technology, material, level, and application. The growth among segments helps the industry to develop strategies to approximate the market and determine the application areas of the different packaging. In 2021, based on the six segments cited above, it was expected that smart packaging, moisture control active packaging, RFID (Radio-frequency identification) tags, plastic materials, primary packaging materials, and the food segment would dominate the market of packaging.

According to the Global Industry Analysts (2021), the active and intelligent packaging market in the United States was estimated at US$ 6.4 billion in the year 2021. China, the world's second-largest economy, is forecast to reach a projected market size of US$ 5.2 billion by the year 2026. Those numbers show the high potential of the active and intelligent packaging in the market and make the industry and the research more focused on the development of new technologies to be able to encompass more and more fresh food, cold chains, and other sectors.

Future trends in the packaging market certainly lie in the development of different types of active packaging that can effectively increase the shelf life of food products; and also in intelligent packaging, especially with the introduction of new inks and biosensors that are able to respond to changes in temperatures, and that can solve problems.

The packaging profile changed, but the industries still bump into some problems related to regulatory agencies in many countries, which mostly have not yet adapted to the innovations that have been taking place in the packaging sector. Regulations and “positive lists” systems are often lacking, or in some cases, the laws are so strict that it is very difficult for packaging to comply with regulations. All of this makes growth in the packaging sector slower than it should.

Conclusion

The concept of packaging as a simple food protector or marketing tool has changed in recent years to meet the needs of consumers and the food industry. New packaging systems are being developed by industries and researchers all over the world. Among these systems are the active and intelligent packaging, which allow greater safety and better traceability of food. This change in the required packaging profile is mainly driven by the consumer’s preference for products with a longer shelf life, ready-to-eat meals, clean label products, and also by the demand for greater safety in relation to the food being purchased and consumed. Sustainability issues associated with the reduction, reuse, and recycling of those packages have also been of great importance in new developments.

Assessing the safety of active and intelligent packaging has been more complex than traditional food packaging. When evaluating traditional packaging, the migration rate of the packaging components into the food and the packaging’s ability to protect the food from external agents are mainly considered. In active and intelligent packaging, more factors must be controlled, as the inclusion of substances in the packaging material can change its permeability, mechanical properties, and sealing capacity, which can lead to changes in the food and consequent loss of safety. The legislation for active packaging is in general unspecific and must be re-evaluated to include new guidelines considering new substances used and technologies developed. Thus, more precise legislation must be available for the new active and intelligent packaging on the market as soon as possible to contribute to the advance of these technologies.

Due to the COVID-19 pandemic, in the years 2020-2021, the active and intelligent packaging market was severely affected, as well as many other industries. Blockage conditions in many parts of the world have affected global trade, and many companies have had to close their manufacturing facilities, disrupting demand and the supply chain. Moreover, research on active packaging was also affected. It is estimated that after the pandemic, when economic conditions return to normal, the growing demand from industries and consumers will once again drive the growth of the active and intelligent packaging market.

References

Abral, H., Ariksa, J., Mahardika, M., Handayani, D., Aminah, I., Sandrawati, N., Pratama, A. b., Fajri, N., Sapuan, S. M. and Ilyas, R. A. (2020). Transparent and antimicrobial cellulose film from ginger nanofiber. *Food Hydrocolloids*, 105266.

Ahvenainen, R. (2003). Active and intelligent packaging: An introduction. In: Ahvenainen, R. *Novel Food Packaging Techniques.* Boca Raton, FL: CRC Press, 5–21.

Alamri, M. S., Qasem, A. A. A., Mohamed, A. A., Hussain, S., Ibraheem, M. A., Shamlan, G., Alqah, H. A. and Qasha, A. S. (2021). Food Packaging's Materials: A Food Safety Perspective. *Saudi Journal of Biological Sciences*, 4490–4499.

Anderson, W. A. C. and Castle, L. (2003). Benzophenone in carton-board packaging materials and the factors that influence its migration into food. *Food Additives and Contaminants*, 607–618.

Arvanitoyannis, I. S. and Kotsanopoulos, K. V. (2014). Migration phenomenon in food packaging. Food–package interactions, mechanisms, types of migrants, testing and relative legislation—a review. *Food and Bioprocess Technology*, 21–36.

Arvanitoyannis, I. S. and Stratakos, A. C. (2011). *Migration from food packaging materials in technologies of preservation of food and food packaging.* Thessaloniki, Hellas: University Studio Press.

Bi, F., Zhang, X., Bai, R., Liu, Y., Liu, J. and Liu, J. (2019). Preparation and characterization of antioxidant and antimicrobial packaging films based on chitosan and proanthocyanidins. *International Journal of Biological Macromolecules*, 11–19.

Braga, L. R., Rangel, E. T., Suarez, P. A. Z. and Machado, F. (2018). Simple synthesis of active films based on PVC incorporated with silver nanoparticles: evaluation of the thermal, structural and antimicrobial properties. *Food Packaging and Shelf Life*, 122–129.

Braga, L. R. and Silva, F. M. (2017). Active packaging: a new approach for food packaging. B*razilian Journal of Food Research*, 170–186.

Coltro, L. and Machado, M. P. (2011). Antioxidant-specific migration from plastic packaging to food. *Polymers,* 390–397.

Comission Regulation (EU). (2011). Regulation No. 10/2011 of 14 January 2011 on Plastic Materials and articles intended to come into contact with food. *Official Journal of The European Union.*

Cushen, M., Kerry, J., Morris, M., Cruz-Romero, M. and Cummins, E. (2012). Nanotechnologies in the food industry - Recent developments, risks and regulation. *Trends in Food Science & Technology,* 30–46.

Cushen, M., Kerry, J., Morris, M., Cruz-Romero, M. and Cummins, E. (2013). Migration and exposure assessment of silver from a PVC nanocomposite. *Food chemistry*, 389–397.

Dainelli, D., Gontard, N., Spyropoulos, D., Zondervan-van den Beuken, E. and Tobback, P. (2008). Active and intelligent food packaging: legal aspects and safety concerns. *Trends in Food Science & Technology,* S103–S112.

Data Bridge Market Research. (2021). *Global Interactive Packaging Market – Industry Trends and Forecast to 2028.* Available at: https://www.databridgemarketresearch.com/reports/global-interactive-packaging-market.

Day, B. (2008). Active packaging of food. In: Kerry J. and Butler P. *Smart packaging technologies for fast moving consumer goods*. Chichester, England: John Wiley & Sons Ltd., 1–18.

Debeaufort, R, Quezada-Gallo, J. A. and Voilley, A. (1998). Edible films and coatings: tomorrow's packaging: a review. *Critical Reviews in Food Science and Nutrition*, 299–313.

Dey, A. and Neogi, S. (2019). Oxygen scavengers for food packaging applications: A review. *Trends in Food Science & technology*, 26–34.

Dobrucka, R. and Cierpiszewski, R. (2014). Active and intelligent packaging food–research and development–a review. *Polish Journal of Food and Nutrition Sciences*, 7–15.

Droval, A. A., Benassi, V. T., Rossa, A., Prudencio, S. H., Paiao, F. G. and Shimokomak, I. M. (2012). Consumer attitudes and preferences regarding pale, soft, and exudative broiler breast meat. *Journal of Applied Poultry Research*, 502–507.

European Parliament, Council of the European Union. (2004). Regulation (EC) No 1935/2004 of the European Parliament and of the Council of 27 October 2004 on materials and articles intended to come into contact with food and repealing Directives 80/590/EEC and 89/109/EEC. *Official Journal of the European Union.*

European Commission. (2009). Commission Regulation (EC) No 450/2009 of 29 May 2009 on active and intelligent materials and articles intended to come into contact with food. *Official Journal of the European Union.*

Fang, Z., Zhao, Y., Warner, R. D. and Johnson, S. K. (2017). Active and intelligent packaging in meat industry. *Trends in Food Science & Technology*, 60–71.

Firouz, M. S., Mohi-Alden, K. and Omid, M. (2021). A critical review on intelligent and active packaging in the food industry: Research and development. *Food Research International,* 1–24.

Food and Agriculture Organization of the United Nations (FAO). (2021). *Food loss and Food Waste.* Available at: https://www.fao.org/food-loss-and-food-waste/flw-data.

Food and Drug Administration (FDA). (2021). Guidance for Industry: Preparation of Food Contact Substance Notifications (Administrative). Available at: https://www.fda.gov/regulatory-information/search-fda-guidance-documents/guidance-industry-preparation-food-contact-substance-notifications-administrative.

Gaikwad, K. K., Singh, S. and Ajji, A. (2019). Moisture absorbers for food packaging applications. *Environmental Chemistry Letters*, 609–628.

Gaikwad, K. K., Singh, S. and Lee, Y. S. (2018). High adsorption of ethylene by alkali-treated halloysite nanotubes for food-packaging applications. *Environmental Chemistry Letters*, 1055–1062.

Gaikwad, K. K., Singh, S. and Negi, Y. S. (2020). Ethylene scavengers for active packaging of fresh food produce. *Environmental Chemistry Letters*, 269–284.

Ganiari, S., Choulitoudi, E. and Oreopoulou, V. (2017). Edible and active films and coatings as carriers of natural antioxidants for lipid food. *Trends in Food Science & Technology*, 70–82.

Garrido-Miranda, K. A., Rivas, B. L., Pérez-Rivera, M. A., Sanfuentes, E. A., and Peña-Farfal, C. (2018). Antioxidant and antifungal effects of eugenol incorporated in bionanocomposites of poly(3-hydroxybutyrate)-thermoplastic starch. *Food Science and Technology – LWT*, 260–267.

Ghanbarzadeh, B., Almasi, H. and Entezami, A. A. (2010). Physical properties of edible modified starch/carboxymethyl cellulose films. *Innovative Food Science & Emerging Technologies*, 697–702.

Gibis, D. and Rieblinger, K. (2011). Oxygen scavenging films for food application. *Procedia Food Science*, 229–234.

Global Industry Analysts. (2021). *Global Active and Intelligent Packaging Market to Reach $33.1 Billion by 2026.* Available at: https://www.prnewswire.com/news-releases/global-active-and-intelligent-packaging-market-to-reach-33-1-billion-by-2026--301355743.html.

Hernandez, R. J. and Giacin, J. R. (1998). Factors affecting permeation, sorption, and migration process in package-product system. In: Taun, I. A. and Singh, R. P. *Food storage stability*. Boca Raton, FL: CRC Press, 269–324.

Jacobsson, A., Nielsen, T. and Sjöholm, I. (2004). Effects of type of packaging material on shelf-life of fresh broccoli by means of changes in weight, colour and texture. *European Food Research and Technology*, 157–163.

Janjarasskul, T. and Suppakul, P. (2018). Active and intelligent packaging: The indication of quality and safety. *Critical Reviews in Food Science and Nutrition*, 808–831.

Karen, A., Barnes, C. and Sinclair, R. (2006). *Chemical migration and food-contact materials*. Boca Raton, FL: CRC Press.

Kuai, L., Liu, F., Chiou, B. S., Avena-Bustillos, R. J., McHugh, T. H., and Zhong, F. (2021). Controlled release of antioxidants from active food packaging: A Review. *Food Hydrocolloids*, 106992.

Kuai, L., Liu, F., Chiou, B-S., Avena-Bustillos, R. J., McHugh, T. H. and Zhong, F. (2021). Controlled release of antioxidants from active food packaging: A review. *Food Hydrocolloids,* 1–14.

Kuswandi, B. and Jumina. (2020). Active and intelligent packaging, safety, and quality controls. In: Siddiqui, M. W. *Fresh-cut fruits and vegetables*. Cambridge, MA: Academic Press, 243–294.

Latou, E., Mexis, S. F., Badeka, A. V. and Kontominas, M. G. (2010). Shelf-life extension of sliced wheat bread using either an ethanol emitter or an ethanol emitter combined with an oxygen absorber as alternatives to chemical preservatives. *Journal of Cereal Science*, 457–465.

Lee, D. S. (2014). Antioxidant packaging system. In: Han J. H. *Innovations in food packaging*. San Diego, CA: Academic, 111–131.

Maboonchuay, T., Chalapat, K. and Siraleartmukul, K. (2016). Humidity sensor using carboxymethyl cellulose hydrogel membrane. In: *2016 13th international conference* *on electrical engineering/electronics, computer, telecommunications and information technology (ECTI-CON)*, 1–3.

Maisanaba, S., Llana-Ruiz-Cabello, M., Gutiérrez-Praena, D., Pichardo, S., Puerto, M., Prieto, A. I., Jos, A. and Camean, A. M. (2017). New advances in active packaging incorporated with essential oils or their main components for food preservation. *Food Reviews International*, 447–515.

Marini, M., Niederhausern, S., Iseppi, R., Bondi, M., Sabia, C., Toselli, M. and Pilati, F. (2007). Antibacterial activity of plastics coated with silver-doped organic-inorganic hybrid coatings prepared by sol-gel processes. *Biomacromolecules*, 1246–1254.

Martins, V. G., Romani, V. P. and Martins, P. C. (2018). Active Compounds Incorporated in Agro-based Films: The Efficacy and Effect in Film Properties for Food Packaging. In: Galaz, A. D. and Bailey, D. S. *Food Packaging and Preservation: Techniques, Applications and Technology*. Hauppauge, NY: Nova Science Publishers.

Menzel, C., González-Martínez, C., Vilaplana, F., Diretto, G. and Chiralt, A. (2020). Incorporation of natural antioxidants from rice straw into renewable starch films. *International Journal of Biological Macromolecules*, 976–986.

Mills, A. (2005). Oxygen indicators and intelligent inks for packaging food. *Chemical Society Reviews,* 1003-1011.

Mills, A., Doyle, G., Peiro, A. M. and Durrant, J. (2006). Demonstration of a novel, flexible, photocatalytic oxygen-scavenging polymer film. *Journal of Photochemistry and Photobiology A: Chemistry*, 328–331.

Miltz, J., Ram, A. and Nir, M. M. (1997). Prospects for application of post-consumer used plastics in food packaging. *Food Additives and Contaminants*, 649–659.

Mordor Intelligence. (2021). *Active and intelligent packaging market - Growth, trends, COVID-19 impact, and forecasts (2021 - 2026).* Available at: https://www.mordorintelligence.com/industry-reports/active-and-intelligent-packaging-market-industry.

Moreirinha, C., Vilela, C., Silva, N. H. C. S., Pinto, R. J. B., Almeida, A., Rocha, M. A. M., Coelho, E., Coimbra, M. A., Silvestre, A. J. D. and Freire, C. S. R. (2020). Antioxidant and antimicrobial films based on brewers spent grain arabinoxylans, nanocellulose and feruloylated compounds for active packaging. *Food Hydrocolloids*, 1–9.

Mousavi, S. M., Desobry, S. and Hardy, J. (1998). Mathematical modelling of migration of volatile compounds into packaged food via package free space, Part II: spherical shaped food. *Journal of Food Engineering*, 473–484.

Müller, P. and Schmid, M. (2019). Intelligent Packaging in the Food Sector: A Brief Overview. *Foods*, 1–12.

Muriel-Galet, V., Cran, M. J., Bigger, S. W., Hernández-Muñoz, P. and Gavara, R. (2015). Antioxidant and antimicrobial properties of ethylene vinyl alcohol copolymer films based on the release of oregano essential oil and green tea extract components. *Journal of Food Engineering*, 9–16.

National Agency of Health Surveillance (ANVISA). (2008). Resolution Nº 17, March 17, 2008. Provides about the Technical Regulation on the Positive List of Additives for Plastic Materials for the Preparation of Packaging and Equipment in Contact with Food. *Official Diary of the Union.*

National Agency of Health Surveillance (ANVISA). (2008). Resolution Nº 17, March 17, 2008. Provides about the Technical Regulation on the Positive List of Additives for Plastic Materials for the Preparation of Packaging and Equipment in Contact with Food. *Official Diary of the Union.*

National Agency of Health Surveillance (ANVISA). (2012). Resolution N° 56, November 16, 2012. Provides about the positive list of monomers, other initiating substances and polymers authorized for the preparation of plastic packaging and equipment in contact with food. *Official Diary of the Union.*

Nikolic, M. V., Vasiljevic, Z. Z., Auger, S., and Vidic, J. (2021). Metal oxide nanoparticles for safe active and intelligent food packaging. *Trends in Food Science & Technology,* 655–668.

Ozer, B. B. P., Uz, M., Oymaci, P. and Altinkaya, S. A. (2016). Development of a novel strategy for controlled release of lysozyme from whey protein isolate based active food packaging films. *Food Hydrocolloids*, 877–886.

Pathak, N., Caleb, O. J., Geyer, M., Herppich, W. B, Rauh, C. and Mahajan, P. V. (2017). Photocatalytic and photochemical oxidation of ethylene: Potential for storage of fresh produce—A review. *Food and Bioprocess Technology*, 982–1001.

Pereira, D. C. S., Teixeira, N. S., Zin, I. M., Gonçalves, R. A. dos S. and de Melo, N. R. (2020). Active and intelligent packaging: Security, legal aspects and global market. *Brazilian Journal of Development*, 61766–61794.

Reis, L. C. B., De Souza, C. O., Da Silva, J. B. A., Martins, A. C., Nunes, I. L. and Druzian, J. I. (2015). Active biocomposites of cassava starch: The effect of yerba mate extract and mango pulp as antioxidant additives on the properties and the stability of a packaged product. *Food and Bioproducts Processing*, 382–391.

Restuccia, D., Spizzirri, U. G., Parisi, O. I., Cirillo, G., Curcio, M., Iemma, F., Puoci, F., Vinci, G. and Picci, N. (2010). New EU regulation aspects and global market of active and intelligent packaging for food industry applications. *Food Control*, 1425–1435.

Robertson, G. L. (2006). *Food packaging principles and practice.* Boca Raton, FL: CRC Press.

Rossi, L. (2000). European Community legislation on materials and articles intended to come into contact with food. In: Piringer, O. G. and Baner, A. L. *Plastic Packaging Materials for Food: Barrier function, mass transport, quality assurance and legislation.* Weinheim: Wiley-VCH, 393–406.

Sanches-Silva, A., Costa, D., Albuquerque, T. G., Buonocore, G. G., Ramos, F., Castilho, M. C., Machado, A. V. and Costa, H. S. (2014). Trends in the use of natural antioxidants in active food packaging: a review. *Food Additives & Contaminants: Part A*, 374–395.

Sarantópoulos, C. I. G. L., Oliveira, L. D., Padula, M., Coltro, L., Alves, R. M. and Garcia, E. E. (2002). *Embalagens plásticas flexíveis: principais polímeros e avaliação de propriedades*. Campinas, SP: CETEA/ITAL.

Sarantópoulos, C. and Cofcewicz, L. S. (2016). Active packaging for perishable products. *Packaging Technology and Development Bulletin* – ITAL, 1–12.

Shin, H. S. and Lee, Y. (2003). Antioxidant-impregnated food packaging materials for inhibition of lipid oxidation. *Food Science Biotechnology*, 737–746.

Silva, A. L. P., Prata, J. C., Walker, T. R., Duarte, A. C., Ouyang, W., Barcelò, D. and Rocha-Santos, T. (2021). Increased plastic pollution due to COVID-19 pandemic: Challenges and recommendations. *Chemical Engineering Journal*, 126683.

Sobeh, M., Mahmoud, M. F., Abdelfattah, M. A. O., Cheng, H., El-Shazly, A. M. and Wink, M. (2018). A proanthocyanidin-rich extract from Cassia abbreviata exhibits antioxidant and hepatoprotective activities *in vivo*. *Journal of Ethnopharmacology*, 38–47.

Song, Y. S. and Hepp, M. A. (2005). US food and drug administration approach to regulating intelligent and active packaging components. In: Han J. H. *Innovations in food packaging*. San Diego, CA: Academic, 475–481.

Sun, M., Yang, X., Zhang, Y., Wang, S., Wong, M. W., Ni, R. and Huang, D. (2019). Rapid and visual detection and quantitation of ethylene released from ripening fruits: The new use of grubbs catalyst. *Journal of Agricultural and Food Chemistry*, 507–513.

Tovar, L., Salafranca, J., Sánchez, C. and Nerín, C. (2005). Migration studies to assess the safety in use of a new antioxidant active packaging. *Journal of Agricultural and Food Chemistry*, 5270–5275.

Triantafyllou, V. I., Akrida-Demertzi, K. and Demertzi, P. G. (2005). Determination of partition behaviour of organic surrogates between paperboard packaging materials and air. *Journal of Chromatography A*, 74–79.

Umaraw, P., Munekata, P. E. S., Verma, A. K., Barba, F. J., Singh, V. P., Kumar, P. and Lorenzo, J. M. (2020). Edible films/coating with tailored properties for active packaging of meat, fish and derived products. *Trends in Food Science & Technology*, 10–24.

Vilela, C., Kurek, M., Hayouka, Z., Röcker, B., Yildirim, S., Antunes, M. D. C., Nygaard, J. N., Pettersen, M. K. and Freire, C. S. R. (2018). A concise guide to active agents for active food packaging. *Trends in Food Science*, 212 – 222.

Wang, H., Liu, H., Chu, C., She, Y., Jiang, S., Zhai, L., Jiang, S. and Li, X. (2015). Diffusion and antibacterial properties of nisin-loaded chitosan/poly (L-lactic acid) towards development of active food packaging film. *Food and Bioprocess Technology*, 1657–1667.

Wang, J., Gardner, D. J., Stark, N. M., Bousfield, D. W., Tajvidi, M. and Cai, Z. (2018). Moisture and oxygen barrier properties of cellulose nanomaterial-based films. *ACS Sustainable Chemistry & Engineering*, 49–70.

Werner, B. G., Koontz, J. L. and Goddard, J. M. (2017). Hurdles to commercial translation of next generation active food packaging technologies. *Current Opinion in Food Science*, 40–48.

Wessling, C., Nielsen, T. and Leufven, A. (2000). The influence of alphatocopherol concentration on the stability of linoleic acid and the properties of low-density polyethylene. *Packaging Technology Science*, 19–28.

Xu, J., Wei, R., Jia, Z. and Song, R. (2020). Characteristics and bioactive functions of chitosan/gelatin-based film incorporated with ε-polylysine and astaxanthin extracts derived from by-products of shrimp (*Litopenaeus vannamei*). *Food Hydrocolloids*, 1-13.

Yildirim, S., Röcker, B., Pettersen, M. K., Nilsen-Nygaard, J., Ayhan, Z.; Rutkaite, R., Radusin, T., Suminska, P., Marcos, B. and Coma, V. (2018). Active packaging applications for food. *Comprehensive Reviews in Food Science and Food Safety*, 165–199.

Chapter 12

Concepts and Implementation of HACCP in Food Packaging

**Nathália Cristina Cirone Silva*
and Bruna Lourenço Crippa**
Department of Food Science and Nutrition, Faculty of Food Engineering (FEA), University of Campinas (UNICAMP), Campinas, SP, Brazil

Abstract

Food packaging represents an important link in the food production chain, as they may or may not be in contact with food. Thus, ensuring that the food after the packaging process is safe for the consumer is essential. Unfortunately, the migration and interaction between food packaging and food could be harmful to consumers. In the previous chapter, our colleagues have highlighted various issues to be considered for quality control in food packaging. In this chapter, we will introduce the use of the HACCP system as a tool for controlling and ensuring the safety of food produced as well as food packaging. By using this system to raise all the critical control points within the production of packaging, the supplier will be able to control the dangers involved in its production, ensuring greater safety for its customers and, consequently, consumers.

1. Introduction

Food packaging is essential for the food industry once allows major production, so the product could be stored and transported around the world,

* Corresponding Author's Email: ncirone@unicamp.br.

In: Food Packaging
Editor: Wing-Fu Lai
ISBN: 979-8-88697-249-8

preventing food waste. Also, increase the preservation and shelf life of the products, gives consumer information about the ingredients, and could be used as marketing. The food industry considers choosing the packaging material the processability, printability, strength, barrier properties, cost-effectiveness, sustainability, and legal requirements (Deshwal et al., 2019).

Even though the packaging pros mentioned, hazards from food packaging are possibly thought the contact with the material of package and the food. The interaction between food and its packaging could be classified into migration, permeation, and sorption. The migration could occur by contaminants as well as additives, being negative or positive for food quality; the permeation can be beneficial for modified atmosphere; and the sorption is associated with aroma and flavor (Alamri et al., 2021). Safe packaging cannot bring hazards to the product, which could be biological, chemical, or physical agents. Food contamination is directly related to human health, once could cause foodborne diseases. In this way, practical measures to prevent and reduce food contaminants during production (Li et al., 2021). In this way, the application of Hazard Analysis and Critical Control Points (HACCP) during the production of food packaging is essential to avoid contamination in the products.

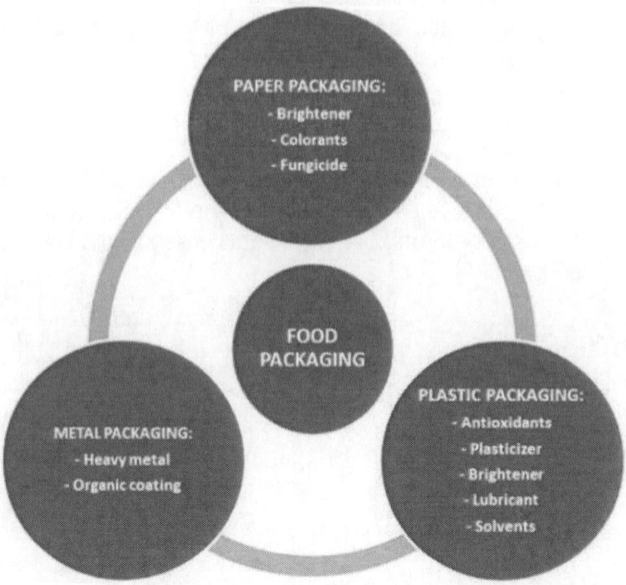

Figure 1. The material components of food packaging and their main contaminants. Source: (Rather et al., 2017, Alamri et al., 2021, Li et al., 2021).

2. Food Packaging As an Important Component in Food Production

Food packaging has an important role in the preservation and quality of food which is a result of the interaction between food, package, and environment. Food packaging have as the common materials glass, plastics, paper and other flexible packages such as coatings and adhesives (Raheem, 2012). Despite food packaging preserving food, the chemicals contamination from it could migrate to the food is a concern. Packaging processes include the use of additives as stabilizers, antioxidants, plasticizers, and slipping agents necessary to the properties of the material. During food storage, the food quality depends on the food type, packaging, and storage conditions. The transference of compounds through the packaging material to food could occur (Guerreiro et al., 2018, Xue et al., 2019). The transference of these substances to the food is a phenomenon called migration (Rather et al., 2017). Also, physical and microbiological contaminations are mainly related to processing, handling, and storage of packing (Bovee et al., 1997). The migration is a phenomenon that occur by substances of low molecular weight which can diffuse to the food, and migration is influenced by the temperature, contact time, nature of pollutant, contact surface size, and others (Xue et al., 2019). Conventional materials used for the package include paper, plastic, and metal. In Figure 1 it is possible to observe the main contaminants in each most common type of material.

2.1. Plastics

Plastic packaging materials are becoming the primary material used for packing food once has an adequate functional property necessary and low costs (Yates et al., 2021). Plastics are derived from synthetic or natural polymers made from different polymers and additives. Additives give the variation of the plastics, allowing to mold the material as well as stability to oxidation and light exposure, and impact resistance (McMillin, 2017). In general, plastics used for food packaging have large molecules that do not migrate into food when it is used properly. But if industry or consumers use these incorrectly, for example, heat-resistant or "sticky" plastics (like cling film) can leach into food during cooking or storage. Other examples are the BPA (bisphenol A) that can be released into food if washed with harsh detergents, or DEHA (diethylhexyl adipate) can migrate into fatty foods such as meat or cheese (NZFSA, 2020). Many studies are showing the toxicity of compounds in plastic packaging and the migration of it to food v. Among the contaminants observed in plastic packaging, we can find plasticizers, thermal and light stabilizers, slip additives, antioxidants, solvents, monomers, acrylonitrile (Alamri et al., 2021).

2.2. Paper

Paper is used in the food industry as a primary package when it is in contact with food; and as a secondary package, when it is used for storage and transportation, but plain paper does not have the functional properties expected. It is used more for dry foods, but coating and o other processes extend its use to wet and fatty food types (Geueke et al., 2018). The paper packages are, the majority, a mixture of paper, some organic substances, and some additives such as aluminum or plastic, and the use of these materials could allow that substances to migrate into food through direct or indirect contact (Deshwal et al., 2019). For paper packaging, some studies observed the migration between the package and the food (Xue et al., 2019). The main contaminants in this type of package are dioxin, benzophenone, nitrosamines, chlorophenols, and chloroanisole (Alamri et al., 2021).

2.3. Metal

The metal is extensively used in packaging, they are a great barrier to light, gas, and moisture can withstand high heating temperatures, etc (Deshwal and Panjagari, 2020). Metal food packaging is extensively used to produce cans, tubes, other containers, closures, films, and barrier layers (Geueke et al., 2018). Two major processes are a concern to healthy: the migration and the interaction from between metal food packing and food. As we see, migration refers to the transformation of components between food and packaging, and Interaction is the physical, chemical, or microbiological reaction of the food package (Deshwal and Panjagari, 2020). The interaction of metal food packing and food are responsible for the contamination, caused mainly by the corrosion, pitting, perforation, and the later migration of metallic ions to food (Amelia et al., 2012). The main contaminants observed are bisphenol A, lead, tin, chromium, and aluminum (Deshwal and Panjagari, 2020, Alamri et al., 2021).

2.4. Other Compounds of Packaging

To produce food packaging, printing inks and adhesives are applied to plastic and paper packaging, and the industry adds substances to these components (Galbiati et al., 2021). Printing inks are composed of one or more of these compounds: colorants, binders, solvents, and several additives (e.g., plasticizer, thickener, slip agent) (Alsing Pedersen et al., 2012). In the adhesives, we can observe coupling agents, epoxy resin, antioxidants, photoinitiator, solvent, and others (Wrona et al., 2017).

3. Hazard Analysis and Critical Control Points (HACCP)

Initially, the Hazard Analysis and Critical Control Points (HACCP) system was developed to ensure the microbiological safety of food production that would be used in space in the early days of the US manned space program. (Mortimore & Wallace, 1998; Wang et al., 2010). Currently, HACCP is an internationally recognized system and is used to ensure food safety. This system covers both the assessment of microbiological risks, as well as the physical and chemical risks associated with food and packaging (Slatter, 2003). HACCP aims to identify the source of potential hazards at all food production points. In addition, through HACCP, it is possible to identify possible risks, so that preventive actions can be inserted, controlling or reducing these risks to safe levels for the consumer's health (Slatter, 2003).

3.1. Principles of the HACCP System

In 1971, HACCP was described for the first time and in more detail at the Conference for Food Protection. The seven HACCP principles were issued in 1989 by the National Advisory Committee on Microbiological Standards for Food (NACMCF). In 1992, the NACMCF system was revised and a new procedure for risk analysis was introduced. Codex Alimentarius then took on a HACCP system similar to NACMCF, but using different and more simplified terms (Fortin et al., 2021). Therefore, to prevent the occurrence of problems, seven basic principles (Table 1) are used in the development of HACCP plans.

Table 1. Seven principles of the HACCP system. Source: (FDA, 2017)

Principle 1	Conduct a hazard analysis.
Principle 2	Determine the critical control points (CCPs).
Principle 3	Establish critical limits.
Principle 4	Establish monitoring procedures.
Principle 5	Establish corrective actions.
Principle 6	Establish verification procedures.
Principle 7	Establish record-keeping and documentation procedures.

CCPs are critical in preventing, eliminating, or reducing risks to acceptable levels. CCP is a step within a process, where food safety risks must be contained. In addition, critical limits (CL) must be established for each CCP, this CL will define which foods/packaging are safe or potentially unsafe, in addition to classifying them into acceptable and unacceptable (Slatter, 2003). For the use of the 7 principles to be satisfactory, Codex Alimentarius recommends following the logical sequence of 12 steps (Table 2), thus enabling the development of a well-structured and objective HACCP plan.

Table 2. Logic sequence for the application of the Codex HACCP principles (Adapted from Codex, 2009)

Step 1	Team Formation;
Step 2	Product Description;
Step 3	Identification of the Intent to Use;
Step 4	Construction of the Process Flowchart;
Step 5	On-site confirmation of the Flowchart;
Step 6	List of all potential hazards Conducting the Hazard Analysis, considering Control Measures;
Step 7	Determination of CCPs;
Step 8	Establishment of Critical Limits for each CCP;
Step 9	Establishment of Monitoring System for each CCP;
Step 10	Establishment of Corrective Actions;
Step 11	Establishment of Verification Procedures;
Step 12	Establishment of Documentation and Record Keeping;

3.2. HACCP in Food-Packaging

In the food packaging industry, the crucial point for implementing the HACCP plan is the determination of CCPs. The packaging supplier must be able to identify as many CCPs as possible within their industry, as packaging is a crucial point within the food production chain. Common CCPs in the food packaging industry are allergens, for example, natural dyes or other chemical products used in printing or making up the packaging. Therefore, this information must be included in the ingredient declaration. Metal is also an important CCP, if there is a risk of contamination at any point, suppliers can use a metal detector at this stage. Unlike the food industry where glass must be avoided in production processes, in the packaging industry it is allowed, but its use must be controlled and inspected if it breaks. Another important critical point in food packaging is pest control since various types of packaging can attract different pests during storage. Although pest control is part of the GMP basic rules, in the packaging industry this point can be a CCP. This control takes place both in external and internal areas, as well as on the trucks that will carry out the delivery logistics (Mortimore & Wallace, 2013).

The CCPs will vary according to each packaging industry and each material, the crucial thing is that they are identified by the suppliers, that they are part of the HACCP plan, and that there are control and monitoring measures. An important point raised by Raaska (2005) is the confusion between aspects related to product quality and safety, which can generate excess CPPs, create an uncontrollable system, and difficult to execute control measures for each CPP. Therefore, risk assessment at the manufacturing level of food or food packaging must be based on scientific knowledge. In addition to being carried out by a team that has the knowledge and experience necessary to carry out this assessment (Raaska, 2005).

Some works in the literature show what were the advances related to the use of the HACCP system in the food packaging industry, for example, Sjöberg et al., 2002 evaluated the application of the HACCP system in five paper and board factories and four converting paper or card intended for contact with food. As a result, they were able to raise the main hazards associated with the production of these packagings, which in these cases were mainly microbiological hazards. In this way, after implementation, it was possible to direct research and hygiene measures to avoid this contamination and microbiological propagation within the production process.

The applicability of the HACCP system was also studied in returnable PET bottles and the authors raised important points to be considered in the implementation of HACCP, for example, the use of bottles for the storage of chemical products by consumers, thus, the HACCP plan should include the detection of volatile contaminants, in addition to a cleaning procedure. The cleaning procedure of these bottles must also be properly controlled since post-cleaning chemical residues are a risk to the consumer's health (Bovee et al., 1997). Recently, Xingyi (2016), studied the design of metal packaging for food using the HACCP system and found that the system was effective in controlling the migration of phenolic substances from packaging to food, making it possible to reduce these indices to acceptable levels.

3.3. Legislations and Norms

In the USA (FDA), Europe, South America, and several other countries and continents, some legislations aim to protect the consumer from contamination by possible materials and packaging. The European framework regulation (EC) No. 1935/2004, for example, deals with materials and articles intended to come into contact with food (European Commission, 2011; Kamboj et al., 2020). The EU/10/2011 Regulation of the European Commission, addresses legal issues on the migration of plastic materials in contact with food (European Commission, 2011; Kamboj et al., 2020), and finally, Regulation (EC) No. 2023/2006 deals with good manufacturing practice (European Commission, 2006; Kamboj et al., 2020).

In the US, through the FDA, several legislations on food packaging protect consumers from possible risks and harm to health, such as the regulations on the components of a food contact material (21 CFR 174-179) and the regulatory 21 CFR 182-186 which contains a list of GRAS notices to ensure that the component used is listed and recognized as safe (GRAS), in addition to other regulations (FDA, 2018). In addition to the legislations that each country has and which include food packaging, there are some standards used in the food industry. These standards are internationally recognized and aim to ensure the implementation of food safety around the world. Among the, standards that include the HACCP plan for food packaging, we can mention FSSC 22000 which is based on ISO 22000 and includes the technical standard ISO TS 22002-2 for the manufacture of packaging, BRC (British Retail Consortium) which is recognized by the Global Food Safety Initiative (GFSI), IFS PACsecure which was created in conjunction with the PAC (Packaging Association of Canada) and incorporated by the IFS (International Featured Standard), also recognized by the GFSI.

3.4. Monitoring, Corrective Actions, Validation, and Verification

Monitoring is a very important step within the HACCP system, it aims to assess whether a CCP is being controlled. This monitoring will generate a written record, which will be used to track the daily operation for verification. As a result of this monitoring, it will be possible to assess trends together with the HACCP system before the validation step. For this monitoring to be effective and reliable, monitors must be trained, records must be made in real-time, and must be accurate (Fortin et al., 2021).

For corrective action, we can define as "any action to be taken when the results of monitoring in the CCP indicate a loss of control." Thus, if after monitoring a deviation is pointed out, corrective action must be taken so that this deviation is controlled again. As it is the last defense device that will protect the consumer from potentially unsafe foods (in case of CCP failure), corrective actions must be effective and must be identified by the HACCP team during their study (Wallace and Mortimore, 2016). To confirm that the HACCP system is working efficiently, the verification of the developed procedures needs to be done. Verification will tell us if the food safety requirements defined in the HACCP plan comply when they are working in practice. Validation, on the other hand, will bring us evidence that the elements used in the HACCP plan are effective and capable of controlling all potential hazards (Wallace and Mortimore, 2016).

4. Challenges in Implementing the HACCP System in Food Packaging

There are several benefits that the HACCP system can provide for the food packaging industry and consequently for food consumers. The system, when well implemented, will contribute to the production of safe packaging that will help maintain food quality. In addition, companies with an implemented HACCP system or companies with ISO certifications that include the HACCP system, increase competitiveness within the domestic and foreign markets, in addition to keeping buyers with this differential (Cantanhede et al., 2018). However, the road to implementation may not be so easy and has its challenges. Cantanhede et al., (2018), analyzed through a questionnaire, 11 manufacturers of plastic packaging for food with FSSC-22000 Packaging certification, and found the "biggest difficulties" of the study and implementation of the HACCP system: resistance and lack of technical knowledge on the part of the employees, and the resistance was mainly linked to the organizational culture of the company; lack of experience with food safety and system implementation; lack of organizational infrastructure, high cost and excessive documentation for implementation were also cited as difficulties.

Another point is that small and medium-sized companies that previously did not have a quality management system with record keeping also find it difficult to develop specific HACCP programs (Mortimore, 2001; Panisello e Quantick, 2001; Raaska, 2005). In this way, food packaging companies that want to overcome these difficulties in implementing the HACCP system can hire companies specializing in the training and implementation of this system, as well as in the training of employees responsible for maintaining the HACCP plan. Seeking implementation before any mandatory need of a supplier, implementing the system more smoothly and without pressure, and seeking to raise awareness of the company's employees in advance, can also contribute to making this process less difficult for everyone involved. The company and its employees must be culturally aware of the importance and benefits that the implementation of this system will bring to everyone who works there.

Conclusion

Food packaging is important to maintain food quality and safety; however, some hazards are associated with the production process of this packaging, which can contaminate the food that will be marketed and then consumed. In this way, the implementation of the HACCP system in the packaging industries aims to identify these potential hazards, monitor them, and outline corrective measures in case there is any failure in the manufacturing process and if contamination occurs. Some challenges may arise in the implementation of this system, due to the high number of documents and records that are part of this tool. In addition, technical and scientific knowledge about the production and the packaging material produced is necessary for the implementation of the HACCP system. In this way, a properly trained and aware team will guarantee an effective implementation, bringing many benefits to this industry and consequently to food safety.

References

Alamri, M. S., A. A. A. Qasem, A. A. Mohamed, S. Hussain, M. A. Ibraheem, G. Shamlan, H. A. Alqah, and A. S. Qasha. 2021. Food packaging's materials: A food safety perspective. *Saudi Journal of Biological Sciences* 28(8):4490-4499.

Alsing Pedersen, G., E. Carlson, S. Ekroth, P. Kostamo, Å. Lagersted Nordström, G. Olafsson, L. Rajakangas, R. Vaz, and B. Fabech. 2012. Food contact materials and articles: Printing Inks: Check lists for compliance in industry and trade and control by food inspection. *TemaNord.* No. 2012:521. Nordic Council of Ministers, Copenhagen.

Amelia, B., G. Gutt, A. Sonia, A. Dabija, and G. Constantinescu. 2012. Study regarding the tin and iron migration from metallic cans. *J. Agroal. Proc. Tech.* 18:299-303.

Bovee, E. H. G., N. de Kruijf, J. Jetten, and A. W. Barendsz. 1997. HACCP approach to ensure the safety and quality of food packaging. *Food Additives & Contaminants* 14(6-7):721-735.

Deshwal, G. K. and N. R. Panjagari. 2020. Review on metal packaging: materials, forms, food applications, safety and recyclability. *Journal of Food Science and Technology* 57(7):2377-2392.

Deshwal, G. K., N. R. Panjagari, and T. Alam. 2019. An overview of paper and paper based food packaging materials: health safety and environmental concerns. Journal of Food Science and Technology 56(10):4391-4403.

Fortin, N. D., C. C. Carr, and J. D. Scheffler. 2021. Chapter 32 - HACCP and other regulatory approaches to prevention of foodborne diseases. Pages 545-561 in *Foodborne Infections and Intoxications (Fifth Edition).* J. G. Morris and D. J. Vugia, ed. Academic Press.

Galbiati, E., L. Jacxsens, and B. De Meulenaer. 2021. Hazard prioritisation of substances in printing inks and adhesives applied to plastic food packaging. *Food Additives & Contaminants: Part A* 38(9):1608-1626.

Guerreiro, T. M., D. N. de Oliveira, C. F. O. R. Melo, E. de Oliveira Lima, and R. R. Catharino. 2018. Migration from plastic packaging into meat. *Food Research International* 109:320-324.

Li, C., C. Li, H. Yu, Y. Cheng, Y. Xie, W. Yao, Y. Guo, and H. Qian. 2021. Chemical food contaminants during food processing: sources and control. *Critical Reviews in Food Science and Nutrition* 61(9):1545-1555.

McMillin, K. W. 2017. Advancements in meat packaging. *Meat Science* 132:153-162.

Rather, I. A., W. Y. Koh, W. K. Paek, and J. Lim. 2017. The Sources of Chemical Contaminants in Food and Their Health Implications. *Frontiers in Pharmacology* 8(830).

Slatter, J. 2003. Hazard Analysis Critical Control Point. Pages 3023-3028 in *Encyclopedia of Food Sciences and Nutrition (Second Edition).* B. Caballero, ed. Academic Press, Oxford.

Wrona, M., C. Nerín, M. J. Alfonso, and M. Á. Caballero. 2017. Antioxidant packaging with encapsulated green tea for fresh minced meat. *Innovative Food Science & Emerging Technologies* 41:307-313.

Xue, M., X. S. Chai, X. Li, and R. Chen. 2019. Migration of organic contaminants into dry powdered food in paper packaging materials and the influencing factors. *Journal of Food Engineering* 262:75-82.

Chapter 13

Reducing Human Error in Quality Control Checking of Fresh Produce Labels

James H. Smith-Spark[1,*], Hillary B. Katz[1], Thomas D. W. Wilcockson[2] and Alexander P. Marchant[1]

[1]London South Bank University, London, UK

[2]Loughborough University, Loughborough, UK

Abstract

Human error in the quality control checking of fresh produce labels results in financial loss, reputational damage, and a significant carbon footprint. This chapter reviews a research project aimed at understanding the reasons for such human error. In the course of the project, observations were taken in a packaging facility, historical error records were studied, key operatives were interviewed, and laboratory-based work was conducted. The in-situ observations highlighted the dynamic environment in which label-checking took place. The interviews revealed that no explicit training was given in label-checking. Respondents also identified a range of cognitive and situational factors likely to contribute to increased human error. Laboratory-based work, using an eye tracker to record eye movements during simulated label-checking tasks, showed that varying strategies were adopted by different quality control professionals. A systematic approach, in which one bit of information was checked at a time, was associated with more accurate performance. Several cognitive abilities were found to predict accurate label-checking performance in both quality control professionals and university

* Corresponding Author's Email: smithspj@lsbu.ac.uk.

In: Food Packaging
Editor: Wing-Fu Lai
ISBN: 979-8-88697-249-8

students. Implications for personnel selection, training, human performance, and task design are identified. The understanding of human quality control checking gained from this chapter can be used to reduce human error and, thus, waste across different manufacturing domains.

1. Introduction

Human error in the quality control checking of the labels which accompany packaged fresh produce has been estimated to cost the UK supermarket industry millions of pounds per year in fines, product recall, repackaging, and replacement (S. Hinks, Product Technical Manager: Fruit and Floral, Sainsbury's Supermarkets Ltd, personal communication, July 2012). Despite the financial, reputational, and environmental costs attached to this reducible if not avoidable waste, human error in the quality control checking has been underexplored area of applied research. The current chapter reviews a research program that was carried out in order to understand the reasons for human error in the quality control checking of the labels which accompany packaged fresh produce and to identify the ways in which performance could be improved. There is a pressing need for such research not just in the UK food industry but wherever similar forms of quality control take place.

In the course of this research program, a range of studies were carried out [1-4]. These employed different methodologies in order to gain a greater understanding of the factors that contribute to human error in quality control label-checking. As a first step, an understanding of the label-checking process was acquired through in situ observations and interviews with the key personnel involved in label-checking. Further to this, the historical error data held by the packaging company were consulted to provide insights into the type of errors that occurred and how these mapped on to shift patterns and times of day. Following this initial work, the research program moved to the laboratory setting, with label-checking behavior being studied under carefully controlled conditions; firstly, to determine whether there was an optimal approach to label-checking and, secondly, to see the extent to which the accuracy of label-checking performance could be predicted on the basis of cognitive abilities and other characteristics.

2. The Process of Packaging Fresh Produce

The company's commercial office receives updates on a weekly basis from the commercial teams of a number of supermarket chains, setting out size changes, promotions, et cetera for the fresh products that they require. Members of the commercial office enter the updates into a product specification sheet. The product specification sheet contains the details of the supermarket orders for processing by the pack-house, with one line of the spreadsheet reflecting one packaging run. For each entry on the product specification sheet, between three and 11 fields of information are displayed (such as the name of the product, its country of origin, its best-before date, and the quantity or weight of the required product). Figure 1 shows a simplified example of a product specification sheet, as used in the experimental work described in this chapter. As is the case with the product specification sheets used in the pack-house, some items have current promotional offers on them, while others do not. The information so entered into the product specification sheet is then checked by two members of staff in the commercial office against the details set out in the email or telephone communication from the supermarket.

Once it has been checked by the commercial office, the product specification sheet is sent to packaging line leaders and label print room staff in the pack-house. The information that it provides is used by the pack-house to make up the orders for shipping across the entire supermarket chain over the course of that week. Further updates to the current product specification sheet are made if necessary. At the end of the week, a new product specification sheet is produced.

PRODUCT	COUNTRY	GROWER	QUANTITY	BB	BARCODE	Promotion Ribbon / Label
WHITE FLESH NECTARINES	FRUITS-SAL, SOUTH AFRICA	SILVER FIRE	X 4 FRUIT	02-Dec	0168 9341	MIX & MATCH Any 2 for £4
YELLOW FLESH NECTARINES	FRUITS-SAL, SOUTH AFRICA	CRIMSON GIO	X 4 FRUIT	01-Dec	0158 5049	MIX & MATCH Any 2 for £2.50
YELLOW FLESH NECTARINES	CAPE-REM, SOUTH AFRICA	MAYGLO	X 4 FRUIT	02-Dec	0158 5049	
STRINGLESS BEANS	MOROCCO	LARA CASTANDA	225G	04-Dec	0103 9429	
STRINGLESS BEANS	HOLLAND	VALSTAR	225g	03-Dec	0103 9429	Half price

Figure 1. An example of a simplified product specification sheet.

Figure 2. An example of a fresh produce label with an accompanying flash label.

In the pack-house, the packaging line leader leads a team of operatives working on a particular packaging line. It is their task to process the orders specified on the product specification sheet. The packaging line leader initiates the process by generating a label check sheet. He or she then checks this check sheet against the size or quantity of the fresh produce being received and the list of growers and varieties approved by the supermarket in question. Following this, the specified quantities or weights of loose fresh produce are transported from warehouse storage to be packaged according to the requirements set out in the product specification sheet (e.g., four apples per pack, 250 g of grapes per punnet). The type of packaging required varies depending on the type of fresh produce, such that the contents may be packed in punnets with a sealed label, netted or sealed in polythene packets.

The printing of labels to accompany the packaged fresh produce is an integral part of the packaging process. The line leader orders the required number of labels from the print room, where staff receive the orders and print the labels (except where fresh produce packets are sealed in film, where dedicated printing machines form a part of the packaging line itself). The line leader then collects the labels and distributes them to his or her team for appending to each unit that is packaged. Depending on the product, the number of fields of information presented on the product label varies but the details presented on it should match those set out on the product specification sheet.

In addition, a further ribbon or sticker may be required to accompany the label. Known as a "flash label," this label highlights any current promotional activity on the product (for example, "Buy one, get one free"). An example of a product label is shown in Figure 2, together with an accompanying flash label (in this case, "Mix & Match. Any 2 for £2").

While errors can originate in other parts of the company (e.g., the sales department, where orders are taken from various supermarket chains), the focus of the research program was on those arising in the pack-house, which was identified by the company's management as being the main source of error. Different packaging technologies had been introduced at different times on different packaging lines in order to meet the demands of packaging fresh produce in different ways (such as single items, e.g., melons, and those packaged in punnets, e.g., berries and grapes)). Given the resultant complexity of the production lines, the company's management did not deem an overarching software solution to the problem to be possible. The quality control process, therefore, is in place to ensure that the information which is displayed on the product packaging matches exactly with the contents of the packaging itself, both in terms of product characteristics and current promotional offers on the product. The task of checking labels involves two main sources of information – the product label and the product specification sheet – in addition to a visual check of the product itself (e.g., to ensure that four apples are indeed contained in the packaging as indicated on the product label). In carrying out a quality control check, the operative must cross-check each of the fields on the fresh produce label against their corresponding entries on the product specification sheet.

It was observed there was a constant flow of labels to be checked throughout the day. Observations of label-checking on the pack-house floor indicated that it took quality control staff 20s on average to complete a label check. The results of the check are recorded on a checklist which has to be signed off by the label-checker. Three or four such label checks are conducted by different operatives before the fresh produce leaves the packaging facility.

3. Insights from Historical Error Data and Interviews with Key Label-Checking Operatives

The vast majority of errors are detected early in the packaging process but there are very considerable costs attached to those label errors which escape detection, both in terms of monetary fines and/or the recall and replacement of otherwise perfectly good fresh produce from supermarket shelves. By way of example, in the month of March 2014, less than 1% of orders contained label errors and, of this 1%, only 5% went undetected during quality control checking.

There are two day shifts and one night shift in the pack-house, with the latter having fewer operatives and fewer packaging lines running. The average shift length was 9.5 hours, including breaks. When the historical error data were examined, more errors were recorded as having occurred during the day shifts (particularly in the morning, between 06:00 and 10:30, and especially so either side of their morning break at 09:00). This increased recorded frequency coincided with the point in the day when quality control staff identified themselves as being at their busiest. Only 15% of errors occurred in the afternoon shift. Fewer lines, typically three, run during the night shift (18:00-06:00) and few errors were recorded as having occurred during this shift. The most frequently recorded errors involved packaging (rather than labelling) errors, involving promotional offers and the wrong stop/start dates for these offers. Spelling errors were also a frequent source of error (e.g., misspelling the name of the grower of the fresh produce or typing "friut" instead of "fruit" since there was no spellchecking facility on the system to detect and correct such errors automatically).

In addition to studying the historical error data held by the packaging facility, structured interviews were conducted with key pack-house staff involved in the label checking process [3]. Three or four errors were estimated to occur in the pack-house per day but the interviewees reported that most of these errors were not recorded since they had been corrected at the time of being noticed. Uncorrected mistakes can be very costly. For example, a packaging error on an order of raspberries was reported as requiring 15 people to work for three hours to repackage 400 punnets of raspberries. Over 60% of the quality control staff interviewed identified the following key contributing factors to label-checking error: feeling unwell, a high workload, a small point size being used in the printed information appearing on either the label or the product specification sheet, more fields of information being presented on labels, and high noise levels occurring in the pack-house environment.

The interviews also revealed that the staff had not received any explicit instruction or training in the task of label-checking, mainly due to there being no information available to staff or management concerning what optimal performance of the task should look like. Indeed, from a management perspective, this was one of the main drivers for their participation in the research program.

4. An Experimental Paradigm to Explore Label-Checking Performance

After gaining insights from the historical data, and interviews with key quality control personnel, the focus of the research program turned to exploring label-checking behavior experimentally under laboratory conditions. This allowed the approaches of naïve and experienced label checkers to be observed and compared under different conditions. This section begins by giving an overview of the simulated label-checking paradigm developed for the laboratory work. From there, the use of the paradigm to understand approaches to label-checking, to identify ways to guide optimal performance, and to determine whether different cognitive abilities might predict performance are considered in turn.

A gaze-contingent paradigm was developed to simulate the real-world task. This required the label-checker to have to shift their gaze alternately from the product specification sheet to the product label itself, thus moving visually between the two sources of information. The label-checking task was programmed in such a way that only one source of information (either the product specification sheet or the product label and any accompanying flash label) was visible on the computer monitor at any one time. If the participant's eyes were fixed on the top half of the display, then only the product specification sheet would be viewable. If the participant's eyes were fixated on the bottom half of the display, then only the product label could be seen. The number of fields of information was held constant at seven (these were product type, country of origin, name of the grower, quantity {either by number of fruit or by weight}, best before date {BB}, barcode, and flash label {promotion ribbon / label}; see Figure 2).

Having established an experimental paradigm with which to explore label-checking, the next phase of the research program involved recording the eye movements of professional label-checkers, employees of the packaging facility itself, who were brought into the lab in order to undertake the simulated task. The data so produced were analyzed to discover whether any individual differences existed in their label-checking strategies [2]. Accuracy of performance was measured, both overall (regardless of whether or not an error was present) and specifically in response to trials where there was a lack of congruence between the information presented on the product specification sheet and that appearing on the label (i.e., a label "error"). Some participants were found to have adopted a systematic approach to the task of label-checking, checking one piece of information at a time in a very serial fashion. Others took what could be described as being a "chunking" approach [5], in which the participants encoded and retained in short-term memory several bits of information from the label before checking them in the same visual pass on against the corresponding entries on the spreadsheet. A final group of participants showed no discernible pattern to their label-checking, being highly idiosyncratic in their behavior. There were, thus, differences in label-checking behavior. A "systematicity index," showing the number of intervening fixations between an individual fixating on a field containing an error and fixating on the field on the spreadsheet that corresponded with it, was calculated. This index indicated how directly the information on the label was compared with information on the spreadsheet and lower scores reflected a more systematic approach to label-checking. It was found that the four label-checkers with the highest error detection scores made significantly fewer intervening fixations than the four least accurate label-checkers. This indicated that a systematic approach to label-checking, in which one field of information on the label was checked at a time against its corresponding entry on the product specification sheet, produced the most accurate levels of performance in detecting labelling errors. In finding that different approaches were taken to label-checking by different professional label-checkers, the experimental findings were consistent with the interviews in which quality control operatives reported that no explicit instruction or training in label-checking has been provided at their place of work. This lack of formal instruction was borne out in the range of approaches adopted by the quality control staff, which varied in their effectiveness. The use of a particular strategy did not correlate with the number of years of experience that an operative had in the quality control checking of fresh produce labels.

Following the identification of different strategic approaches to label-checking, the next step was to see whether this understanding could be used to shape the behavior of individuals when carrying out the task. Some preliminary research was, therefore, conducted on 58 university students, who were naïve to the quality control checking processes involved in the checking of product labels [4]. Two blocks of label-checking trials were presented using a gaze-contingent eye-tracking paradigm. Depending on whether their gaze was centered on the top or bottom half of the monitor screen, the participants would see either the product label or the product specification sheet. This approach was taken in order to simulate the need for the label-checker to orientate visually between the label and the product specification sheet when performing a label check. In the first block of trials, the participants were allowed to follow their own personally determined method of label-checking. For the second block of trials, the participants were allocated randomly to one of three conditions. The first condition acted as a control condition, in which the participants were allowed to continue label-checking using their own idiosyncratic method. The remaining two conditions provided differing levels of computerized support to the participants. The support given by the first of these computer-guided conditions was unimodal, with visual highlighting of the current field of information which required checking. The second condition was bimodal with visual highlighting of the current field and a pre-recorded voice reading out the information presented in the field currently required to be checked. Improvements were seen in the accuracy of performance across all three conditions when moving from the first to second block of trials. However, significant differences were found in the extent of this improvement depending upon the experimental condition. Relative to the control condition, the bimodal support condition resulted in a significantly greater level of improvement (5% compared with 0.6%), while there were no significant differences between the control condition and the unimodal support condition nor between the control condition and the bimodal support condition. While a 5% improvement may seem small in absolute terms, this would translate to considerable savings in both financial and environmental terms. The empirical demonstration of the importance of releasing information to label checkers in a strictly serial manner was fed into the development of a commercially released software app called Greenlight Label Check. A three-month onsite trial of the app demonstrated the success of the approach, with the percentage of undetected label errors falling to 0%.

A further aim of the research program was to determine whether the accuracy of label error detection could be predicted by performance on standard laboratory measures of cognition. Findings in this area could potentially be used to guide personnel selection for the quality control role or to highlight the aspects of the cognition in which individual label-checkers would need to be supported. In broad terms, the literature with which the task of label-checking has greatest affinity is that of visual inspection, which also requires the checking products by eye for deviations from an accepted norm or correctness in one or more dimensions. The ergonomics literature has considered a range of different visual inspection tasks, such as those involved in X-ray security inspections at airports [6], aircraft inspection [7], contact lens inspection [8], international nuclear safeguard inspection [9,10], telecommunications [11], the manufacture of consumer products [12], and the production of pharmaceutical products [13]. However, unlike most visual inspection tasks, in which a manufactured item is compared against a mental representation of an exemplar stored in long-term memory, the quality control check of fresh produce labels requires two sources of information to be compared, both of which vary in informational content from check to check. There are, therefore, important differences in the structure of each task and associated cognitive demands, especially with respect to the demands made on short-term memory. In label-checking, unlike other types of visual inspection task, information from one source needs to be encoded and retained in short-term memory while it is compared with the information presented on the other source.

Alongside a label-checking task, Katz et al. [1] presented a battery of basic cognitive and perceptual measures to 51 university students naïve to the process of quality control checking fresh produce labels. Due to the time and resource constraints involved in testing a relatively large number of participants, the label and the product specification sheet were presented simultaneously on the monitor screen rather than using the gaze-contingent paradigm described previously. A battery of basic cognitive measures was also presented to the participants to measure information processing speed, attention, verbal and visuospatial short-term memory and working memory, and the propensity to mind wander. Of these measures, only verbal short-term memory was found to be a significant predictor of label-checking accuracy. The ability to recall a greater number of digits in the same order as they were presented showed a weak to moderate positive correlation with label-checking accuracy.

The observations and interviews described previously had highlighted the dynamic and complex pack-house environment. These findings suggested that label-checkers might draw upon higher-order cognition (or executive functions; e.g., [14, 15]) in order to maintain task performance in the face of interruptions and distractions. Therefore, Smith-Spark et al. [3] administered executive functioning measures (the "core" executive functions of inhibition, updating, and set shifting; e.g., [15]) as well as a measure of the speed with which information could be processed to a mixed sample of professional label-checkers and naïve university students. Inhibition relates to the ability to prevent habitual responses in favor of responses that are relevant to task at hand and current conditions. Updating describes the ability to refresh the contents of working memory in the light of new information becoming available in the environment. Set shifting reflects cognitive flexibility and the ability to move fluently between different cognitive operations or cognitive sets. Information processing speed was measured in terms of the rate at which the participants could convert one form of information to another. Level of alertness was also measured using a subjective self-report measure which required the participants to rate their current level of sleepiness. The same gaze-contingent paradigm as described previously in this chapter was used to assess label-checking performance. It was found that better abilities on two of their battery of measures, namely information processing speed and inhibition, were related to better label-checking performance. Overall accuracy of label-checking performance (i.e., collapsed across trial types) was positively predicted by inhibitory ability, while the ability to detect label errors was also positively predicted by information processing speed.

5. Methodological Considerations and Future Research Directions

While the verisimilitude of the simulated label-checking task to the task conducted on the pack-house floor was praised by the professional quality control staff taking part in the study, it should be noted that potentially important differences exist. Firstly, blocks of successive trials were presented. While the mean time to complete each label check was similar to that observed in situ, the presentation of label after label in an unbroken sequence was not reflective of performance in the pack-house, where quality control checkers would be engaged in other quality assurance work between label checks. While several labels might need to be checked in succession, a block of 50 label checks would be highly unlikely to occur.

Further to this, the number of fields presented on the label remained constant throughout the experiment rather than varying as it would in the pack-house. Work by Gallwey and Drury [16] has indicated that having more information to inspect results in reduced levels of performance. To add even greater verisimilitude to the task (and to explore the effect that the number of fields of information to be checked might have on performance), further research should explore this factor experimentally. In addition, the number of fields of information that could contain potential errors was limited to one. Gallwey and Drury [16] have found that visual inspection performance is lower when multiple fields contain an error rather than an error being present in only one field (see [16] for a review of the potential impact of this on performance). In the pack-house itself, however, each and every field printed on a label could, potentially, be incorrect, albeit with differing likelihoods of error occurring, and it would be the task of the label-checker to identify every error appearing on the label and not just the first error that they encountered.

The current experimental paradigm is not able to account for strategic behavior relating to participants exiting the label check after having found a single error. It would also be informative to explore the effects of distractors (such as unpredictable noises) on performance, given the difference between the controlled laboratory setting in which label-checking was investigated and the noisy and dynamic environment in which quality control checking takes place. The ability to maintain the goal of label-checking under these conditions would be likely to link to executive function and, as a result, it might prove fruitful to look at the association between performance under such dynamic conditions and measures of executive function, perhaps finding stronger links than those described in the previous section.

Harris [17] has highlighted the influence of the frequency of defects on visual inspection, with inspection accuracy decreasing as the rate of errors decreases. In the label-checking experimental paradigm, the percentage of trials containing a label error was 20%. The rate of errors had to be considerably higher than in the pack-house due to time and resource constraints on both the professional label checkers and researchers.

Owing to the complexity of label checking in terms of the cognitive and situational factors involved, laboratory studies can only go so far in teasing out key determinants of performance. The insights obtained from the current program of research depended on experimental design choices having to be made. These maximized the degree of similarity between the laboratory task and the real-world situation that it was designed to replicate but it was not possible to copy or analyze all in situ elements. Thus, differences exist in that the laboratory task involved the presentation of large numbers of trials in succession, constraints on the number of fields of information presented on the fresh produce labels, the presence of errors in solely one field, and the frequency rate of errors. These and other factors remain to be explored. The need to explore environmental influences on performance has already been highlighted, together with suggestions for further extending the investigation of cognitive factors contributing to label-checking performance. Further research might explore the contribution of individual differences in characteristics such as conscientiousness and motivation, to label-checking performance.

Conclusion

Overall, the data gained from a range of methodological approaches have indicated the need for standardized training for operatives involved in the quality control checking process in order to avoid idiosyncratic and often less effective label-checking strategies from being adopted. The research has also identified several cognitive abilities which may predispose some workers to be better suited to the role of label-checking than others. Higher levels of verbal short-term memory, information processing speed (involving the rate at which information could be converted from one form to another), and inhibition have all been found to be positive predictors of label-checking accuracy. Performance on measures tapping into these areas of cognition may indicate the degree of fit between an individual worker and the demands posed by structure and constraints of the label-checking task. Finally, the research program has demonstrated that the process of labels checking can be guided in ways that make it more effective, suggesting a software application-based approach to label-checking in which human label-checkers are pushed towards a serial and systematic approach to the checking of label information fields by the controlled release of information over time.

Through the study of behavior both in situ and in the controlled laboratory setting, this research program has advanced the understanding of factors that influence the performance of humans in detecting errors during quality control checks of fresh produce labels. The research described in this chapter is broadly applicable to any type of product label which needs to be checked by humans during the manufacturing process and/or distribution of packaged products to retail points and thereby helping to reduce wastage and the carbon footprint generated by manufacturing.

Funding

The research reported in this chapter was funded by Innovate UK, grant number 101393.

References

[1] H. B. Katz, J. H. Smith-Spark, T. D. W. Wilcockson, A. Marchant, A. (2015). Cognitive predictors of accuracy in quality control checking, in G. Airenti, B. G. Bara, G. Sandini (Eds.), *Proceedings of the EuroAsianPacific Joint Conference on Cognitive Science, CEUR Workshop Proceedings,* Aachen, Germany, 2015, pp. 750-755. http://ceur-ws.org/Vol-1419/.

[2] J. H. Smith-Spark, H. B. Katz, T. D. W. Wilcockson, A. P. Marchant, Optimal approaches to the quality control checking of product labels, *Int. J. Ind. Ergon.* 68 (2018) 118-124. https://doi.org/10.1016/j.ergon.2018.07.003.

[3] J. H. Smith-Spark, H. B. Katz, T. D. W. Wilcockson, A. P. Marchant, Factors affecting accuracy in the quality control checking of fresh produce labels: A situational and laboratory-based exploration, *Hum. Factor. Ergon. Man.* 29 (2019) 447-458. https://doi.org/10.1002/hfm.20806.

[4] J. H. Smith-Spark, H. B. Katz, A. Marchant, T. D. W. Wilcockson (2016), Reducing quality control errors by guiding behavior, in: *Book of Proceedings of the 6th International Ergonomics Conference Ergonomics 2016 - Focus on synergy,* Zagreb, Croatia: Croatian Ergonomics Society, 2016, pp. 315-322.

[5] G. A. Miller, The magical number seven, plus or minus two: Some limits on our capacity for processing information, *Psychol. Rev.* 63 (1956) 81-97. https://psycnet.apa.org/doi/10.1037/0033-295X.101.2.343.

[6] K. M. Ghylin, C. G. Drury, A. Schwaninger, A., Two-component model of security inspection: Application and findings, in: *Proceedings of the 16th World Congress of Ergonomics,* Maastricht, The Netherlands: International Ergonomics Association, 2006.

[7] C. G. Drury, Exploring search strategies in aircraft inspection, in: D. Brogan, A. Gale, K. Carr (Eds.), *Visual Search 2*, Taylor & Francis, London, UK, 1993, pp. 101-112.

[8] P. Rao, S. R. Bowling, M. T. Khasawneh, A. K. Gramopadhye, B. J. Melloy, Impact of training standard complexity on inspection performance, *Hum. Factor. Ergon. Man.* 16 (2006) 109-132. https://doi.org/10.1002/hfm.20045.

[9] L. E. Matzen, M. C. Stites, H. A. Smartt, Z. N. Gastelum, The impact of information presentation on visual inspection performance in the international nuclear safeguards domain, in: S. Yamamoto, H. Mori, (Eds.), Human interface and the management of information. Visual information and knowledge management. HCII 2019. *Lecture Notes in Computer Science, Volume 11569*, Springer, Cham, Switzerland. 2019, pp. 56-75.

[10] M. C. Stites, L. E. Matzen, H. A. Smartt, Z. N. Gastelum, Effects of note-taking method on knowledge transfer in inspection tasks, in: S. Yamamoto, H. Mori (Eds.), Human interface and the management of information. Visual information and knowledge management. HCII 2019. *Lecture Notes in Computer Science*, Volume 11569, Springer, Cham, Switzerland, 2019, pp. 594-612.

[11] G. H. Jameeson, Inspection in the telecommunications industry: A field study of age arid other performance variables, *Ergonomics* 9 (1966) 297-303.

[12] F. C. H. Lee, S. S. Man, A. H. S. Chan, Effects of magnification modes and location cues on visual inspection performance, *PLoS ONE* 14 (2019) e0213805. https://doi.org/10.1371/journal.pone.0213805.

[13] J. A. Melchore, Sound practices for consistent visual inspection, *AAPS PharmSciTech*, 12 (2011) 215-221. https://doi.org/10.1208/s12249-010-9577-7.

[14] A. Diamond, Executive functions, *Ann. Rev. Psychol.* 64 (2013) 135-168. https://doi.org/10.1146/annurev-psych-113011-143750.

[15] A. Miyake, N. P. Friedman, M. J. Emerson, A. H. Witzki, A. Howerter, T. D. Wager, The unity and diversity of executive functions, and their contributions to complex "frontal lobe" tasks: A latent variable analysis, *Cognitive Psychol.* 41 (2000) 49-100. https://doi.org/10.1006/cogp.1999.0734.

[16] T. J. Gallwey, C. G. Drury, Task complexity in visual inspection, *Hum. Factors* 28 (1986) 595-606. https://doi.org/10.1177%2F001872088602800509.

[17] D. H. Harris, Effect of defect rate on inspection accuracy, *J. Appl. Psychol.* 52 (1968) 377-379. https://doi.org/10.1037/h0026241.

About the Editor

Wing-Fu Lai, PhD
Department of Applied Biology and Chemical Technology,
The Hong Kong Polytechnic University, Hong Kong
Email: rori0610@graduate.hku.hk

Wing-Fu Lai received his MSc degree in Materials Engineering and Nanotechnology from the City University of Hong Kong, and earned his PhD in Chemistry from the University of Hong Kong. He has been accredited as Shenzhen Municipal "Overseas High-caliber Personnel". Professionally, he is a Registered Nutritionist in the UK, and a Fellow of the UK Higher Education Academy. He has also received his Certified Food Scientist credential from the International Food Science Certification Commission in the US. His research focus lies in the development and engineering of polymeric materials for food and pharmaceutical applications.

Index